EVOLUTION IN VICTORIAN BRITAIN

Nineteenth Century Science, Technology and Medicine Sources

Series Editors: Piers J. Hale and Meegan Kennedy

Available:

Science and Sound in Nineteenth-Century Britain
Edited by Edward J. Gillin
4 volume set

Forthcoming:

Women in Medicine in the Long Nineteenth Century
Edited by Claire Brock
4 volume set

Scientific Methodology in Nineteenth-Century Britain
Edited by Charles H. Pence
4 volume set

Electric Power in Victorian Britain
Edited by Nathan Kapoor
4 volume set

Insects in Victorian Culture
Edited by John Clark
4 volume set

Evolution in Victorian Culture
Edited by Piers J. Hale and Caden C. Testa
5 volume set

Nineteenth-Century British Microscopy
Edited by Meegan Kennedy
4 volume set

Meteorology in Nineteenth-Century Society
Edited by Aitor Anduaga
4 volume set

Women, Gender and Science in Victorian Culture
Edited by Leila McNeill & Anna N. Reser
4 volume set

EVOLUTION IN VICTORIAN BRITAIN

Edited by
Caden C. Testa and Piers J. Hale

Volume I
Evolution Before Darwin

Routledge
Taylor & Francis Group
LONDON AND NEW YORK

First published 2025
by Routledge
4 Park Square, Milton Park, Abingdon, Oxon OX14 4RN

and by Routledge
605 Third Avenue, New York, NY 10158

Routledge is an imprint of the Taylor & Francis Group, an informa business

British Library Cataloguing-in-Publication Data
A catalogue record for this book is available from the British Library

Library of Congress Cataloging-in-Publication Data
Names: Testa, Caden C., editor. | Hale, Piers J., editor.
Title: Evolution in Victorian Britain. Volume 1, Evolution before Darwin / edited by Caden C. Testa and Piers J. Hale.
Description: Abingdon, Oxon ; New York, NY : Routledge, 2025. | Includes bibliographical references and index.
Identifiers: LCCN 2024016495 (print) | LCCN 2024016496 (ebook) | ISBN 9781032791128 (v. 1 ; hardback) | ISBN 9781003490548 (v. 1 ; ebook)
Subjects: LCSH: Evolution (Biology)—Great Britain—History—19th century. | Transmutation of animals—Great Britain—History—19th century. | Darwin, Charles, 1809–1882—Contemporaries.
Classification: LCC QH361 .E924 2024 (print) | LCC QH361 (ebook) | DDC 576.8/20941—dc23/eng/20240417
LC record available at https://lccn.loc.gov/2024016495
LC ebook record available at https://lccn.loc.gov/2024016496

ISBN: 978-1-032-16844-9 (set)
ISBN: 978-1-032-79112-8 hbk (vol I)
ISBN: 978-1-003-49054-8 ebk (vol I)

DOI: 10.4324/9781003490548

Typeset in Times New Roman
by Apex CoVantage, LLC

CONTENTS

VOLUME I EVOLUTION BEFORE DARWIN

ACKNOWLEDGEMENTS

We would like to thank Dr. Kerry Magruder and Dr. JoAnn Palmeri at the History of Science Collections at the University of Oklahoma for their help both in locating and providing good readable copies of some of the primary sources contained in this volume.

This research was supported in part by a grant from the Research Council of the University of Oklahoma Norman Campus.

SERIES PREFACE

Nineteenth-Century Science, Technology and Medicine: Sources and Documents

This volume is part of the Routledge Historical Resources primary source collection, "Nineteenth-Century Science, Technology and Medicine: Sources and Documents." This series offers a fresh look at the historical development of science, technology, and medicine through the eyes of subject experts and specialists from around the world. The series provides curated access to significant primary-source documents on these subjects, collating and contextualizing these valuable resources so that they may ground not just scholarly research but undergraduate and graduate research and teaching as well. As well as being published in print and ebook formats, the contents of this and all other volumes in this series will be available through the Routledge Historical Resources database, where they will be presented in a fully integrated and searchable format, together with supplemental secondary sources and images and newly-commissioned scholarly written and video essays.

Our editors have framed their volumes around themes they consider most pertinent to understanding nineteenth-century developments in their chosen field and their significance to broader changes in nineteenth-century British culture and society. Often, the editors have chosen to provide short extracts to illustrate particular points or themes; in other cases, they provide entire essays, book chapters, or periodical articles as appropriate. Each volume is prefaced by an introductory essay in which the editors review their topic and discuss the significance of their sources. They also frame each entry or collection of related entries with explanatory headnotes. Thus, each volume or set of volumes will provide a useful resource of easy-to-navigate, well-contextualized sources curated by an expert eye, which will serve as a useful introduction for those unfamiliar with the field. Since the editors build each volume to represent a unique, deeply considered scholarly perspective on the foundational primary sources for a particular subject, the series should also prove valuable to established scholars, even those with deep experience in the field. Indeed, given the range of topics covered and sources collated in this series, we anticipate that this project will foster an exciting and insightful interdisciplinary perspective on the histories and cultures of nineteenth-century British science, technology, and medicine. We, as general editors, have learned a lot through our engagement with the editors who have contributed to this project, and we believe that wider audiences will do so, too.

Throughout the project, we have asked editors to comment on the state of the existing historiographical debate and to specify where their own work fits in. Some editors have chosen to do this through the provision of recommendations for further reading; others do so through explicit commentary on this topic in their introductory essays and in the headnotes and footnotes of each of the sources they have selected.

The histories of science, technology, and medicine are wide and often disparate fields, even though they are often encountered by students in the context of a single university department. Although they have much in common, scholars of science, technology, and medicine also propose field-specific questions and approaches; these are represented in the project through the disparate interests of the range of scholars contributing to the series. We have also worked to bring the insights of literary scholars as well as historians to the project.

We have made every effort to cover major nineteenth-century themes and topics in the scientific and technological disciplines as well as to give space to specialist topics. Of course, as is inevitably the case in a project of this nature, which must rely on the availability of expert scholars in particular fields, there will remain gaps. We have attempted to address these where we can in the form of introductory essays and video essays in the database. We do believe, though, that the reader and researcher will find that this project presents the major themes, as well as many specialist topics, in detail and with scholarly rigour.

We are very grateful to Routledge for their invitation to take on this project, and we would especially like to thank the commissioning editor, Rachel Douglas, her successor, Laura Pilsworth, and Simon Alexander. Most of all, however, we are grateful to the extensive list of scholars from around the world who have agreed to contribute and who make this series what it is.

Piers J. Hale and Meegan Kennedy

GENERAL INTRODUCTION

The idea that all life, including ourselves, has evolved from other earlier forms has proven to be one of the most controversial in the history of Western thought. It challenged people to reflect on the very question of what it meant to be human and to think deeply about what a common ancestry with other and lower forms might mean for how we think about ourselves not only as biological but also as social, cultural, and moral beings. Most readily associated with the work of Charles Darwin (1809–1882) – both his *On the Origin of Species,* published in 1859,[1] and *Descent of Man*, which was published in 1871[2] – evolutionary ideas effected such a reframing of the Western sense of self that scholars have long referred to the period that followed the publication of *Origin* as "the Darwinian Revolution" to indicate the extent of this transformation in both science and society.[3]

As we show in the first of our volumes in *Evolution in Victorian Britain,* though, despite the subsequent emphasis on Darwin and his proposed mechanisms of natural and sexual selection, "transmutation," or "the developmental hypothesis," as evolution was often referred to for the first half of the nineteenth century, has a longer history.[4] Ideas of species development can be found in antiquity, but we can trace the origins of modern evolutionary thought to the Enlightenment and to both the natural history and the political and social philosophy of eighteenth-century Europe. The idea that the development of science and learning might allow humanity not only to attain sufficiency and thus end poverty and inequality, but also to lay the foundations for continual improvement in all things, helped to establish a progressive ideology within which older notions of natural barriers and boundaries were challenged and, in many instances, overthrown. The Enlightenment and its social and political consequences thus provided a new lens through which to ask – and to answer – questions about human nature; about the meaning of sexual, racial, and class differences; and about the implications of a natural origin and development of mankind for religious belief and social organization. As we show here, Enlightenment aspirations for the freedom, equality, and brotherhood of all men had significance across Europe and, following the French Revolution, took on a deeply threatening aspect to those who supported monarchist and other hierarchical forms of government. Indeed, the shadow of the French Revolution hung over British social and scientific debates throughout the nineteenth century. It is for this reason that we consider works from a number of prominent French thinkers in our collection. Their works were translated and widely read in Britain. In addition, of course, we also include the more significant British Enlightenment and radical thinkers; most notable for our purposes are William Godwin (1756–1836) and Erasmus Darwin (1731–1802). Godwin was arguably Britain's most famous radical figure; his *Enquiry Concerning Political Justice* was published in 1793,[5] following on the heels of his wife Mary Wollstonecraft's (1759–1797) *Vindication of the Rights of Woman* (1792).[6]

THE 'SPECIES QUESTION' IN THE NINETEENTH CENTURY

Such political ideas did not evaporate when, by the nineteenth century, the area of focus was narrowed – or perhaps it would be more accurate to say broadened – to address the question of the origin of the new species that were being unearthed from the fossil record as the world's first industrial nations dug into the earth in search of ores and minerals, or in cutting rail and canal networks across the land. It was this revelation in European science that framed what was called 'the species question' as a prominent one in the early years of the nineteenth century. It was, as Darwin noted first in his 1839 account of his voyage on the *Beagle*, but later and more famously in the first pages of *On the Origin of Species*, what the internationally renowned philosopher of science and scientific practitioner Sir John Herschel (1792–1871) had identified as the "mystery of mysteries" in modern science.[7] The emergence of 'the species question' in England in the early years of the nineteenth century thus had its origins in the new science of geology rather than emerging from eighteenth-century natural history. Despite this, though, it was clear that the discovery of new fossil forms asked questions that demanded zoological answers, and in consequence, the term 'palaeontology' was coined to describe the study of fossil forms and their implications for our understanding of the connections between what we now call the earth and the life sciences.[8] The significant debates in this period in the history of geology have been discussed extensively elsewhere, most notably by the historian of science Martin Rudwick.[9] However, for our purposes, it is relevant to note here just two main aspects of these debates: First, that they raised the question not only of the origin of new species but of what scientific practitioners meant by the word species. Were species 'real' in nature, or was it simply a word that naturalists and systematists used to define a group of organisms that were similar in some particular way? Second, these discoveries pressed questions about the tempo and mode of change in Earth's history. Most notable, at least in the debates as they unfolded in England, were the French geologist and anatomist Georges Cuvier (1769–1832) and the Scottish geologist Charles Lyell (1797–1875). While both men agreed that species were fixed and immutable, they disagreed vehemently over the tempo and mode of change that characterized Earth's history. Their disagreement was over how to read the rocks and, most fundamentally, over the meaning of the layered strata that geological investigation revealed. Lyell, in his *Principles of Geology* (1830–1833),[10] followed the Scottish geologist James Hutton (1726–1797) in arguing that earth history had been one of slow and gradual change over many millions of years, while Cuvier, although he admitted that the earth was surely many hundreds of thousands of years old, held back from Lyell's extravagant claims about the age of the earth, on the basis that he thought that such a hypothesis was both unnecessary and inconsistent with the evidence. Instead, he suggested that the often stark differences between one stratum and the next indicated that there had been frequent major "revolutions."[11] These, he thought, had been caused by great floods which had both caused extinctions and laid down new layers of silt and sediment, often entrapping the remains of those animals now found as fossils. This much, he argued, fit with the geological evidence blanketing the earth. It was the nineteenth-century historian and philosopher of science William Whewell (1794–1866) who coined the terms 'uniformitarianism' and 'catastrophism' to describe the two schools of thought. These differences regarding belief in gradual and uniform change over time versus more stochastic and saltationary change in Earth's history had implications for similar debates about the tempo and mode of evolutionary change that lasted well beyond the nineteenth century.

DARWIN, MALTHUS, AND THE POLITICS OF EVOLUTION

There remains debate among historians as to exactly when Charles Darwin came to believe in the mutability of species. It is possible that this was during the years of his medical education in Edinburgh, for as is now widely appreciated, transmutationist ideas were widely discussed and debated in the lecture halls of the Edinburgh anatomy teachers as well as in the pages of the major scientific and philosophical journals published in the city.[12] Darwin met and was clearly influenced by a number of men who had deep sympathies for the French naturalists and transmutationists Jean-Baptiste Lamarck (1744–1829), Étienne Geoffroy Saint-Hilaire (1722–1844), and others whose works were clearly embedded in an evolutionary framework. Further, as a number of historians have shown, for all that the 'species question' might have been prompted by fossil finds, in truth, it remained deeply connected to debates about human improvement and the effects that changes in the environment, diet, and climate might have upon human as well as animal species.[13] Thus, as the study of these debates in Edinburgh shows, the consideration of human natural history was always to the fore, and this inevitably included the question of the nature of human racial variety, as well as the natural historical and geographical relations between them. Enlightenment experiments and dissertations on human and animal breeding only added to the controversial nature of the question.[14] To put it bluntly, evolution was never only about fossils and finches; it was always about us and about what implications it might have for society – questions about sex and race were central, but so too, and especially in England, were questions about class. In this way, we can see that the Enlightenment ideas and ideals about how best to organize society to ensure human betterment – whether they highlighted liberté, égalité, and fraternité or the beheading of aristocrats and monarchs – were always implicit and sometimes explicit in debates about evolution.

It was for this reason that early transmutationist ideas were controversial as much as for the notion that they might undermine belief in a young earth or a religious conception of creation, although in practice, of course, religious conservativism tended to map onto political conservatism.[15] The Francophile and politically radical associations of Erasmus Darwin, Lamarck, and Geoffroy's ideas tainted later expressions of transmutationism. This was so in the case of the anonymously published and deeply evolutionary work *The Vestiges of the Natural History of Creation* (1844).[16] As the historian of science James Secord has shown, the publication and dissemination of this book was a veritable sensation across all sectors of society. In it, the anonymous author, the publisher Robert Chambers (1802–1871), argued that the universe was subject to general laws of progressive development. Even though he tried to frame his evolutionary ideas in the context of a divine law-giver, his work was still read as deeply threatening to conventional morals and politics.[17]

Darwin paid close attention to the reception and reviews of *Vestiges* and was mindful to distance his own efforts from those of Chambers, as well as from the earlier radical and Enlightenment thinkers who had tied their ideas of progressive change to a radical political agenda.[18] There has been debate about exactly why Darwin waited so long to publish his theory. Many scholars have argued that he delayed because he was aware of the religious connotations his work would have, and there are some good reasons for accepting this line of argument.[19] Other scholars have noted that Darwin was not the angst-ridden theorist he has often been portrayed to have been. Rather, he was simply busy publishing his geological work and working hard in the time he had to find answers to problems he imagined future readers might have with this developing theory of natural selection.[20] Although the parties to this debate see these accounts of Darwin's delay as mutually exclusive, it is, in fact, quite possible that there is truth to both sides of this story, although we might add concern over the political connotations as well as religious concerns.

It was in the context of the political arguments that followed the 1832 reform act and specifically, about the reform of the poor law that Darwin was led to pick up and read *An Essay on the Principle of Population* by the minister and political economist Thomas Robert Malthus (1766–1834). Malthus had originally written his argument against the radical political and social aspirations of Godwin, Nicolas de Condorcet (1743–1794), and others. The fact that there would always be more mouths than resources meant that there would always be inequality and poverty, and thus, the notion of perpetual improvement was contrary to natural laws. People needed to eat, people had an instinct to reproduce, and the population would always expand faster than the ability of society to produce food, and the rest was inevitable.

By the late 1830s, though, liberals were reading Malthus quite differently. Instead of seeing Malthus as making an argument against positive change, they read him as setting up the necessary conditions for social improvement. If one was chaste and worked hard, then each individual could advance their position.[21] The poor law, which would segregate the poor by sex into workhouses, would ensure that the less moral and less useful members of society no longer increased their numbers.

Darwin read Malthus and noted, as he recalled in his autobiography, "Here, then, I had at last got a theory by which to work."[22] Cutting away the moral restraint that might benefit liberals, Darwin saw that in nature, overpopulation would mean an intense struggle for existence, and then – his original insight – that under these conditions, any organism that had a variation that was in any way beneficial to it would have an advantage, and would likely survive to reproduce. Any individual that had a deleterious variation would just as assuredly succumb. If these variations were to any extent heritable, then the logical conclusion would be that species would change over time, even though it might take many, many generations. As Darwin noted in *Origin*, his theory was "the doctrine of Malthus applied with manifold force to the whole animal and vegetable kingdoms."[23]

A number of Darwin's contemporaries instantly recognized the political significance of Darwin's theory. Far from the Francophile radicalism that led to equality and socialism, with Malthusian theory at its heart, Darwin's work served to naturalize the individualism and competition of liberal capitalism. The comparative anatomist Thomas Henry Huxley (1825–1895) welcomed the publication of *Origin* as "a veritable Whitworth gun in the armoury of liberalism."[24] Karl Marx (1818–1883) famously wrote to his comrade and friend, Friedrich Engels (1820–1895), "It is remarkable how Darwin rediscovered, among the beasts and plants, the society of England with its division of labour, competition, opening up of new markets, 'inventions,' and Malthusian 'struggle for existence.' It is Hobbes' *bellum omnium contra omnes*" – Thomas Hobbes's war of each against all.[25] The fact that liberal theologians like Charles Kingsley (1819–1875) also welcomed Darwin's work did a lot to allay the religious concerns of many of Darwin's readers.[26]

DARWINISM AND *DESCENT OF MAN*

1860 was not 1845. The kind of lawful evolutionary ideas framed anonymously in *Vestiges* were published in a Britain that was still politically unstable. The 1832 Reform Act had not opened the floodgates to revolution, as some conservatives had feared, but it had enfranchised and legitimized the growing middle class of capitalists, entrepreneurs, and industrialists.[27] By 1860, many of these former radicals were now a part of the establishment themselves, and young liberals were gaining an entry into science and certainly were in the business of writing about it. The ready acceptance of Darwin's evolutionary ideas among this group is a testament to the extent to which they saw nature in their own guise and welcomed a work that underpinned their own methods and mode of action. Of course, and as we shall show across this series of source collections, this was not the only way of reading *Origin*, and the moral meaning of Darwin's works was very much open to interpretation. This was even more the case following his 1871 publication of *The Descent of Man and Selection in Relation to Sex*. It was in this book that Darwin ventured into print on the evolution and development of humankind. While by this time, many other authors had written about the implications of evolutionary ideas for humanity, Darwin was, for obvious reasons, the most authoritative voice on the topic. It was in *Descent* that Darwin had given, amongst much else, an evolutionary account of the human mind and morals.[28] He had been pressed to address the topic by the defection of his colleague Alfred Russel Wallace (1823–1913) from the belief that natural selection could account for these most important aspects of humankind. And, as Darwin might have imagined, these were precisely the most controversial issues, and they upset the delicate religious compromise he had established by excluding discussion of man from *Origin* and by openly acknowledging the role of "the Creator" from the 1860 second edition onwards.[29]

By the late 1860s, Darwin had clearly affected a religious compromise. Influential theologians like Charles Kingsley, the American botanist and devout Presbyterian Asa Gray (1810–1888), and the Unitarian Frances Power Cobbe (1822–1904) were not alone in welcoming *Origin* as a scientific theory that gave what they took to be a sound answer to the species question while allowing that a God of law lay behind the unfolding of the laws that pertained to His creation. In *Descent*, though, Darwin gave a natural historical and thoroughly contingent account of the development of the mind and morals. Conscience, morality, ethics: all of these things, if Darwin was right, were simply ideas that had served the good of the community in its struggle for existence but were otherwise quite meaningless. Cobbe and Kingsley each publicly distanced themselves from Darwin's position, Cobbe in her review of *Descent* published in the *Theological Review*, "Darwinism in Morals," and Kingsley in his 1871 Presidential Address to the Devonshire Association for the Advancement of Science, Literature, and the Arts.[30] However, while they rejected Darwin's opinion on this particular point, they did not walk back their long-established support for evolution. Rather, they doubled down on their own theistic interpretations of Darwin's work and of the necessary place of God at the heart of the evolutionary process. This set the ground for a further range of theistic interpretations of evolution that went on in the context of broader debates about the increasingly apparent turn towards scientific naturalism and away from theism within the broader scientific community.[31] It is clear that although the publication of *Origin of Species* had done away with the old idea of an ever-creating God, it did not undermine the idea of an ever-acting God. These disputes went on, and are worth following in greater detail, in the Metaphysical Society, a debating club established in 1869 to debate exactly these issues. Its members met monthly until it was disbanded in 1880.[32]

MALTHUS OR MUTUALISM

Descent of Man was not only read as having theological and metaphysical implications but also as having political import. Whereas, in *Origin,* Darwin had emphasized competition between individuals in a bitter struggle for existence, in *Descent* he emphasized instead the cooperation and mutualism that had allowed one society to best another. It was in this context that he sought to explain the evolution of mankind's social and moral characteristics.[33] It is, therefore, unsurprising that anarchists and socialists read the moral and political meaning of Darwin's work to be quite different from the ways in which the likes of the liberal comparative anatomist Thomas Huxley had read it. Noting that Darwin had acknowledged in *Origin* that he used the term the struggle for existence "in a broad and metaphoric sense" to include the struggle against the environment and reliance of one being upon another, the anarchist Peter Kropotkin (1842–1921) wrote a series of essays in which he sought to reclaim the politics of evolution for anarchist-socialism.[34] They were originally published in response to Huxley's essay "The Struggle for Existence: A Programme," in the journal *Nineteenth Century* from 1889 through the 1890s, and were brought together in book form under the title *Mutual Aid. A Factor in Darwinism* in 1902.[35] As we show in our third collection of sources, although Kropotkin's *Mutual Aid* was influential, it was far from the only socialist interpretation of evolution.[36]

EVOLUTION AND THE 'WOMAN QUESTION'

In *Descent,* Darwin elaborated on the theory of sexual selection that he had already outlined briefly in *Origin of Species*.[37] It was through sexual selection that Darwin sought to explain the sexual dimorphism that was evident across so many species. Darwin described sexual selection as having two distinct aspects: one was male-male conflict in order to gain access to fertile females, which he called "the law of battle," and the other was the selection of the most attractive males by the females. Darwin called this "female choice." In *Descent,* Darwin suggested that it was through these processes that the human male had become "more courageous, pugnacious, and energetic than women, and has a more inventive genius."[38] The human female, by contrast, had become more maternal, physically weaker than man, less intellectually developed, and a more emotional, intuitive, and moral being.[39]

Further, although female choice continued to play a role in human sexual selection, as Darwin noted in *Descent,* because the male had become "more powerful in body and mind than woman" during the course of the evolution of the human species, he had at quite an early stage "gained the power of selection." Male preference had thus come to play a dominant role in the moulding of the female form and character.[40] More often than not, husbands and brothers chose whom the single women of their family were to marry. In this respect, Darwin's theory did nothing to challenge or change the conventional bourgeois sexual division of labour.[41]

Others wrote more extensively on this topic; among them was the essayist, ardent Darwinian, and psychologist Henry Maudsley (1835–1918). Maudsley wrote on the science of mind and mental development, as well as on mental pathology. He had experience working in asylums for lunatics and the insane but turned quickly to writing.[42] His works became foundational in the developing field of mental science. They influenced Darwin in the preparation of his 1872 book *The Expression of Emotion in Man and Animals*, but he was also deeply influenced by Darwin in turn. Maudsley adopted Darwin's ideas regarding sexual dimorphism and held a view of women that was quite in accord with those we have already outlined in Darwin's thinking on the matter. If anything, though, Maudsley's writing on women was even more demeaning, and he was challenged on this in public by the physician and suffragist Elizabeth Garret Anderson (1836–1917).[43]

Indeed, if we look at the writings of advocates of women's rights, whether they were suffragists or socialists of one kind or another, following the publication of *Descent,* it was almost routine for them to appeal to Darwin's theory in one way or another to support their own claims.[44] In some cases, they accepted his account that it was the relatively recent turn in 'civilized society' to refuse women's choice that had indeed resulted in women's evolution towards biological and, thus, social inferiority. It was thus necessary, they argued, to change the prevailing sexual and social arrangements so that women's true nature might be developed and equality might be achieved.

DARWINISM, COLONIALISM, IMPERIALISM, AND RACE

Darwin appealed to sexual selection not only to account for sexual differences but also the differences between human racial groups. Different groups, he surmised, had developed different standards of beauty, and the selection of mates had resulted in these preferences being carved into the bodies of both the male and the female of each racial group across many generations of selection.[45]

As recent scholarship has shown, although Darwin's family was deeply opposed to slavery and was active in the abolitionist movement, in many ways, Darwin was fairly conventional in his general beliefs about racial differences.[46] Like many other white, middle-class Europeans, he thought that his own class and culture were better than those of other nations and races. Thus, even though he was committed to a monogenist account of the development of all humans (that is, he believed that all humans shared a common ancestor – a view that was in contrast to the polygenism that predominated in the Anthropological Society of London), he thought that the non-European races were, on the whole, less evolved than those of northern Europe.

Where Darwin had appealed to the competition between tribes and communities to account for the development of human social and moral qualities, others chose to read this as implying that competition and war between racial, ethnic, and national groups was a good thing for the further development of society. As such, Darwin's name and ideas were often appealed to in ways that sought to justify colonial and imperial endeavours, and to underpin long-held beliefs about white supremacy.[47]

This series of primary source volumes on *Evolution in Nineteenth Century Britain* aims to show just how ubiquitous the ideas and language of evolution became throughout Britain and across the nineteenth century, but we should acknowledge here, too, that they had a much wider influence as well. As we aim to show, however, although scholars have long and quite rightly referred to a "Darwinian Revolution" following the publication and popularization of Charles Darwin's *On the Origin of Species* in November 1859, discussion of "transmutationist" or "developmental" ideas was rife throughout the nineteenth century. Indeed, and as we shall further explore through the first volume in this set *Evolution Before Darwin*, the discussion of evolution before Darwin was much wider than has conventionally been understood.[48] It has not been unusual, for instance, for histories of this period to acknowledge that Darwin had his forerunners. Buffon (1707–1788), Erasmus Darwin, Jean-Baptiste Lamarck, Herbert Spencer (1820–1903), and the anonymous author of the book *The Vestiges of the Natural History of Creation* are routinely discussed in this light. However, more recently, historians such as Adrian Desmond, James Secord, Paul Elliot, and Pietro Corsi, amongst others, have been working to show that discussion of evolutionary ideas was a much more widespread and popular phenomenon than this list of admittedly significant individuals alone suggests. Indeed, it is fair to say that Darwin's *On the Origin of Species* needs to be viewed merely as Darwin's contribution to an already ongoing and highly controversial social, cultural, and political, as well as scientific, debate. This is an important volume in this series, for rather than just providing an introduction to the discussion of developmental ideas before Darwin, it sets up the social and cultural framework within which and through which people encountered and read Darwin. The idea of transmutation already had a range of political and cultural meanings prior to 1859, and thus, we cannot really understand the debates that followed the publication of *Origin* without recognizing and appreciating this fact.

Our second and subsequent volumes turn more explicitly to Darwin and the publication, reception, and popularization of his work. And we reiterate that following the publication of *Origin,* it really does make sense to talk about a 'Darwinian revolution.' What will become clear is that even though the meaning of Darwinism was and remained very much contested throughout the nineteenth century, Darwin had been successful in capturing the ground from earlier theorists to the extent that his name became synonymous with the very idea of evolution. 'Darwinism' had been born! Further, even though Darwin's theories of natural and sexual selection each remained subject to criticism and question among scientific practitioners, particularly as to whether they were adequate to account for speciation, both natural selection and sexual selection – as well as the idea of a natural and inevitable "struggle for existence" and "progressive development" – came to dominate all aspects of Victorian culture, politics, and society. As a number of scholars have shown, this was so not only across literature but also in art, architecture, dance, and more.[49]

As readers will expect, in our second volume, *Evolution and Religion in Nineteenth-Century Britain*, we turn first to the perceived theological implications of Darwin's theory of evolution by natural selection.[50] In our introduction to that collection, we reiterate that we have to keep in mind that readers read and judged the religious implications of Darwin's contribution to the species question in light of their preconceptions of the implications of developmentalism more broadly. What we shall recognize, though, is that one of Darwin's more significant achievements was in managing to convert to his way of thinking a number of prominent and influential liberal theologians, who were able to help smooth the way for a wider acceptance of evolutionary ideas as quite compatible with religious belief in a way that earlier theorists just were not able to do. This cemented a compromise that brought an admittedly revised orthodoxy into line with the latest science at the, by this time, somewhat small cost of the idea of 'special creation.' We will chart the ways in which this compromise was undermined by Darwin's account of the evolution of both the moral sense and moral standards, and the debates that developed from this, in this second volume.

In our third and following volumes, we turn more overtly to the perceived political implications of evolutionary and "Darwinian" ideas. Volume three, *Evolution and Socialism in Nineteenth-Century Britain,* is dedicated to socialist Darwinism.[51] While there have been many studies on social Darwinism and the association that some of Darwin's contemporaries drew between Darwin's individualistic and competitive theory of natural selection and laissez-faire capitalist economics, we want to highlight that there was a wider diversity of political responses. Certainly, Marx and Engels were of the opinion that with *Origin,* Darwin had made a great contribution to their own political perspective, but following the publication of *Descent of Man* in particular, socialists of all shades argued that the theory of evolution gave a natural ground for socialism over capitalism and collectivism over individualism. As we show, however, socialists found that there were as many different ways of reading Darwin's work as there were varieties of socialism. It was not simply the case, though, that people bent Darwin to their own ends – although this is certainly a part of the story; rather, it was frequently the case that in embracing Darwin, socialists often found that they had to alter their political beliefs to fit with the biology they had committed themselves to.

Our fourth volume, *Evolution, Sex, Gender, and the Woman Question in Nineteenth-Century Britain*, expands on this topic to cover the impact that evolutionary ideas had on the discussion of women's place in society.[52] The "woman question," as it was termed, dominated Victorian political debate in the second half of the nineteenth century. The Industrial Revolution drove many social and demographic changes and the influx of many young women into rapidly expanding cities in search of work and, increasingly, education, political representation, and an active place in society. It was clear that with the formation of new classes, both proletarian and bourgeois, conventional ideas about gender and conventional gender roles could not stand. Evolutionary ideas about transformation, development, sex, and progress were thus a rich resource for those who sought to make both conservative as well as progressive claims about the place of women in society. As we shall see, not only does this volume include the voices of those who used biological arguments to make the case for female inferiority and thus substantiate existing social and sexual hierarchies, but also a number of prominent voices who argued the contrary. Perhaps predictably, we find the voices of capitalists and socialists, conservatives and reformers, and men and women on both sides of this question. Each appealed to evolutionary arguments to frame and defend their position. Here, we want to emphasize the voices of those, especially of those women, who appealed to Darwin and his ideas to find a political space for their own liberation. These sources are less well known and thus, we see this aspect of this collection in particular as a vital and valuable contribution to existing scholarship.

The fifth volume, *Evolution, Race, and Colonialism in Nineteenth-Century Britain*, engages with the ways in which evolutionary ideas were invoked to describe, categorize, and marshal the concept of race across the nineteenth century, and we see this collection as complementary to other collections in this series that speak specifically to the science of race in nineteenth-century Britain.[53] The fact that evolutionism often drew linear – even if sometimes branching – theories of species development meant that it was taken to speak to contemporary concerns and anxieties about the relationship between different racial, national, and ethnic groups of humans. In Britain, in particular, participants in the debates about the morality and legality of slavery in the first half of the century and about the rights and wrongs of imperialism and colonialism in the second routinely appealed to and invoked evolutionary ideas to make their case. Thus, in this volume, we look at Darwin's views on race, the earlier history of race science from the 1820s through to the 1850s, as well as the ways in which Darwin's work was later appealed to, appropriated from, and applied to re-create and reinforce biological categories of race from the 1870s onwards.

The nineteenth-century playwright, sometime Fabian socialist, and social commentator George Bernard Shaw (1856–1950) once remarked that what made Darwin so popular and Darwinism so ubiquitous was that "he had the luck to please everybody who had an axe to grind."[54] And, as we hope to have shown through these volumes, in a sense, Shaw was right. But beyond merely suggesting that Darwin's work was open enough to appeal to anyone with a cause, we also hope to have shown that the discourse of development, transmutation, and evolution became so ubiquitous that no one with a social or political agenda could afford not to embrace at least some aspects of Darwin's work, even when they reframed, muted, or denied others. "Darwin" and "Darwinism" became social and political signifiers that allowed their users (Darwin included) to speak with the authority of nature and natural law. In the context of social, cultural, and political arguments about how we might best live, how we might ensure social progress (and what that might look like, and for whom), and how we might guard against social degeneration (and what that might mean, and to whom), evolutionary language was at once irresistible and indispensable. By the third quarter of the century, it was certainly the case that both Darwin's name and the broad concept of 'Darwinism' had become an integral and necessary part of British nineteenth-century culture.

Notes

1 Charles Darwin, *On the Origin of Species by Means of Natural Selection; or the Preservation of the Favoured Races in the Struggle for Life* (London: John Murray, 1859).
2 Charles Darwin, *The Descent of Man and Selection in Relation to* Sex (London: John Murray, 1871).
3 Michael Ruse, *The Darwinian Revolution. Science Red in Tooth and Claw* (Chicago: University of Chicago Press, 1977) is perhaps the most classic example.
4 Caden Testa and Piers J. Hale, *Evolution Before Darwin* (London: Routledge, 2025).
5 William Godwin, *An Enquiry Concerning Political Justice and Its influence on General Virtue and Happiness*, 2 vols. (London: G.G. J & J Robinson, 1793).
6 Mary Wollstonecraft, *A Vindication of the Rights of Woman with Strictures on Political and Moral Subjects* (Boston: Peter Edes 1792).
7 Charles Darwin, *Journal of Researches into the Natural History and Geology of the Countries Visited During the Voyage of H.M. S. Beagle Round the World, Under the Command of Capt. Fitz Roy, R.N.* (New York: Harper & Brothers, 1846), pp. 145–6; Charles Darwin, *On the Origin of Species by means of Natural Selection, or the preservation of the favoured races in the Struggle for Life*, (London: John Murray, 1859), p. 1.
8 Henri Marie Ducrotay de Blainville, "Analyse des principaux travaux dans les sciences physiques, publiés dans l'année 1821," *Journal de phyique*, tome XCIV, p. 54. De Blainville coined the term specifically to describe the work of Georges Cuvier in this area of study.
9 Martin J.S. Rudwick, *Bursting the Limits of Time: The Reconstruction of Geohistory in the Age of Revolution* (Chicago: University of Chicago Press, 2005); Martin J.S. Rudwick, *Worlds Before Adam. The Reconstruction of Geohistory in the Age of Reform* (Chicago: University of Chicago Press, 2008).
10 Charles Lyell, *Principles of Geology*. 3 vols. (London: John Murray, 1830–33).
11 Georges Cuvier, *Essay on The Theory of the Earth* (Edinburgh: William Blackwood, 1827).
12 Bill Jenkins, *Evolution Before Darwin: Theories of the Transmutation of Species in Edinburgh, 1804–1834* (Edinburgh: Edinburgh University Press, 2021).
13 Bill Jenkins, "Race before Darwin: Variation, adaptation and the natural history of man in post-Enlightenment Edinburgh, 1790–1835" *British Journal for the History of Science*, (2020), 53 (3): 333–350. Evelleen Richards, "The 'Moral Anatomy' of Robert Knox. The Interplay Between Biological and Social Thought in Victorian Scientific Naturalism," *Journal of the History of Biology* (1989), 22, (3): pp. 373–436. B. Ricardo Brown, *Until Darwin. Science, Human Variety, and the Origins of Race* (London: Pickering and Chatto, 2010).
14 Caden C. Testa and Piers J. Hale, "Mapping evolutionary ideas in Britain in the first half of the nineteenth century," in Caden C. Testa and Piers J. Hale, (eds.), *Evolution Before Darwin* (London: Routledge, 2025).
15 Adrian Desmond, *The Politics of Evolution. Medicine, Morphology, and Reform in Radical London* (Chicago: University of Chicago Press, 1989), pp. 213–20; Piers J.

Hale, *Political Descent. Malthus, Mutualism and the Politics of Evolution in Victorian England* (Chicago: University of Chicago Press, 2014).

16 [Robert Chambers], *The Vestiges of the Natural History of Creation* (London: John Churchill, 1844).

17 James Secord, *Victorian Sensation. The Extraordinary Publication, Reception, and Secret Authorship of The Vestiges of the Natural History of Creation* (Chicago: University of Chicago Press, 2000).

18 Secord, *Victorian Sensation;* Janet Browne, *Charles Darwin. Voyaging. A Biography* (Princeton N.J., Princeton University Press, 1995), pp. 462–65.

19 Adrian Desmond and James Moore, *Darwin. The Life of a Tormented Evolutionist* London & New York: W. W. Norton & Co., 1991).

20 Robert J. Richards, *Darwin and the Emergence of Evolutionary Theories of Mind and Behavior*, Chicago: University of Chicago Press, 1987; John van Wyhe, "Mind the Gap: Did Darwin Avoid Publishing his Theory for Many Years?" *Notes and Records of the Royal Society* (May 2007), 61, (2): pp. 177–205.

21 John Stuart Mill, *Autobiography of John Stuart Mill* (New York: New American Library, 1964), p. xii.

22 Nora Barlow, (ed.), *The Autobiography of Charles Darwin 1809–1882* (London: W. W. Norton & Co., 1958), p. 120.

23 Darwin, *Origin*, 1859, p. 62–3.

24 Thomas Henry Huxley, "Darwin On the Origin of Species," *Westminster Review*, 73, no.144 (1860), pp. 541–70.

25 Karl Marx to Friedrich Engels, June 18, 1862, in Karl Marx and Friedrich Engels, *Karl Marx and Friedrich Engels: Collected Works*, 50 vols. (New York: International Publishers, 1975–2004), 41:381.

26 Piers J. Hale, "Darwin's Other Bulldog: Charles Kingsley and the Popularisation of Evolution in Victorian Britain," *Science and Education*, (2012), 21, pp. 977–1013.

27 Eric J. Evans, *The Great Reform Act of 1832* (London: Routledge, 1994).

28 Piers J. Hale, "Charles Darwin, sexual selection, and the evolution of other-regarding ethics" *British Journal of the History of Science, Themes*, 6, (2021), pp 157–177.

29 Darwin, *On the Origin of Species by Means of Natural Selection, or the Preservation of the Favored Races in the Struggle for Life* (London: John Murray, 1860), p. 490.

30 Frances Power Cobbe, "Darwinism in Morals," *Theological Review: A Quarterly Journal of Religious Thought and Life* 8, no.33 (1871), pp. 167–192.

Charles Kingsley, "President's Address," *Report and Transactions of the Devonshire Association for the Advancement of Science, Literature and the Arts* 4, no.1, (1871), p. 383.

For further commentary see Piers J. Hale, "Darwin, Design, and the Honey Bee. The Morality of the Hive," *American Bee Journal*, (Oct. 2020), pp. 1103–1107.

31 Bernard Lightman, "Darwin and the Popularization of Evolution," *Notes and Records of the Royal Society* 64 (2010): pp. 5–24.

32 For more on the Metaphysical Society see Alan Willard Brown, *The Metaphysical Society. Victorian Minds in Crisis, 1869–1880* (New York: Columbia University Press, 1947); and more recently Catherine Marshall, Bernard Lightman, & Richard England, (eds.), *The Metaphysical Society (1869–1880). Intellectual Life in Mid-Victorian England* (Oxford: Oxford University Press, 2019); also see Caden C. Testa and Piers J. Hale, *Evolution and Religion* (London: Routledge, 2024).

33 Darwin, *Descent*, (1871), chapter 3.

34 Piers J. Hale, *Political Descent. Malthus, Mutualism, and the Politics of Evolution in Victorian England (*Chicago: University of Chicago Press, 2014), chapter 5.

35 Peter Kropotkin, *Mutual Aid. A Factor in Evolution*, (London: Freedom Press, 1902).

36 Caden C. Testa and Piers J. Hale, (eds.), *Evolution and Socialism* (London: Routledge, 2025).
37 Darwin, *Origin*, 1859, pp. 87–90.
38 Darwin, *Descent*, 1871, vol. 2, p. 316.
39 Darwin, *Descent*, 1871, vol. 2, pp. 326–9.
40 Darwin, *Descent*, 1871, vol.2, p. 371.
41 For a thorough study of the social, cultural and scientific development of Darwin's theory of sexual selection see Evelleen Richards, *Darwin and the Making of Sexual Selection* (Chicago: University of Chicago Press, 2017).
42 For more on Maudsley see Daniel Pick, *Faces of Degeneration. A European Disorder, c.1848–1918* (Cambridge: Cambridge University Press, 1989); Elaine Showalter, *The Female Malady. Women, Madness and English Culture 1830–1980* (London: Virago, 1987), see chapter 5 especially.
43 See Caden C. Testa and Piers J. Hale, *Evolution and the Woman Question* (London: Routledge, 2025).
44 See Testa and Hale, *Evolution and the Woman Question.*
45 Darwin, *Descent*, 1871, vol.2, pp. 343–354.
46 Darwin, *Descent*, 1871, vol. 1., pp. 214–250; Adrian Desmond and James Moore, *Darwin's Sacred Cause: How a Hatreds of Slavery Shaped Darwin's views on Human Evolution* (New York: Houghton Mifflin Harcourt, 2009).
47 Caden C. Testa and Piers J. Hale, (eds.), *Evolution, Race, and Colonialism in Nineteenth-Century Britain* (London: Routledge, 2025).
48 Caden C. Testa and Piers J. Hale, *Evolution Before Darwin* (London: Routledge, 2025).
49 Bernard Lightman and Bennett Zon, (eds.), *Evolution and Victorian Culture* (Cambridge: Cambridge University Press, 2014).
50 Caden C. Testa and Piers J. Hale (eds), *Evolution and Religion in Nineteenth-Century Britain* (London: Routledge, 2025).
51 Caden C. Testa and Piers J. Hale (eds.), *Evolution and Socialism in Nineteenth-Century Britain* (London: Routledge, 2025).
52 Caden C. Testa and Piers J. Hale (eds.), *Evolution, Sex, Gender, and the Woman Question in Nineteenth-Century Britain* (London: Routledge 2025).
53 Caden C. Testa and Piers J. Hale, (eds.), *Evolution, Race, and Colonialism in Nineteenth-Century Britain* (London: Routledge, 2025); see also, for example, the volumes on race in nineteenth-century science edited by B. Ricardo Brown.
54 George Bernard Shaw, 'Preface' to *Back to Methuselah*: *A Metabiological Pentateuch* (London: Constable and Company Ltd., 1921).

INTRODUCTION TO VOLUME I: EVOLUTION BEFORE DARWIN

Mapping Evolutionary Ideas in Britain in the First Half of the Nineteenth Century

The history of the development of evolutionary ideas has focused largely on Charles Darwin for a good reason. His theories of natural and sexual selection, which he developed across both *On the Origin of Species* (1859) and *The Descent of Man* (1871), have become central to modern biology. Historians frequently refer to the period that followed the publication of *Origin* as 'the Darwinian Revolution' precisely because Darwin's evolutionary ideas had a lasting impact not only upon science but on so many other aspects of Western culture and society. Indeed, it has frequently been claimed that the extent of the change that Darwin's work effected in the dominant Western conception of what it meant to be human is arguably unmatched in the history of Western thought and is akin to the scale of change in worldview brought about by Copernicus.

However, despite this emphasis, historians have long known that Darwin was far from the only person to suggest that life had evolved, and neither was he the first to do so. This fact clearly undermines the view that is commonly held by the wider public that before Darwin published *On the Origin of Species* in 1859, everyone thought that each species existed just as it had been since the Creation. Indeed, in light of what geologists already knew about the earth before 1800, the notion that before Darwin, everyone believed that the earth was only 6,000–10,000 years old (depending on how one counted the various 'begats' recorded in *Genesis*) is also untenable. By the early 1800s, geologists were already proposing that the earth must be many hundreds of thousands of years old, if not older, and in the first volume of his three-volume *Principles of Geology* (1830–1833), the Scottish geologist Charles Lyell (1797–1875) revived the uniformitarian and gradualist ideas of his countryman James Hutton (1726–1797) to argue that the age of the earth should be counted in millions and not simply in hundreds of thousands of years.[1] Further, though, and undermining the historical focus on Darwin, the historian of science Peter J. Bowler has long argued that it is not as certain as is often presumed that even after 1859, Darwin's ideas about the origin of new species were the most influential of the various evolutionary schemes that had already been proposed. It is in light of this that Bowler has claimed that we can better understand the post-Origin years in terms of a "non-Darwinian Revolution." Naturalists did embrace evolution but often remained skeptical of natural selection.[2]

While not all historians agree with Bowler's "non-Darwinian" terminology, or all he means it to imply, it is certainly the case that elements of pre-Darwinian theories of development persisted long after Darwin published his own work – indeed, aspects of Darwin's own argument contributed to this phenomenon.[3] Thus, there is good reason to revisit these earlier theories not just to contextualize Darwin's later achievement – although consideration of these earlier theories and theorists is essential if we are to appreciate the deeply moral and political implications that transmutation was taken to have across the nineteenth century – but also in order to really get an appreciation of the range of developmentalist ideas that were theorized, published, and debated in the decades before Darwin.

It is this last point that we want to emphasize here, for it informs the methodological approach we want to emphasize in this collection and to advocate for among our readers. We read these sources not simply as the publications of significant precursors to Darwin but rather as writers whose texts were not just published but popularized and became popular. In saying this, we recur to the important 1994 article by Roger Cooter and Stephen Pumfrey in which they appealed to their colleagues in the history of science to take the study of the popularization of science seriously and to think too about what it might mean to study science in, and science as, popular culture.[4] Since the publication of Cooter and Pumfrey's essay, there has been a whole swath of scholars who have turned to the study of science popularization but also to the study of science in and as popular culture, and we read these texts in light of this historiographic approach. We are thus indebted to the decades of work by scholars who have developed these approaches to our discipline. In following their lead, we gain a much richer account of the history of evolution which reveals the depth of the social, political, and cultural importance of evolutionary ideas across this period.

Through this introduction and the headnotes provided for the sources in this collection, we thus seek to put these texts into context. We recognize that they were each written as a contribution to an ongoing debate about the nature of the human condition, and reconstructing this debate – and its social, political, and cultural implications – is a necessary part of making sense of them. Further, such contextualization allows us to see more fully what was at stake for the writers of these selections as well as for those they wrote for, and against. We also recognize that the study of the popularization of the ideas in these texts calls for more than thinking in terms of the mere dissemination of the arguments and ideas contained therein, but rather often in their appropriation, reframing, and rearticulation, often in ways over which the original author had little or no control.

We have divided this work into largely chronological sections, but before proceeding on our broad historical narrative, by way of a preface, we introduce this collection with excerpts from Alfred Lord Tennyson's poem *In Memoriam*. We do so for several reasons. First, Tennyson's poem, which preceded Darwin's *Origin* by almost a decade, was rife with evolutionary ideas. More to the point, though, Tennyson gave voice to the cultural, social, and metaphysical anxieties that came along with a full acceptance of a thoroughly natural history of mankind and the world we inhabit. The anonymous *Vestiges of the Natural History of Creation* combined with the deep time of Lyell's *Principles of Geology* provided the context in which an evolving nature cared nothing for the individuals, or indeed the species, that came into being and passed away across the long march of geological time. As the historian of science James Secord has suggested, "for thousands of readers, Tennyson's *In Memoriam* offered the most profound integration of the evolutionary narrative into everyday experience,"[5] and this is something we need to keep in mind as we read the sources provided here. Although the majority of the texts we have provided here are written as non-fiction and framed more as what modern readers might more readily identity as 'science' than as 'literature,' we need to look too for evidence of evolution's impact in wider social, political, and literary culture if we are to grasp the importance that transmutationist or evolutionary ideas had upon Victorian society. Following our preface, we turn our attention to transmutationist ideas in Enlightenment thought and then on to how these ideas were received, interpreted, and often appropriated in Edinburgh in the 1820s, a particularly vibrant time in the intellectual discussion and debate in Scotland's capital city. In our third section, we turn to the articulation of, and debate over, the 'developmental hypothesis' in mid-Victorian England. After the 1832 Reform Act, British politics and society changed significantly, and so too did the debate about transmutation. Finally, in a fourth and shorter section, we include a selection of authors who only later came to notice as having anticipated Darwin. Darwin acknowledged them in the 'historical sketch' that he wrote as a preface to the third and subsequent editions of *Origin*. We close this essay with a further brief commentary on our methodological approach.

EVOLUTIONISM IN THE ENLIGHTENMENT

Although the focus of much of the literature on the history of evolutionary thought has almost always been on Darwin, even the earliest scholarly accounts of this history have sought to contextualise Darwin's work by including at least brief discussion of those who framed earlier evolutionary accounts of life's history. Indeed, the move during the eighteenth-century Enlightenment to champion the almost limitless possibilities of human reason and to pursue an empirical investigation of the natural world led to theories of organic change over half a century before Darwin's work was published. As a number of scholars have pointed out, the contemporary discoveries and theories of mineralogists and geologists regarding a dynamic, rather than a static, earth history seemed to imply the necessity of a dynamic history of life as a matter of logic.[6] However, as Peter Bowler has argued, the materialist philosophies propounded by Enlightenment thinkers were often constrained by their Cartesian insistence that animals were mere machines, as well as by their rejection of what they supposed to be unsubstantiated and unscientific appeals to supernatural 'vital forces' of generation. Indeed, even when the crucial question of reproduction or 'generation' was considered, it was almost always invoked as the mechanism that perpetuated stability rather than as a dynamic driving force of biological novelty and change, and Bowler thus contends that their thinking was of quite a different and limited character when compared to later developmentalist thinking.[7]

As we shall see in the works presented here, though, there were thinkers who pressed beyond this framework. The work of the French naturalist Georges Louis Leclerc, the Comte de Buffon (1707–1788), was clearly suggestive of some belief in the natural development of species through adaptation to changed environmental circumstances, for instance. Across the channel in England, the British naturalist, philosopher, and physician Erasmus Darwin (1731–1802) framed reproduction as a driving and productive force of speciation and development. In works such as *The Botanic Garden* (1789; 1791), *Zoonomia; or The Laws of Organic Life* (1794), and *The Temple of Nature* (1803), he developed and laid out his transmutationist theory. He wrote racy poetry about the sex lives of plants in defence of the Linnean system of classification and speculated upon an evolutionary origin and development of life and society.[8]

Other Enlightenment thinkers who invoked the idea of species change did appeal to a vital life force as the motive power that drove natural development, although they were clear in trying to distance their own ideas from discredited earlier vitalist thinking. Here, the German physician, physiologist, and anatomist Johann Friedrich Blumenbach (1752–1840), the French invertebrate specialist and botanist Jean-Baptiste Lamarck (1744–1829), and his colleague the anatomist Étienne Geoffroy Saint-Hilaire (1772–1844) merit our consideration. Blumenbach framed the idea of a "*Bildungstrieb*," a developmental drive to explain the developmentalism seen in embryology and across the natural world more broadly, whereas Lamarck appealed to a "*sentiment intérieur*" as the animating force of development and adaptation.[9] Geoffroy supported and expanded Lamarck's theory but gave emphasis to the divergent development in embryo.

Of the three, Lamarck has generally been given the most attention, especially in the context of the reception of evolutionary or transmutationist ideas in England, for he not only articulated a protracted argument in which he suggested that present-day species had descended from earlier kinds and clearly explained the mechanism by which he believed they had done so, but his work was also widely commented on and popularized in England, a point to which we shall return shortly.[10] Neither Blumenbach nor Geoffroy, though, was without influence.[11]

Lamarck devised his evolutionary or 'transmutationist' theory in the 1790s and brought it together most clearly in his 1809 book *Zoological Philosophy*. His views have frequently been summarized to emphasize his belief in the 'inheritance of acquired characters' – the idea that adaptations acquired in an individual organism's lifetime could then be transmitted to their offspring; however, as scholars have shown, and as we shall indicate here, his ideas were more complicated than this. Indeed, this was very much a secondary element of his understanding of transmutation. Rather, he believed that once organisms had come into being through a spontaneous generation, they were subject to an 'internal sentiment' (*sentiment interieur*) that drove their development from their simple beginning to an ever-greater complexity. This generally linear development was then subject to direct adaptation resulting from the contingencies of environmental circumstances, and these acquired characteristics might then be inherited by their offspring.[12] As Caden Testa has shown, although historians of evolutionary ideas have often tended to look at Lamarck's theory of development in isolation from his other work and overall aims, his transmutationist ideas, along with his ideas about health, agriculture, and meteorology, were in fact a part of Lamarck's broader contribution to ongoing debates about how best to regenerate the health and prosperity of the French nation.[13] In the context of these debates, it had been quite common to propose means of biologically improving or perfecting human bodies as a way of enacting various systematic reforms that would address broad social and political problems in France. Such hopes of improving the human species were deeply tied to Enlightenment-era ideas about the progress and perfectibility of humanity, as well as suggestions that biological interventions might provide the answers to the political upheaval France faced throughout the decline of the Old Regime and the revolutionary years. Some of Lamarck's direct intellectual influences, including both Buffon and Condorcet, had also made such suggestions about the relationship between biology and politics, arguing that interventions ranging from encouraging public health and hygiene to selectively 'breeding' human beings would improve the species and ultimately allow for the creation of a better and more enlightened nation.

Other and earlier French philosophes, such as the ardent materialist Julien Offray de La Mettrie (1709–1751) and Pierre Louis Moreau de Maupertuis (1698–1759), also argued for developmentalism of a sort. Maupertuis's 1745 work *Venus physique* was particularly important in this regard, as was his later *Système de la nature* (1751). He not only advanced a theory of embryological development that suggested it as a source of novelty and creative development, but applied his ideas to the origin and development of humans and of the human races. Important here too were the writings of the French philosopher and mathematician Nicolas de Condorcet (1743–1794). Condorcet's *Sketch of a Historical View of the Progress of the Human Spirit* (1795) quickly became one of the most influential works of the French Enlightenment period and underpinned a prevailing faith in human progress as a result of science, learning, and reason. Written in 1794 while Condorcet was in hiding amidst the turmoil of the revolution, this work was published both in England and in France in 1795, achieving wide circulation. Alongside William Godwin's *Inquiry Concerning Political Justice* (1793), it quickly became a major reference point in English political radicalism. Like Condorcet, William Godwin (1756–1836) argued that the application of science and learning to the human condition would ensure progress and end poverty and injustice. It was this optimism that underpinned his faith that humankind could win sufficient resources from nature to ensure that no one went hungry and that mankind could live under social and political conditions of equality – an idea that Godwin's partner, Mary Wollstonecraft (1759–1797), had made clear in her 1792 work *A Vindication of the Rights of Woman* also applied to women.[14]

The radical political connections to transmutation and developmentalism need to be stressed because they were not only a central part of the French calls for *liberté*, *egalité*, and *fraternité*, but, as the historian of science Adrian Desmond has pointed out, also of the ambitions and rationale of a significant faction of the English political radical movement.[15] Such was the popularity and influence of both Condorcet and Godwin that Thomas Robert Malthus was prompted to put pen to paper to counter their arguments in his 1798 work *An Essay on the Principle of Population*. It was the 6th edition of this book that Darwin famously read in September 1838, and that immediately prompted him to recognize that the incommensurable ratio between population growth and the increase in available resources would lead to a struggle for existence that would have telling consequences for the different varieties of a species. Our concern here, though, is to point out that Malthus's work – and thus Darwin's theory, in which he vociferously acknowledged his debt to Malthus – had its own deeply political associations.[16]

EDINBURGH TRANSMUTATIONISTS

Both in his Introduction to *Origin* and even more so following its publication in November 1859, Darwin was keen to distance his own theory of evolution by means of natural selection from earlier ideas and from those of Lamarck in particular. This was a line of argument he maintained in his autobiography, asserting that the works of his grandfather, of Lamarck, and of other theorists had little impact upon his own thinking.[17] Historians have been reluctant to accept Darwin's dismissal of these earlier thinkers quite so readily, however, and for good reason. As we make clear here, the ideas not only of Erasmus Darwin and Lamarck, but also of Lamarck's colleague Étienne Geoffroy Saint-Hilaire and of Blumenbach, were widely circulated and discussed in Edinburgh throughout the 1820s and 30s, exactly when Darwin was in the city studying medicine. Not only was Darwin's mentor, the naturalist and political radical Robert Grant (1793–1874), deeply impressed by Lamarck and Erasmus Darwin's ideas, but so too, it seems, was the geologist Robert Jameson (1774–1854), under whom Darwin also studied.[18] Further, radical, materialist, and developmentalist phrenology was also discussed – and condemned – at the meetings of the Plinian Society, the scientific society at which Darwin read his first research paper.[19] Edinburgh was also home to the anatomist Robert Knox (1791–1862). Knox was the most popular tutor for Edinburgh students seeking classes in anatomy. He had studied extensively in Europe, including with both Georges Cuvier and Geoffroy Saint-Hilaire, and he brought the insights of transcendental anatomy back into his Edinburgh classroom and into his published research. He later spelt out the main thrust of his ideas on this topic in his "Contributions to the Philosophy of Zoology," which appeared in *The Lancet* in 1855.[20] Following Geoffroy, he saw the origins of new species to lie in embryological developments and the adaptation of organisms to a changed climate, alterations in diet, and crossing of varieties. Knox was at one time reckoned to be the most famous transmutationist in Britain, although he later became more widely known for his involvement in the infamous Burke and Hare murders. Knox, like many of the French Enlightenment thinkers, was interested in the question of the origin and mutability of the human races. He presented papers on this topic from the 1820s in which he argued that all the human races were of a common origin and had descended from the original Caucasian race.[21] Later, he wrote The *Races of Men* (1850, second revised edition 1862), in which he wrote extensively on the topic, and described what he took to be a biological racial hierarchy.[22]

Scholars have long recognized the deeply evolutionary bent of Edinburgh's scientific community. Such ideas were not uncontroversial, of course, as Darwin saw first-hand at the Plinian Society (indicated above). But there were also a number of papers published in the *Edinburgh New Philosophical Journal* in the 1820s that clearly lent support to French evolutionary ideas. For many years, historians thought that the most explicit of these was anonymously authored by Jameson; however, recent research by Pietro Corsi has now established that this was not the case.[23] Rather, the essay in question, "Observations on the Nature and Importance of Geology," which favourably presented Lamarck's transmuationist ideas, was a translation of a report of a recent scientific meeting in France.[24] This is suggestive of the extent to which Edinburgh was in touch with debates about anatomy, physiology, and transmutation in Europe as they were unfolding. As Corsi points out, this translation appeared only days after the original French account of the meeting had appeared.[25]

THE 'DEVELOPMENTAL HYPOTHESIS' IN MID-VICTORIAN ENGLAND

It is a common assertion amongst scholars of English nineteenth-century history that the spectre of the French Revolution hung over English politics and society throughout the century, and transmutationist ideas were intimately associated with Francophile revolutionary politics. This was certainly the case in the 1820s but remained so even after the 1832 Reform Act extended the franchise to male heads of households with a significant income.[26] The social historian E. P. Thompson has long since argued that it was the passage of the 1832 Act that was the last stage in the 'making of the English working class,' and although this claim has been much debated, it is certainly clear that it did usher in the political making of the English middle class.[27]

As James Secord has argued in his study of the anonymous publication and reception of the evolutionary pot-boiler *The Vestiges of the Natural History of Creation*, by the 1840s, writers were framing transmutationist ideas in such a way as to appeal to the political and cultural preferences and prejudices of this newly enfranchised and increasingly hegemonic middle class.[28] *Vestiges*, published in 1844 by the London medical publisher John Parker, was the work of the Scottish publisher and editor of *Chambers Journal* Robert Chambers (1802–1871). As Secord has shown, the book made transmutation the talk of the town a decade and a half before *On the Origin of Species* was published. As Secord notes, we can look to *Vestiges* as an example of the importance of our understanding of the production and reception of printed works in the period we are interested in. The middle years of the nineteenth century not only saw a new industrial middle class gain political representation through the 1832 Reform Act, but as it came to economic prominence, it also established its own unique culture, a middle-class ideology that claimed to be universal and natural. As scholars such as Jürgen Habermas have long noted, the bourgeois public sphere was centred upon the written word and its reading.[29] Increasing literacy rates across this expanding social class coincided with and drove developments in print production that saw cheaper paper and new methods for bookbinding, as well as new and faster modes of distribution. The expansion of the rail network brought news, literature, fashions, and ideas from London and the continent to the provincial towns across the nation, and lending libraries originated in this period, as did the middle-class soiree and conversazione – social society gatherings at which the great and the good (and the aspiring) came together, and at which the discussion and demonstration of knowledge and opinion on current literature was the mode of social discourse. As Secord describes, *Vestiges of the Natural History of Creation* was one of the main topics of social debate for more than one season. As we have mentioned, the book was published anonymously, and this fact only enhanced its popularity. The mystery of the identity of the author itself became a major element of discussion of the book. What it also did was ensure that the topic of species change was well debated and extensively popularized across the nation throughout the 1840s. Echoes of Chambers' work found their way into contemporary literature as well as other aspects of Victorian culture. The very idea of progress, which had been so challenging when uttered by French materialist radicals and their English allies, was quickly appropriated to fit with the progressive development of middle-class ambition, industrialization, and the framing of liberal capitalist economics. *Vestiges* was met with an immense amount of criticism from the conservative and religious scientific establishment, some of the most notable of whom – William Whewell and Adam Sedgwick, for instance – had taught Darwin at Cambridge once he had left Edinburgh and the study of medicine behind, and moved south to study with the intent of becoming a country parson. We should bear in mind not only that Darwin, who had begun working out his own evolutionary ideas in the late 1830s after returning from his voyage on H.M.S. *Beagle*, witnessed this response with interest and read reviews of *Vestiges* carefully, but also that all of the history we have described up to this point coloured how Darwin's readers approached his book when he did eventually publish in 1859.

While it would be interesting to include more detailed commentary on those who were critical of *Vestiges*, their arguments, and their motivations – it is notable, for instance, that Thomas Huxley was deeply critical of the book – there is not the space to do so. It would also be useful to give greater consideration to those who were sympathetic to the book. The physiologist William Benjamin Carpenter not only wrote an (anonymous) favourable review of the book but actually had a hand in helping to revise and improve later editions of the text.[30] Further, though, we should also be mindful of the broad impact of the work, not only on middle-class readers, but also among artisans and working men and women. As Roger Cooter and Stephen Pumfrey long ago noted, this makes us have to think about the full range of political and moral meanings that popular science could have.[31] *Vestiges* made many converts to evolution, and exactly how these readers interpreted, appropriated, and re-conceptualized Chambers' work is a vital part of the social history of the history of science that we need to try to recapture.

Vestiges was not the last major evolutionary work published before *Origin* appeared, though, even if it was immensely popular. Across the mid-1840s and into the 1850s, the young middle-class radical journalist and philosopher Herbert Spencer (1820–1903) put pen to paper to publish his own ideas about the significance of evolution for humankind. Spencer, whose name has subsequently (although not unproblematically) become associated with the philosophy of 'social-Darwinism,' published numerous essays and books on the topic of evolution, or 'the developmental hypothesis.' Spencer understood the mechanism of speciation to be very much akin to that outlined variously by Erasmus Darwin and by Lamarck in particular; he, too, emphasized adaptation to the environment and the subsequent inheritance of these acquired adaptations to circumstance. After 1860, he incorporated Darwin's theory of natural selection into his own work but did so in such a way as not to unsettle his own already-developed ideas. Spencer is very much worth our attention given that, although it was Darwin's name that became synonymous with evolution following the publication of *Origin*, after 1860, Spencer quickly became much more influential, not only in Britain but around the globe. Indeed, the extent of Spencer's international influence in the reception of 'Darwinism' has only relatively recently begun to be appreciated.[32]

Spencer was first led to think seriously about the origin of species as a result of his early work as an engineer on the expanding railways that were being cut through the hills and vales of the British countryside throughout this period. He became interested in the fossils that came to light as he worked, and picked up a copy of Charles Lyell's recently published *Principles of Geology*. Lyell's work, as we know, was significant in that he argued that one could best understand Earth's history by accepting the premise that the evidence left to us in the strata of the Earth had been laid down by forces that were consistent with those that we see in operation today – i.e., that natural laws were consistent in their operation. This impressed Spencer not only in terms of its implications for how he read the mineral and fossil evidence he uncovered, but also impressed on him the belief that the whole universe must, of necessity, be subject to uniform, overarching natural laws. In the second volume of *Principles of Geology*, though, Lyell had done his best to reject the transmutationist arguments of Geoffroy and Lamarck and had likely felt the need to do so because his own studies clearly lent themselves to reading earth history as one of progressive development – a belief in the uniformity of law would imply that the fossil record could be read as a record of progressive development. Such conclusions were too troubling for Lyell to entertain because of what he saw as their disturbing theological implications. He therefore resisted the idea that earth history was progressive, contending instead that it was cyclical (he was subsequently mocked for the implications of his thinking on this topic as suggesting that one day in the future, dinosaurs would return once again to the earth).

Ironically, it was because Lyell did not want his readers to be led towards transmutationist conclusions that he included such a detailed account of Lamarck's argument, and as a result, it became one of the most widely read and fullest accounts of Lamarck's work available to English readers. Spencer was far from alone in developing a 'decided leaning' towards Lamarck's theory as a result of reading Lyell – indeed, Darwin had a similar experience. Further, as the historian Anne Dewitt has recently noted, Lyell's account of Lamarck was embraced and rehashed by freethinkers and published across the atheist press.[33]

Reading Lyell was not Spencer's first encounter with the developmental hypothesis, though, for as he noted in his autobiography, he had "during previous years been cognizant of the hypothesis that the human race has been developed from some lower race."[34] As the historian Paul Elliott has suggested, this was doubtless a result of the many hours Spencer had spent in the library of the Derby Philosophical Society, which held not only works by Erasmus Darwin but also works by – and which contained accounts of the work of – a number of transmutationist continental anatomists.[35] It is also important for us to note that Spencer was interested in transmutation as it applied to humans from the first. This was the case for Darwin, too, of course, but unlike Darwin, Spencer made this central to his writing, where Darwin hedged and delayed engaging with the implications of evolution for humans until he published *Descent of Man* in 1871.[36] Even though Darwin's transmutation notebooks reveal that he too was thinking of the implications of evolution for humankind as the central question, in *Origin,* he stated only that in light of his theory, "Light will be thrown on the origin of man and his history," and this in the very last pages of the book.[37]

It is worth saying a little more about Spencer here because he has so frequently been mischaracterized. Despite some decades of solid scholarship on Spencer, the perception of him as a social Darwinist – in the sense that he took the competition at the heart of Darwin's theory and applied it to society – remains prevalent. This is in large part a result of the immense popular success of Richard Hofstadter's 1944 book *Social Darwinism in American Thought*, which in all sold more than 200,000 copies.[38] As we can already see, though, the chronology of Spencer's evolutionism makes this interpretation problematic, and, as we have noted here, even when Spencer did incorporate Darwin's theory of natural selection into his own, it remained very much subordinate to his own very Lamarckian understanding of adaptation to environment. Spencer's evolutionism was certainly very much in service to his political views, but rather than being the simple endorsement of unrestrained capitalist competition that Hofstadter suggests, Spencer believed that natural adaptation needed to be left to its own devices in order to achieve a state of natural and social equilibrium. The competition and struggle central to Darwin's theory was, in Spencer's view, the consequence of maladaptation, rather than the mechanism that drove adaptation. It was this theory of biological development that underpinned Spencer's anti-statist political radicalism.

Spencer first published his political views in a series of letters published in the radical journal *The Nonconformist* in the early 1840s. He later brought these together and published them as a pamphlet in 1843 under the title *The Proper Sphere of Government*. His first major work was *Social Statics*, published by John Chapman in 1851, and he first spoke out explicitly on his embrace of transmutationist ideas in "The Developmental Hypothesis," published in *The Leader* the following year. Other essays on the "Social Organism" and the book *Principles of Psychology* (1855) followed. Here, he more explicitly applied his ideas of evolutionary development to the human mind and society. Indeed, rather than endeavoring to establish a natural argument for competition, the historian Robert J. Richards has argued that Spencer was, in fact, trying to answer the question of how an ethical and moral society could best be developed.[39] As Richards has shown, however, although Spencer's early works, *Social Statics* and *Principles of Psychology*, impressed the small group of intellectuals who surrounded the published John Chapman's *Westminster Review*, it was only after the publication of *Origin* that his work became truly popular and gained a mass audience.[40]

Between 1862 and 1893, Spencer went on to write his massive *A System of Synthetic Philosophy*, a series of works in which he applied his ideas about the implications of his theory of a universal gradual development to biology, sociology, psychology, and ethics. This contributed to his international reputation, and he sold well over a million books during his lifetime. His ideas were popularized and influenced many of his contemporaries around the globe as well as in Britain, not only across the sciences but in politics and literature as well.

It was in the context of the wide and enduring debate about the developmental hypothesis and its implications for humanity, religious belief, and politics that the theologian and mathematician Baden Powell published *Unity of Worlds*. The book was originally published in 1855, and in it, Powell took issue with the standard Christian narrative pertaining to both the age of the earth and the fixity of species. Although Powell maintained that science and theology were quite separate endeavours, it is clear too that he believed that they could and should be consistent. Accepting the basis of Lyell's claims in the first volume of his *Principles of Geology*, he not only concluded that any claim to a young earth must be erroneous but so too, in light of Lyell's arguments about the uniformity of natural law, he controversially rejected the very idea of a miracle as inconsistent with sound theology. In contrast to Lyell on the mutability of species, though, Powell was clearly open to the idea, and he was certainly opposed to the prevailing orthodoxy of the fixity of species. Although in *Unity of Worlds* Powell neither proposed an evolutionary theory of his own nor actively defended any previous transmutationist theory, he did go as far as to suggest the superiority of both Lamarck's and Geoffroy's ideas to those of the French naturalist Frédéric Cuvier, who denied transmutation. The historian Curtis N. Johnson is quite right to say that in the context of the middle 1850s, Baden

Powell's book could only have been read as a positive endorsement of a natural and developmental understanding of earth and life history.[41]

A perhaps somewhat controversial inclusion in this collection of those who arrived at evolutionary conclusions before Darwin is Alfred Russel Wallace (1823–1913). Wallace is regarded today as the co-discoverer of evolution by natural selection because he independently came to very similar conclusions to Darwin, and sent an essay outlining his views to Darwin prior to the latter's publication of *Origin*. Darwin was dismayed that he was about to be forestalled after having worked carefully on developing his theory for some twenty years, but the intervention of his friends and colleagues, Lyell and the botanist Joseph Dalton Hooker, resulted in the presentation of Darwin's and Wallace's work together at the Linnean Society of London in 1858. Extracts from Darwin's notes and from a letter he had written to the American naturalist Asa Gray in 1857 were presented alongside Wallace's essay, allowing Darwin to maintain priority. It should be noted here that although this was an arrangement that was undertaken without Wallace's input (he was half a world away at the time), he was later quite happy with it and always referred to natural selection as Darwin's idea.[42] Wallace's essay "On the Tendency of Varieties to Depart Indefinitely from the Original Type" was preceded by another that had been published in 1855 in the *Annals and Magazine of Natural History*. In this essay, "On the Law Which has Regulated the Introduction of New Species," Wallace discussed the implications of geology for the distribution of species, noting that "the present condition of the organic world is clearly derived by a natural process of gradual extinction and creation of species" and acknowledged that

> to discover how the extinct species have from time to time been replaced by new ones down to the very latest geological period, is the most difficult, and at the same time the most interesting problem in the natural history of the earth.[43]

While Wallace did not go so far as to lay out a clear answer to this problem in this essay, he did state that he hoped his readers would see it as a step in the right direction towards its solution.

The last of the sources we have included in this collection are included almost as an appendix. These authors anticipated Darwin and Wallace's theory of natural selection, but although their contributions were written prior to the publication of *Origin*, they were not widely known until afterwards when either the authors themselves or others brought them to Darwin's attention.

DARWIN'S 'HISTORICAL SKETCH'

It seems likely that Darwin had originally contemplated a historical survey of the treatment of transmutation prior to his own work on the subject as a preface to *Origin*, although he did not include one until the third edition, which was published in April 1861. This was an addition that was clearly prompted by the numerous authors who wrote to Darwin about their own earlier work or who criticized Darwin for not acknowledging the work of earlier theorists. As historian Curtis N. Johnson pointed out, in this 'historical sketch,' Darwin acknowledged some eighteen predecessors; however, in subsequent editions, he expanded this number to thirty-five. Some of these were well-known and widely commented on, and we have covered a few of these already, including Lamarck, Geoffroy, and Baden Powell, for instance. We close out the volume, though, with extracts from the American physician W. C. Wells; the grain merchant, forester, and horticulturalist Patrick Matthew (1790–1874); and the plant breeder and hybridiser William Herbert (1778–1847), who had suggested that it was a distinct likelihood that new species might be produced through hybridization in nature.[44]

CONCLUDING REMARKS: PUTTING TEXTS IN CONTEXT

A primary source collection necessarily brings to the foreground the works of individuals whose work has been considered by historians to be significant in one way or another. And this is the crux. While most historians of evolution would likely agree that the texts included in this volume represent important contributions to the early history of modern evolutionism, it is important for those reading these works to think about why they are important. In early historical works, many of these authors have been framed simply as 'forerunners' of Darwin: those who vaguely theorized or speculated, but failed to fully elaborate a system of evolution in a truly scientific manner. This style of 'Whig' history is now only rarely to be found among scholars, but it remains prevalent among public readers and specialists outside the field of the history of science. We hope to have made clear that our aim here is different, and we hope to encourage the consideration of these and other sources in the context of their creation and not in the context of events or people that came only afterwards. Thus, a study of Buffon, Lamarck, Erasmus Darwin, Spencer, Chambers, or any of the other evolutionary thinkers needs to proceed with a focus on these individuals and their ideas, and how their ideas were received, interpreted, appropriated, and often reframed, developed, and ultimately acted upon. That is, we must see these texts as a potential entry point into the study of a social and cultural history of science. Thus, for instance, while we do want to ensure that our readers have the opportunity to read often extended extracts from a range of early-nineteenth-century evolutionary theorists, we also want to press readers to go beyond the texts in search of context to recognize that there may be several and even competing contexts to which any particular text is relevant. As Desmond, Secord, Jenkins, Corsi, and Testa have shown, Lamarck can be read in the context of Enlightenment France or in the context of Edinburgh or London. In each place, his texts had (often several) different and often highly contested meanings. So too, as Paul Elliott has shown, the work of Erasmus Darwin was widely popularized and debated in the English midlands, and James Secord has documented the many and various ways in which *Vestiges of the Natural History of Creation* was read, each highly dependent upon local as well as national politics and culture. We might usefully apply the methods of these scholars to our study of other transmutationist and evolutionary theorists, whether they were established men of science; political radicals, socialist and atheist freethinkers in London or Edinburgh; or genteel ladies who read *Vestiges* by candlelight and thereafter saw the world, and their place in it, quite differently.[45]

Notes

1 Charles Lyell, *Principles of Geology* (3 vols.), facsimile reprint, (Chicago: University of Chicago Press, 1990; 1991); Martin J.S. Rudwick, *Bursting the Limits of Time: The Reconstruction of Geohistory in the Age of Revolution* (Chicago: University of Chicago Press, 2005); Martin J.S. Rudwick, *Worlds Before Adam. The Reconstruction of Geohistory in the Age of Reform* (Chicago: University of Chicago Press, 2008).

2 Peter J. Bowler, *The Non-Darwinian Revolution: Reinterpreting a Historical Myth* (Baltimore: Johns Hopkins University Press, 1988), p. ix.

3 Robert J. Richards, *The Meaning of Evolution*; Piers J. Hale, "Rejecting the Myth of the non-Darwinian Revolution,"; Mark Largent, "The So-Called Eclipse of Darwinism," *Transaction of the American Philosophical Society*, n.s., 99, no.1 (2009), pp. 3–21.

4 Roger Cooter and Stephen Pumfrey, "Separate Spheres and Public Places. Reflections on the History of Science Popularization and Science in Popular Culture," *History of Science*, vol. 32, Iss. 3, (1994), pp. 237–267,

5 James Secord, *Victorian Sensation. The Extraordinary Publication, Reception, and Secret Authorship of The Vestiges of the Natural History of Creation* (Chicago: University of Chicago Press, 2000), p. 530.

6 Martin J.S. Rudwick, *Worlds Before Adam. The Reconstruction of Geohistory in the Age of Reform* (Chicago: University of Chicago Press, 2008); Peter J. Bowler, *Evolution. The History of an Idea*, (Berkley: University of California Press, 1989).

7 Bowler, *Evolution*, pp. 50–52; 57.

8 Janet Browne, "Botany of Gentlemen: Erasmus Darwin and 'The Loves of Plants,'" *Isis* 80, no.4, (1989), pp. 593–621.

9 Johann Friedrich Blumenbach, "Über den Bildungstrieb (Nisus formativus) und seinen Einfluß auf die Generation und Reproduction" in *Göttingisches Magazin der Wissenschaften und Litteratur*, vol. 1 (1780), pp. 247–266; Jean-Baptiste Lamarck; *Philosophie Zoologique*, (Paris: Musée d'Histoire Naturelle, 1809).

10 Andrés Galera, "The Impact of Lamarck's Theory of Evolution Before Darwin's Theory," *Journal of the History of Biology*, (Feb. 2017), 50 (1): 53–70.

11 See Bill Jenkins, "Race before Darwin: Variation, adaptation and the natural history of man in post-Enlightenment Edinburgh, 1790–1835" *British Journal for the History of Science*, (2020), 53 (3): 333–350, on p. 339. Blumenbach, Geoffroy and others were also acknowledged in the work of Robert Knox and James Cowles Prichard, amongst others. See for example, Robert Knox, "Contributions to the Philosophy of Zoology with Special Reference to the Natural History of Man," *The Lancet*, July 14th 1855, pp. 24–26; James Cowles Pritchard, *A Review of the Doctrine of a Vital Principle: As Maintained by Some Writers on Physiology with Observations on the Causes of Physical & Animal Life*, (London: John and Arthur Arch, 1829).

12 Biologists are wont to refer to the mechanism of the inheritance of acquired characters as "Lamarckism," in order to contrast it to the "Darwinian" mechanism of natural selection. However, as historians have long pointed out, Darwin also included the

inheritance of acquired characters as a significant element in the formation of new species. While historians thus generally have a better appreciation for the complexity of both Darwin's and Lamarck's ideas, they also frequently use the label "Lamarckism" and "Lamarckian" in this way. See Piers J. Hale, "Rejecting the Myth of the Non-Darwinian Revolution", *Victorian Review*, vol. 41, no.2, 2015, pp. 13–18. While there has long been an extensive "Darwin Industry" among academic scholars, Lamarck has received much less attention. Among those who have made Lamarck their focus some of the most notable are Pietro Corsi, Snait Gissis, and Caden Testa.

13 Caden Camille Testa, "Species Transformation and Social Reform: The Role of the Will in Jean-Baptiste Lamarck's Transformist Theory," *Journal of the History of Biology*, (March 2023), 56 (1):125–151. On Lamarck's meteorology and his efforts to establish a wider scientific discipline dedicated to the topic see Valentine Delrue, "Le spectacle du ciel: nut en schronheid als drijfveren van de meteorologie (1799–1816)," *Jaarboek de Achttiende Eeuw*, (2022), 54 (1), pp. 148–165.

14 Mary Wollstonecraft, *A Vindication of the Rights of Woman with strictures on Political and Moral Subjects* (Boston: Peter Edes 1792).

15 Adrian Desmond, *The Politics of Evolution. Morphology, Medicine, and Reform in Radical London* (Chicago: University of Chicago Press, 1990).

16 As one of us has written elsewhere, the political associations of Malthus's work developed and changed across the first decades of the nineteenth century; see Piers J. Hale, *Political Descent. Malthus, Mutualism, and the Politics of Evolution in Victorian England* (Chicago: University of Chicago Press, 2014).

17 Nora Barlow (ed.), *The Autobiography of Charles Darwin* 1809–1882 (London: Norton, 1958), p. 49, p. 153.

18 Bill Jenkins, *Evolution Before Darwin: Theories of the Transmutation of Species in Edinburgh, 1804–1834* (Edinburgh: University of Edinburgh Press, 2021).

19 Adrian Desmond and James Moore, *Darwin. The Life of a Tormented Evolutionist* (London: W. W. Norton, 1991).

20 Robert Knox, "Contributions to the Philosophy of Zoology, with special reference to the Natural History of Man," *The Lancet*, July 14th 1855, pp. 24–6. This essay is reprinted in full in this collection.

21 Robert Knox, "Inquiry into the Origin and Characteristic Difference of the Native Races inhabiting the Extra-Tropical part of Southern Africa," *Memoirs of the Wernerian Natural History Society*, v.5, (Edinburgh: 1824), pp. 206–219.

22 Robert Knox, *The Races of Men. A Fragment*, (Philadelphia, Lea & Blanchard, 1850)

23 Pietro Corsi, "Edinburgh Lamarckians? The Authorship of Three Anonymous Papers (1826–1829)." *Journal of the History of Biology* 54 (2021): 345–374.

24 Both Adrian Desmond in *The Politics of Evolution*, and Bill Jenkins in his *Evolution before Darwin* attributed this paper to Jameson; for Corsi's argument for a revision of this interpretation see Corsi, "Edinburgh Lamarckians?"

25 Pietro Corsi, "Edinburgh Lamarckians? The Authorship of Three Anonymous Papers (1826–1829)." *Journal of the History of Biology* 54 (2021): 345–374.

26 On the significance of the 1832 Reform Act see Eric Evans, *The Great Reform Act of 1832* (London: Routledge 1994).

27 E. P. Thompson, *The Making of the English Working Class*, (London: Penguin, 1963); for just one contrary claim see Patrick Joyce, *Visions of the People. Industrial England and the Question of Class, 1848–1914* (Cambridge: Cambridge University Press, 1994).

28 James Secord, *Victorian Sensation. The Extraordinary Publication, Reception, and Secret Authorship of The Vestiges of the Natural History of Creation* (Chicago: University of Chicago Press, 2000).

29 Jürgen Habermas, *The Structural Transformation of the Public Sphere: An Inquiry in a Category of Bourgeois Society* (Cambridge: Polity, 1962, trans. 1989).

30 Secord, *Victorian Sensation*, p. 143.
31 Roger Cooter and Stephen Pumfrey, "Separate Spheres and Public Places: Reflections on the History of Science Popularization and Science in Popular Culture," *History of Science* xxxii, (1994), pp. 237–267, on 246.
32 Bernard Lightman, "The 'Greatest Living Philosopher' and the Useful Biologist. How Spencer and Darwin Viewed Each Other's Contributions to Evolutionary Theory," in Ian Hesketh (ed.), *Imagining the Darwinian Revolution. Historical Narratives of Evolution from the Nineteenth-Century to the Present*, (Pittsburgh: University of Pittsburgh Press, 2022), pp. 37–57. Bernard Lightman, (ed.), *Global Spencerism. The Communication and Appropriation of a British Evolutionist* (Leiden: Brill, 2015); Greta Jones and Robert A. Peel (eds.), *Herbert Spencer. The Intellectual Legacy* (London: Galton Institute, 2004).
33 Anne Dewitt, "Evolution and Political Revolution in Blackwood's Periodical Poetry," *Victorian Periodicals Review*, Vol. 56, no.2 (Summer 2023), pp. 158–181, on p. 164. See [William Chilton], "Theory of Regular Gradation XLV," *Oracle of Reason, September 30, 1843 p. 332–34.* As James Secord notes, in *Victorian Sensation*, p. 313, n.33, Lyell's geology had been appropriated to undermine biblical literalism by other atheist and freethinking writers, such as in Henry Hetherington's *The Freethinker's Information for the People*, 1, no.1–2 (1842): pp. 1–16. The series "The Theory of Regular Gradation" also later appropriated the *Vestiges of the Natural History of Creation* to the atheist cause.
34 Herbert Spencer, *An Autobiography* 2 vols. (New York: D. Appleton, 1904), vol.1, pp. 200–201.
35 Paul A. Elliott, "'The Derbyshire Darwinians': The Persistence of Erasmus Darwin's Influence on a British provincial Library and Scientific Community, c.1780–1850." In C.U.M. Smith and Robert Arnot (eds.), *The Genius of Erasmus Darwin* (Aldershot: Ashgate, 2005); Paul A. Elliott, "Erasmus Darwin, Herbert Spencer, and the Origins of the Evolutionary Worldview in British Provincial Scientific Culture, 1770–1850," *ISIS*, 94, no.1 (2003), pp. 1–29.
36 See volume two of this series, Caden C. Testa and Piers J. Hale, *Evolution and Religion in Nineteenth Century Britain* (London: Routledge, 2025), for more details on this.
37 Darwin, *Origin*, (1859), p. 488.
38 Richard Hofstadter, *Social Darwinism in American Thought* (Boston: Beacon Press, 1992).
39 Robert J. Richards, *Darwin and the Emergence of Evolutionary Theories of Mind and Behavior*, Chicago: University of Chicago Press, 1987).
40 Robert J. Richards, "The Relation of Spencer's Evolutionary Theory to Darwin's," in Greta Jones and Robert A. Peel, (eds.), *Herbert Spencer. The Intellectual Legacy.* (London: The Galton Institute, 2004), pp. 17–36, on p. 19.
41 Curtis N. Johnson, *Darwin's Historical Sketch. An Examination of the 'Preface' to the Origin of Species* (Oxford: Oxford University Press, 2020), p. 6. Johnson's work on the 'historical sketch' that Darwin inserted into later editions of the *Origin of Species* explores the importance and influence of Baden Powell's work at greater length.
42 Martin Fichman, *An Elusive Victorian. The Evolution of Alfred Russel Wallace*, (Chicago: University of Chicago Press, 2004).
43 Alfred Russel Wallace, "On the Law Which has Regulated the Introduction of New Species" *Annals and Magazine of Natural History*, September 1855, reprinted in A. R. Wallace, *Contributions to the Theory of Natural Selection. A Series of Essays*, (London: Macmillan & Co. 1871), pp. 1–25, on pp. 3, 14.
44 Johnson, *Darwin's Historical Sketch*, p. 83.
45 Secord, *Victorian Sensation*, pp. 15–17.

Editorial Headnote

1. Alfred, Lord Tennyson, "In Memoriam A. A. H." (London: Edward Moxon, 1850), cantos LIV, LV, & LVI

Alfred, Lord Tennyson (1909–1892) wrote this poem lamenting the death of his friend Arthur Hallam, who died at the age of twenty-two in 1833. The poem *In Memoriam A.A.H.* was published in 1850, although Tennyson revised it over the following decade. It reflects not only Tennyson's grief over the loss of his friend but also a number of pressing metaphysical and theological concerns that were raised by developments in contemporary science. Tennyson not only invoked ideas about evolution that had been widely popularized by the then quite recently published *Vestiges of the Natural History of Creation* (1844), but also reflected on the evident cruelty and carelessness of nature and the demands that this view of life placed on conventional religious belief about a purposeful, beneficent designer.[1] As will be evident from the many readings in this volume,[2] Tennyson wrote in light not only of *Vestiges* but of developments in geology. The geologist Charles Lyell had argued in his *Principles of Geology* (1830–1833) that the earth was much older than many of his contemporaries in geology had acknowledged (and even they had concluded that the earth was vastly older than a literal reading of "Genesis" would allow), and also that it did not exhibit signs of progression; rather, life on earth was subject to an endless, and apparently purposeless, cyclical development. It was following the publication of *In Memoriam* that Tennyson was appointed poet Laureate, succeeding Wordsworth. As the historian James Secord has noted, "For thousands of readers, Tennyson's *In Memoriam* offered the most profound integration of the evolutionary narrative into everyday experience."[3]

Tennyson remained deeply interested in the relationship between science and religion and the threat that he believed the growth of scientific naturalism posed to orthodox religious belief. As a result, he was instrumental in establishing the Metaphysical Society, a cross-denominational discussion group that included many of the foremost theologians and men of science of that period.[4]

Notes

1 For further discussion of Tennyson in this context, see Michael Ruse, *Darwinism as Religion. What Literature Tells Us About Evolution* (Oxford: Oxford University Press, 2017), pp. 33–35.

2 Caden C. Testa and Piers J. Hale, *Evolution Before Darwin* (London: Routledge, 2025).

3 James Secord, *Victorian Sensation. The Extraordinary Publication, Reception, and Secret Authorship of The Vestiges of the Natural History of Creation* (Chicago: University of Chicago Press, 2000), p. 530.

4 For more on the Metaphysical Society, see Alan Willard Brown, *The Metaphysical Society. Victorian Minds in Crisis, 1869–1880* (New York: Columbia University Press, 1947); and more recently Catherine Marshall, Bernard Lightman, & Richard England, (eds.), *The Metaphysical Society (1869–1880). Intellectual Life in Mid-Victorian England* (Oxford: Oxford University Press, 2019); also see Caden C. Testa and Piers J. Hale, *Evolution and Religion* (London: Routledge, 2025).

1

"IN MEMORIAM A. A. H."

Alfred, Lord Tennyson

Source: Alfred, Lord Tennyson, "In Memoriam A. A. H." (London: Edward Moxon, 1850), cantos LIV, LV, & LVI

LIV

Oh yet we trust that somehow good
Will be the final goal of ill,
To pangs of nature, sins of will,
Defects of doubt, and taints of blood;

That nothing walks with aimless feet;
That not one life shall be destroy'd,
Or cast as rubbish to the void,
When God hath made the pile complete;

That not a worm is cloven in vain;
That not a moth with vain desire
Is shrivell'd in a fruitless fire,
Or but subserves another's gain.

Behold, we know not anything;
I can but trust that good shall fall
At last – far off – at last, to all,
And every winter change to spring.

So runs my dream: but what am I?
An infant crying in the night:
An infant crying for the light:
And with no language but a cry.

LV

The wish, that of the living whole
No life may fail beyond the grave,

DOI: 10.4324/9781003490548-1

Derives it not from what we have
The likest God within the soul?

Are God and Nature then at strife,
That Nature lends such evil dreams?
So careful of the type she seems,
So careless of the single life;

That I, considering everywhere
Her secret meaning in her deeds,
And finding that of fifty seeds
She often brings but one to bear,

I falter where I firmly trod,
And falling with my weight of cares
Upon the great world's altar-stairs
That slope thro' darkness up to God,

I stretch lame hands of faith, and grope,
And gather dust and chaff, and call
To what I feel is Lord of all,
And faintly trust the larger hope.

LVI

'So careful of the type?' but no.
From scarped cliff and quarried stone
She cries, 'A thousand types are gone:
I care for nothing, all shall go.

'Thou makest thine appeal to me:
I bring to life, I bring to death:
The spirit does but mean the breath:
I know no more.' And he, shall he,

Man, her last work, who seem'd so fair,
Such splendid purpose in his eyes,
Who roll'd the psalm to wintry skies,
Who built him fanes of fruitless prayer,

Who trusted God was love indeed
And love Creation's final law –
Tho' Nature, red in tooth and claw
With ravine, shriek'd against his creed –

Who loved, who suffer'd countless ills,
Who battled for the True, the Just,
Be blown about the desert dust,
Or seal'd within the iron hills?

No more? A monster then, a dream,
A discord. Dragons of the prime,
That tare each other in their slime,
Were mellow music match'd with him.

O life as futile, then, as frail!
O for thy voice to soothe and bless!
What hope of answer, or redress?
Behind the veil, behind the veil.

Part 1

THE ENLIGHTENMENT

Editorial Headnote

2. Georges Louis Leclerc, the Comte de Buffon, "Of the Degeneration of Animals" (1766) reprinted in *Buffon's Natural History containing A Theory of the Earth, A General History of Man, of Brute Creation, and of Vegetables, Minerals &c. From the French with notes by the Translator* (London: J.S. Barr, 1792).

Georges Louis Leclerc, the Comte de Buffon (1707–1788), was one of the most prominent naturalists of the French Enlightenment. Educated initially in law, he probably studied medicine, botany, and mathematics at the University of Angers. In 1732, he settled in Paris, where he wrote on mathematics, botany and forestry, and animal reproduction. He also published early studies of timber at the request of the minister of the navy, who, in June 1739, secured Buffon the position of *intendant* (head) at the Jardin du Roi – a post that Buffon held until his death. Over the course of these five decades, he expanded the Jardin's land holdings as well as its specimen collections, making it not just a prized property for the French crown but a leading European institute for research in natural history. In the meantime, he oversaw his estates in Montbard, continued publishing his own research, and amassed considerable personal wealth. His 36-volume encyclopedia *Histoire naturelle* (1749–1788) was read widely throughout Europe and even America.

Buffon argued that living matter was highly plastic and continually in flux; however, naturalists and historians through the nineteenth and twentieth centuries debated whether his name truly belonged in a history of evolutionary ideas.[1] More recently, scholars of the French Enlightenment have emphasized Buffon's interest in human phenotypical differences, especially his belief that humans could be selectively 'bred' in order to alter, preserve, or enhance such features.[2]

In the extract from the *Histoire* provided here, Buffon discusses the effect of the climate on human phenotypes, particularly racialized features, and he argues that selectively breeding humans could change such features faster than nature could. In this way, Buffon presented the human 'races' as arising from a natural, historical process – one that could be accelerated, altered, or reversed by human intervention. He does not go so far as to suggest that such a process could cross the species barrier, but the extract is indicative of the extent to which his discussions of biological plasticity were concerned with ideas of race, a concern that we shall see remained present in later overtly evolutionary arguments.

Notes

1 See, e.g. Arthur O. Lovejoy, "Buffon and the Problem of Species," in Forerunners of Darwin: 1745–1859, ed. Bentley Glass, Owsei Temkin, & William L. Straus, Jr. Baltimore: Johns Hopkins University Press (1959), pp. 84–113.

2 See, e.g. William Max Nelson, "Making Men: Enlightenment Ideas of Racial Engineering," *American Historical Review* 15 no. 5 (Dec. 2010), pp 1364–1394; David Allen Harvey, *The French Enlightenment and Its Others*, New York: Palgrave Macmillan (2012), pp. 138–146.

2

"OF THE DEGENERATION OF ANIMALS"

Georges Louis Leclerc, the Comte de Buffon

Source: Georges Louis Leclerc, the Comte de Buffon, "Of the Degeneration of Animals" (1766) reprinted in *Buffon's Natural History containing A Theory of the Earth, A General History of Man, of Brute Creation, and of Vegetables, Minerals &c. From the French with notes buy the Translator* (London: J.S. Barr, 1792)

OF THE DEGENERATION OF ANIMALS.

002 "Buffon's illustration of an ape standing erect: Histoire naturelle, générale et particuliére, avec la description du cabinet du roi [Georges Louis Leclerc Buffon]."

Wellcome Collection. Image in the public domain.

DOI: 10.4324/9781003490548-3

WHEN man began to disperse himself from climate to climate, his nature underwent several alterations; in the temperate countries which we suppose to be near where he was originally produced, these alterations have been but slight; but they increased in proportion as the distance was greater; and after many centuries had passed away, continents had been traversed, and generations degenerated by the influence of different climates, he ventured to the extremes, and being habituated to the scorching heats of the south, and the frozen regions of the north, the changes have become so great that there is room for imagining that the Negro, the Laplander, and the White, were different species, if, on the one hand, we were not certain that there was but one man originally created, and, on the other, that the White, Laplander, and Negro, can unite and propagate the great family of the human kind. Thus their colours are not original, their disresemblances being only external and superficial. It is the same being which is tinctured with black under the torrid zone, and rendered tawny, with contracted limbs by the rigour of the cold under the polar circle. This fact is alone sufficient to demonstrate that there is more strength, extent, and flexibility, in man than in any other being; for vegetables, and almost every animal, are confined to particular soils and climates. This extension of our nature depends less on the properties of our bodies than those of our minds. By the last man has been enabled to seek those things which are necessary for the delicacy of his body; by that he has found out the means of bearing the inclemencies of the weather, and to conquer the barrenness of the earth. He may be said to have subdued the elements: by a single ray of his intellect he produced the element of fire, which before did not exist on the surface of the earth: he has cloathed, sheltered, and lodged himself, thus providing against every external attack: he has compensated by his reason for every deficiency; and although not so strong, so large, nor so robust, as many animals, yet he has found means to conquer, subdue, enslave, and deprive them of those spaces which Nature seems to have resigned them for their use.

The earth is divided into two great continents: which division is more ancient than all human structures and monuments, and yet man is more ancient than that division, for he is found the same in both. The Asiatic, the European, and the Negro, alike propagate with the American. Nothing can stronger prove them to be the issue of one source than the facility with which they re-unite with the common stock. The blood is different, but the gem is the same. The skin, the hair, the features, and the size, have varied, but the internal form has not changed. The type is general and common, and if it should ever happen, by any unforeseen revolution, that man should be obliged to forsake those climates which he has possessed himself of, and return to his native country, he would in time resume his original features, his primitive size, and his natural colour. But the mixture of races would produce the same effect in a much shorter time. The conjunction of a white male with a black female, or a black male with a white female, equally produce a Mulatto, whose colour is brown, that is, a mixture of black and white. This mulatto intermixing with a white, produces a second mulatto not so brown as the former; and if this second mulatto unites with a white the third mulatto will have no more than a slight tincture of the brown, which will entirely disappear in succeeding generations. Thus, by this mixture with a white, one hundred and fifty, or two hundred years is sufficient to bleach the skin of the Negro; but it would, perhaps, require many centuries to produce this effect by the influence of the climate alone. Since the Negroes were transported to America, which is about two hundred years, not the smallest shade of difference is perceivable to have taken place in the colour of those families which have preserved themselves from mixture. It is true the climate of South America being hot enough to give the natives a brown tint, we ought not to be astonished at the Negroes retaining their colour in that part of the world. Indeed, to make a proper experiment of the change of colour in the human species, some individuals of this black race should be transported from Senegal to Denmark, where the people have generally fair skins, golden locks, and blue eyes; and where the difference of blood, and opposition of colour, are the greatest. We must keep these Negroes with their females apart from the inhabitants, and scrupulously prevent all crossing of their breed. This is the only method of learning how much time it would require to change a Negro into a White, or the White into a Black, by the influence of climate.

This is the greatest alteration that the atmosphere has made on man, and yet this is only superficial. The colour of the skin, hair, and eyes, varies solely by the influence of climate. The other changes, such as that of size, features, and the quality of the hair, do not seem to depend on this cause alone, for among the Negro race, the greatest part of whom have frizzled wool on their heads, a flat nose, and thick lips, we meet with whole nations with long and real hair, and regular features. Again, if we compare, among the white race, the Dane with the Calmuck Tartar, or only the Finlander with the Laplander, who belong to the neighbouring nations, we shall find as much difference between them, with respect to size and features, as there is among the Negroes; consequently we must subjoin some other cause to that of the climate to account for these alterations, which are stronger than the former. The most general and direct cause is the quality of the food, for it is principally by the aliments that man receives the influence of the soil which he inhabits, the air and atmosphere acting more superficially. While the latter alters the external surface by changing the colour of the skin, food acts upon the internal form by its properties, which are constantly relative to those of the earth by which it is produced. Even in the same country we find strong differences between men who occupy the high lands, and those who live in the low. The inhabitants of a mountain are always better made, more spirited, and handsomer than those of a valley: therefore, in countries far distant from the original climates, where herbage, fruit, grain, and the flesh of animals, differ both in quality and substance, the men who feed on them must undergo greater changes. These impressions are not suddenly made. Time is required for man to receive the tincture of the atmosphere, and still more for the earth to transmit him its qualities. Ages, joined to a constant use of the same nutriment, is necessary to influence the form of the features, the size of the body, the substance of the hair, and to produce those alterations which, when, perpetuated by generation, becomes general and constant characters, by which the races, and even the different nations, which compose the human race, are distinguished.

In brute animals these effects are quicker and greater; for, partaking more of the nature of the soil than man, and their food being more uniform and unprepared, the quality is more decisive, and, consequently, its influence stronger; and because as the animals cannot clothe nor shelter themselves, nor make use of the element of fire, they remain constantly exposed to the impressions of the air, and inclemencies of the climate. For this reason every animal has chosen its zone and country according to its nature; for the same reason they remain there, and instead of extending or dispersing themselves, like the human race, they generally continue in those places which are most agreeable to their constitutions. When driven by man, or carried away, or forced by any revolution of the globe to forsake their native country, their nature undergoes such great and strong alterations that they are no longer to be known, but by having recourse to attentive inspection, and even to experiment and analogy. If to these natural causes of alteration in free animals we add that of the empire of man over those which he has reduced to slavery, we shall be surprised to see how far tyranny is able to degrade and disfigure Nature; we shall perceive the stigmas of captivity, and the impressions of her fetters; we shall find that those wounds are deeper, and more incurable, in proportion to their antiquity; and that in the state wherein we have reduced domestic animals it would perhaps be impossible to reinstate them in their primitive form, and to restore those other natural attributes which we have deprived them of.

Thus, the temperature of the climate, the quality of the food, and the evils arising from slavery, are the three causes of the changes and degeneration of animals. The effects of each deserve to be particularly considered, and their relations, when viewed in detail, will present a tablet on which we shall see Nature such as she is at present, and in a more remote degree, what she was before her degradation.

Editorial Headnote

3. William Godwin, *An Enquiry Concerning Political Justice and its Influence on General Virtue and Happiness*, (2 vols.) (Dublin: Luke White, 1793).

William Godwin (1756–1836) was one of the most prominent political radicals of the eighteenth century. He wrote a number of novels in which he explored his political ideas, but far and away his most influential work was his 1793 *Enquiry Concerning Political Justice*. His lover and later his wife, the writer and philosopher Mary Wollstonecraft, was also a politically influential figure. In her *Vindication of the Rights of Woman* (1792), she advocated for women's rights on the basis of their equality with men. Their daughter, Mary Shelley (1797–1851), wrote *Frankenstein*.

Although Godwin, like many other radicals, argued that humanity was subject to infinite "perfectibility," by this, Godwin sought to indicate "the progressive nature of man, in knowledge, in virtuous propensities, and in social institutions," rather than biological development.[1] Indeed, in the extracts provided here, we can see that, in contrast to many of his contemporaries in the Enlightenment, he was deeply sceptical of the role of heredity in transforming humanity. Environment was important, but, he argued, humans came into the world "with no innate principles" – as a blank slate, as it were. Thus, although others might draw analogies between human improvement and animal breeding, Godwin denied the relevance of such a comparison.

The timing of the publication of Godwin's work was all-important for how it was read. The French Revolution of 1789, the establishment of the First Republic, and the onset of the 'Reign of Terror' meant that any radical text could not be read without the threat of violence. The fact that Godwin's work was sold at a price that kept it out of the hands of a popular audience meant that Godwin avoided the political persecution that some likeminded authors experienced, but his work and ideas were reproduced and summarized in more accessible radical tracts and, even though Godwin did not accept the hereditary nature of environmental influence, many of his readers saw things differently, and adapted his ideas about perfectibility to imply transmutationist conclusions.

Given the later importance of Thomas Robert Malthus's 1798 *Essay on the Principle of Population*, both in its own right as a political essay and as the inspiration for Darwin's thinking about natural selection, it is important for us to recognize that Malthus wrote his essay to counter the popularity of both Godwin and the French philosopher and humanist Marie-Jean-Antoine-Nicolas de Caritat, the Marquis de Condorcet (1743–1794).[2]

The 1793 first edition of Godwin's work was printed in instalments. In the second edition, which appeared in 1796, Godwin revised the text to eliminate unnecessary repetition. A third edition was published in 1798 in which Godwin drew back from some of his claims, such as his embrace of some of Rousseau's primitivist ideas and some of his more radical ideas about property.[3]

Notes

1 Quoted in William St. Clair, *The Godwins and the Shelleys: The Biography of a Family* (London: Faber & Faber, 1989), p. 73.

2 Robert J. Mayhew, *Malthus. The Life and Legacies of an Untimely Prophet* (Harvard: Belknap, 2014), pp. 65–74; Claeys, "Malthus and Godwin" p. 52–73.

3 Claeys, "Malthus and Godwin," p. 52. An anarchist and advocate of mutual aid, Peter Kropotkin did more than anyone to popularize Godwin among late-nineteenth century readers, although Kropotkin thought that Godwin had weakened his position on property in later editions.

3

AN ENQUIRY CONCERNING POLITICAL JUSTICE AND ITS INFLUENCE ON GENERAL VIRTUE AND HAPPINESS

William Godwin

Source: William Godwin, *An Enquiry Concerning Political Justice and Its influence on General Virtue and Happiness*, (2 vols.) (Dublin: Luke White, 1793)

THE MORAL CHARACTERS OF MEN ORIGINATE IN THEIR PERCEPTIONS.

WE bring into the world with us no innate principles: consequently we are neither virtuous nor vicious as we first come into existence. No truth can be more evident than this, to any man who will yield the subject an impartial consideration. Every principle is a proposition. Every proposition consists in the connection of at least two distinct ideas, which are affirmed to agree or disagree with each other. If therefore the principles be innate, the ideas must be so too. But nothing can be more incontrovertible, than that we do not bring pre-established ideas into the world with us.

[...]

Hence it inevitably follows, that physical causes, though of some consequence in the history of man, sink into nothing, when compared with the great and inexpressible operations of reflection. They are the prejudices we conceive or the judgments we form, our apprehensions of truth and falshood, that constitute the true basis of distinction between man and man. The difference between savage and savage indeed, in the first generation of the human species and in perfect solitude, can only be ascribed to the different impressions made upon their senses. But this difference would be almost imperceptible. The ideas of wisdom and folly would never have entered the human mind, if men, like beasts, derived neither good nor evil from the reflections and discoveries of their companions and ancestors.

Hence we are furnished with an answer to the analogical argument from the considerable effects that physical causes appear to produce upon brutes. "Breed for example appears to be of unquestionable importance to the character and

DOI: 10.4324/9781003490548-4

qualifications of horses and dogs; why should we not suppose this or certain other brute and occult causes to be equally efficacious in the case of men? How comes it that the races of animals perhaps never degenerate, if carefully cultivated; at the same time that we have no security against the wisest philosopher's begetting a dunce?"

I answer, that the existence of physical causes cannot be controverted. In the case of man their efficacy is swallowed up in the superior importance of reflection and science. In animals on the contrary they are left almost alone. If a race of negroes were taken, and maintained each man from his infancy, except so far as was necessary for the propagation of the species, in solitude; or even if they were excluded from an acquaintance with the improvements and imaginations of their ancestors, though permitted the society of each other, the operation of breed might perhaps be rendered as conspicuous among them, as in the different classes of horses and dogs. But the ideas they would otherwise receive from their parents and civilised or half-civilised neighbours would be innumerable: and, if the precautions above mentioned were unobserved, all parallel between the two cases would cease.

Such is the character of man considered as an individual. He is operated upon by exterior causes immediately, producing certain effects upon him independently of the exercise of reason; and he is operated upon by exterior causes mediately, their impressions furnishing him with materials for reflection, and assuming the form of motives to act or to refrain from acting. But the latter of these, at least so far as relates to man in a civilised state, may stand for the whole. He that would change the character of the individual, would miserably misapply his efforts, if he principally sought to effect this purpose by the operations of heat and cold, dryness and moisture upon the animal frame, The true instruments of moral influence, are desire and aversion, punishment and reward, the exhibition of general truth, and the development of those punishments and rewards, which wisdom and error by the very nature of the thing constantly bring along with them.

[...]

OF THE EQUALITY OF MANKIND.

THE equality of mankind is either physical or moral. Their physical equality may be considered either as it relates to the strength of the body or the faculties of the mind.

This part of the subject has been exposed to cavil and objection. It has been said, "that the reverse of this equality is the result of our experience. Among the individuals of our species we actually find that there are not two alike. One man is strong and another weak. One man is wise and another foolish. All that exists in the world of the inequality of conditions is to be traced to this as their source. The strong man possesses power to subdue, and the weak stands in need of an ally to protect. The consequence is inevitable: the equality of conditions is a chimerical assumption, neither possible to be reduced into practice, nor desirable if it could be so reduced."

Upon this statement two observations are to be made. First, this inequality was in its origin infinitely less than it is at present. In the uncultivated state of man diseases, effeminacy and luxury were little known, and of consequence the strength of every one much more nearly approached to the strength of his neighbour. In the uncultivated state of man the understandings of all were limited, their wants, their ideas and their views nearly upon a level. It was to be expected that in their first departure from this state great irregularities would introduce themselves; and it is the object of subsequent wisdom and improvement to mitigate these irregularities.

Secondly, notwithstanding the incroachments that have been made upon the equality of mankind, a great and substantial equality remains. There is no such disparity among the human race as to enable one man to hold several other men in subjection, except so far as they are willing to be subject. All government is founded in opinion. Men at present live under any particular form, because they conceive it their interest to do so. One part indeed of a community or empire may be held in subjection by force; but this cannot be the personal force of their despot; it must be the force of another part of the community, who are of opinion that it is their interest to support his authority. Destroy this opinion, and the fabric which is built upon it falls to the ground. It follows therefore that all men are essentially independent. – So much for the physical equality.

The moral equality is still less open to reasonable exception. By moral equality I understand the propriety of applying one unalterable rule of justice to every case that may arise. This cannot be questioned but upon arguments that would subvert the very nature of virtue. "Equality," it has been affirmed, "will always be an unintelligible fiction, so long as the capacities of men shall be unequal, and their pretended claims have neither guarantee nor sanction by which they can be inforced[a]." But surely justice is sufficiently intelligible in its own nature, abstracted from the consideration whether it be or be not reduced into practice. Justice has relation to beings endowed with perception, and capable of pleasure and pain. Now it immediately results from the nature of such beings, independently of any arbitrary constitution, that pleasure is agreeable and pain odious, pleasure to be desired and pain to be obviated. It is therefore just and reasonable that such beings should contribute, so far as it lies in their power, to the pleasure and benefit of each other. Among pleasures some are more exquisite, more unalloyed and less precarious than others. It is just that these should be preferred.

From these simple principles we may deduce the moral equality of mankind. We are partakers of a common nature, and the same causes that contribute to the benefit of one contribute to the benefit of another. Our senses and faculties are of the same denomination. Our pleasures and pains will therefore be the same. We are all of us endowed with reason, able to compare, to judge and to infer. The improvement therefore which is to be desired for the one is to be desired for the other. We shall be provident for ourselves and useful to each other, in proportion as we rise above the atmosphere of prejudice. The same independence, the same freedom from any such restraint, as should prevent us from giving the reins to our own understanding, or from uttering upon all occasions whatever we think to be true, will conduce to the improvement of all. There are certain opportunities and a certain situation most advantageous to every human being, and it is just that these should be communicated to all, as nearly at least as the general economy will permit.

There is indeed one species of moral inequality parallel to the physical inequality that has been already described. The treatment to which men are entitled is to be measured by their merits and their virtues. That country would not be the seat of wisdom and reason, where the benefactor of his species was considered in the same point of view as their enemy. But in reality this distinction, so far from being adverse to equality in any tenable sense, is friendly to it, and is accordingly known by the appellation of equity, a term derived from the same origin. Though in some sense an exception, it tends to the same purpose to which the principle itself is indebted for its value. It is calculated to infuse into every bosom an emulation of excellence. The thing really to be desired is the removing as much as possible arbitrary distinctions, and leaving to talents and virtue the field of exertion unimpaired. We should endeavour to afford to all the same opportunities and the same encouragement, and to render justice the common interest and choice.

Note

a "*On a dit – que nous avions tous les mêmes droits. J'ignore ce que c'est que les mêmes droits, où il y a inégalité de talens ou de force, & nulle garrantie, nulle sanction.*" *Raynal, Revolution a'Amerique,* p. 34.

Editorial Headnote

4. Nicolas de Condorcet, "*Outlines of an Historical View of the Progress of the Human Mind. Being a Posthumous Work of the Late M. DE CONDORCET*" (London: J. Johnson, 1795).

Marie-Jean-Antoine-Nicolas de Caritat, Marquis de Condorcet (1743–1794), was a French Enlightenment philosopher and humanist who was known for his progressive views, advocating for causes including educational reform, women's suffrage, and the abolition of slavery. Condorcet argued that the human race was highly perfectible and continually progressing, both mentally and physically. A moderate republican, he believed the French Revolution of 1789 would bring increased freedom, equality, and prosperity to the French people, and represented Paris in the Legislative Assembly. He was later arrested by the Jacobin party, which advocated a more radical position than the centrist Girondins whom Condorcet represented. While in captivity, he wrote *Sketch for a Historical Picture of the Progress of the Human Mind* (1794), in which he explicated his view of human progress toward enlightenment and perfection, including the physical improvement of individuals.

In the extract provided from his *Sketch*, Condorcet argues that physical and moral qualities are inherited traits that can be perfected through social interventions, especially proper education. With the right political and social reforms, Condorcet believed that the human species would continue to progress, both biologically and socially; positive changes to physiology as well as to the moral and intellectual faculties would be passed from parent to offspring, leading each subsequent generation to even greater perfection. The extract shows that Condorcet, like many of his contemporaries, saw reproduction as a critical part of human biological improvement, even drawing analogies between human reproduction and studies of stock breeding. It also shows that for Condorcet, social and political reform was a way of biologically altering the human species – in other words, a means by which humans could consciously and rationally direct their own development.

4

OUTLINES OF AN HISTORICAL VIEW OF THE PROGRESS OF THE HUMAN MIND

Nicolas de Condorcet

Source: Nicolas de Condorcet, *Outlines of an Historical View of the Progress of the Human Mind. Being a Posthumous Work of the Late M. DE CONDORCET* (London: J. Johnson, 1795), pp. 370–1

But may not our physical faculties, the force, the sagacity, the acuteness of the senses, be numbered among the qualities, the individual improvement of which it will be practicable to transmit? An attention to the different breeds of domestic animals must lead us to adopt the affirmative of this question, and a direct observation of the human species itself will be found to strengthen the opinion.

Lastly, may we not include in the same circle the intellectual and moral faculties? May not our parents, who transmit to us the advantages or defects of their conformation, and from whom we receive our features and shape, as well as our propensities to certain physical affections, transmit to us also that part of organization upon which intellect, strength of understanding, energy of soul or moral sensibility depend? Is it not probable that education, by improving these qualities, will at the same time have an influence upon, will modify and improve this organization itself? Analogy, an investigation of the human faculties, and even some facts, appear to authorise these conjectures, and thereby to enlarge the boundary of our hopes.

Such are the questions with which we shall terminate the last division of our work. And how admirably calculated is this view of the human race, emancipated from its chains, released alike from the dominion of chance, as well as from that of the enemies of its progress, and advancing with a firm and indeviate step in the paths of truth, to console the philosopher lamenting the errors, the flagrant acts of injustice, the crimes with which the earth is still polluted? It is the contemplation of this prospect that rewards him for all his efforts to assist the progress of reason and the establishment of liberty. He dares to regard these efforts as a part of the eternal chain of the destiny of mankind; and in this persuasion he finds the true delight of virtue, the pleasure of having performed a durable service, which no vicissitude will ever destroy in a fatal operation calculated to restore the reign of prejudice and slavery.

DOI: 10.4324/9781003490548-5

DOI: 10.4324/9781003495[illegible]-5

Editorial Headnote

5. Thomas Robert Malthus, *An Essay on the Principle of Population*, (London, J. Johnson, 1798), pp. 1, 3–9, 49–54, 110, 111–4.

The Reverend Thomas Robert Malthus (1766–1832), who was known as Robert Malthus, wrote *An Essay on the Principle of Population* in 1798 as a critical response to and rejection of the popular radical ideas of the Enlightenment political theorists William Godwin and Nicolas de Condorcet.[1] Godwin and Condorcet, like many political radicals, had argued that, under the right conditions, humanity was capable of infinite improvement. They further believed that if society were properly organized, then no one would go hungry and equality would be the basis of a just future polity. Malthus, to the contrary, argued that the basic laws of nature that governed population meant that, however desirable such ideas might be, they were simply impossible. Humanity, he argued, was subject to two necessary laws: they would always continue to reproduce, and they would always need to eat. Further, since the population would increase exponentially (geometrically), while the available resources would only increase in a linear ration (arithmetically), there would always be insufficient provisions for all. The poor would always be with us. In the second and subsequent editions, Malthus ameliorated the harsh message of the first edition to emphasize the possibility that moral restraint might alleviate the worst effects of this arrangement. Malthus was a theologian, and although some might have asked why a beneficent God would have created a world subject to such laws, to Malthus, this was evidence that God wanted humanity to embrace a certain form of moral behaviour: to work hard in order to provide for themselves and their families and to exercise moral restraint.

Although Malthus wrote his theory as an argument against the kind of improvement in society and humanity that Godwin, Condorcet, and others sought to realize, by the 1830s, liberal 'philosophical radicals' like John Stuart Mill and Harriet Martineau read Malthus to very different ends – it was through hard work and moral restraint that progress was possible.[2] It was in this context that the young Charles Darwin read Malthus in September 1838 and declared, "Here, then, I had at last got a theory by which to work."[3]

Notes

1 Robert J. Mayhew, *Malthus. The Life and Legacies of an Untimely Prophet* (Harvard: Belknap, 2014), pp. 65–74; Gregory Claeys, "Malthus and Godwin: rights, utility, and productivity," in Robert J. Mayhew (ed.), *New Perspectives on* Malthus (Cambridge: Cambridge University Press, 2016): p. 52–73.

2 John Stuart Mill, *Autobiography of John Stuart Mill* (New York: New American Library, 1964), p. xii.

3 Nora Barlow, (ed.), *The Autobiography of Charles Darwin 1809–1882* (London: W. W. Norton & Co., 1958), p. 120.

5

AN ESSAY ON THE PRINCIPLE OF POPULATION

Thomas Robert Malthus

Source: Thomas Robert Malthus, *An Essay on the Principle of Population*, (London, J. Johnson, 1798), pp. 1, 3–9, 49–54, 110, 111–4

CHAPTER 1

Question stated – Little prospect of a determination of it, from the enmity of the opposing parties – The principal argument against the perfectibility of man and of society has never been fairly answered – Nature of the difficulty arising from population – Outline of the principal argument of the Essay

THE GREAT AND UNLOOKED FOR DISCOVERIES that have taken place of late years in natural philosophy, the increasing diffusion of general knowledge from the extension of the art of printing, the ardent and unshackled spirit of inquiry that prevails throughout the lettered and even unlettered world, the new and extraordinary lights that have been thrown on political subjects which dazzle and astonish the understanding, and particularly that tremendous phenomenon in the political horizon, the French Revolution, which, like a blazing comet, seems destined either to inspire with fresh life and vigour, or to scorch up and destroy the shrinking inhabitants of the earth, have all concurred to lead many able men into the opinion that we were touching on a period big with the most important changes, changes that would in some measure be decisive of the future fate of mankind.

[...]

I have read some of the speculations on the perfectibility of man and of society with great pleasure. I have been warmed and delighted with the enchanting picture which they hold forth. I ardently wish for such happy improvements. But I see great, and, to my understanding, unconquerable difficulties in the way to them. These difficulties it is my present purpose to state, declaring, at the same time, that so far from exulting in them, as a cause of triumph over the friends of innovation, nothing would give me greater pleasure than to see them completely removed.

The most important argument that I shall adduce is certainly not new. The principles on which it depends have been explained in part by Hume, and more at

DOI: 10.4324/9781003490548-6

large by Dr Adam Smith. It has been advanced and applied to the present subject, though not with its proper weight, or in the most forcible point of view, by Mr Wallace, and it may probably have been stated by many writers that I have never met with. I should certainly therefore not think of advancing it again, though I mean to place it in a point of view in some degree different from any that I have hitherto seen, if it had ever been fairly and satisfactorily answered.

The cause of this neglect on the part of the advocates for the perfectibility of mankind is not easily accounted for. I cannot doubt the talents of such men as Godwin and Condorcet. I am unwilling to doubt their candour. To my understanding, and probably to that of most others, the difficulty appears insurmountable. Yet these men of acknowledged ability and penetration scarcely deign to notice it, and hold on their course in such speculations with unabated ardour and undiminished confidence. I have certainly no right to say that they purposely shut their eyes to such arguments. I ought rather to doubt the validity of them, when neglected by such men, however forcibly their truth may strike my own mind. Yet in this respect it must be acknowledged that we are all of us too prone to err. If I saw a glass of wine repeatedly presented to a man, and he took no notice of it, I should be apt to think that he was blind or uncivil. A juster philosophy might teach me rather to think that my eyes deceived me and that the offer was not really what I conceived it to be.

In entering upon the argument I must premise that I put out of the question, at present, all mere conjectures, that is, all suppositions, the probable realization of which cannot be inferred upon any just philosophical grounds. A writer may tell me that he thinks man will ultimately become an ostrich. I cannot properly contradict him. But before he can expect to bring any reasonable person over to his opinion, he ought to shew that the necks of mankind have been gradually elongating, that the lips have grown harder and more prominent, that the legs and feet are daily altering their shape, and that the hair is beginning to change into stubs of feathers. And till the probability of so wonderful a conversion can be shewn, it is surely lost time and lost eloquence to expatiate on the happiness of man in such a state; to describe his powers, both of running and flying, to paint him in a condition where all narrow luxuries would be contemned, where he would be employed only in collecting the necessaries of life, and where, consequently, each man's share of labour would be light, and his portion of leisure ample.

I think I may fairly make two postulata.

First, That food is necessary to the existence of man.

Secondly, That the passion between the sexes is necessary and will remain nearly in its present state.

These two laws, ever since we have had any knowledge of mankind, appear to have been fixed laws of our nature, and, as we have not hitherto seen any alteration in them, we have no right to conclude that they will ever cease to be what they now are, without an immediate act of power in that Being who first arranged the system of the universe, and for the advantage of his creatures, still executes, according to fixed laws, all its various operations.

I do not know that any writer has supposed that on this earth man will ultimately be able to live without food. But Mr Godwin has conjectured that the passion between the sexes may in time be extinguished. As, however, he calls this part of his work a deviation into the land of conjecture, I will not dwell longer upon it at present than to say that the best arguments for the perfectibility of man are drawn from a contemplation of the great progress that he has already made from the savage state and the difficulty of saying where he is to stop. But towards the extinction of the passion between the sexes, no progress whatever has hitherto been made. It appears to exist in as much force at present as it did two thousand or four thousand years ago. There are individual exceptions now as there always have been. But, as these exceptions do not appear to increase in number, it would surely be a very unphilosophical mode of arguing to infer, merely from the existence of an exception, that the exception would, in time, become the rule, and the rule the exception.

Assuming then my postulata as granted, I say, that the power of population is indefinitely greater than the power in the earth to produce subsistence for man.

Population, when unchecked, increases in a geometrical ratio. Subsistence increases only in an arithmetical ratio. A slight acquaintance with numbers will shew the immensity of the first power in comparison of the second.

By that law of our nature which makes food necessary to the life of man, the effects of these two unequal powers must be kept equal.

This implies a strong and constantly operating check on population from the difficulty of subsistence. This difficulty must fall somewhere and must necessarily be severely felt by a large portion of mankind.

Through the animal and vegetable kingdoms, nature has scattered the seeds of life abroad with the most profuse and liberal hand. She has been comparatively sparing in the room and the nourishment necessary to rear them. The germs of existence contained in this spot of earth, with ample food, and ample room to expand in, would fill millions of worlds in the course of a few thousand years. Necessity, that imperious all pervading law of nature, restrains them within the prescribed bounds. The race of plants and the race of animals shrink under this great restrictive law. And the race of man cannot, by any efforts of reason, escape from it. Among plants and animals its effects are waste of seed, sickness, and premature death. Among mankind, misery and vice. The former, misery, is an absolutely necessary consequence of it. Vice is a highly probable consequence, and we therefore see it abundantly prevail, but it ought not, perhaps, to be called an absolutely necessary consequence. The ordeal of virtue is to resist all temptation to evil.

This natural inequality of the two powers of population and of production in the earth, and that great law of our nature which must constantly keep their effects equal, form the great difficulty that to me appears insurmountable in the way to the perfectibility of society. All other arguments are of slight and subordinate consideration in comparison of this. I see no way by which man can escape from the weight of this law which pervades all animated nature. No fancied equality, no agrarian regulations in their utmost extent, could remove the pressure of it even for a single century. And it appears, therefore, to be decisive against the possible existence of a society, all the members of which should live in ease, happiness, and comparative leisure; and feel no anxiety about providing the means of subsistence for themselves and families.

Consequently, if the premises are just, the argument is conclusive against the perfectibility of the mass of mankind.

I have thus sketched the general outline of the argument, but I will examine it more particularly, and I think it will be found that experience, the true source and foundation of all knowledge, invariably confirms its truth.

CHAPTER 2

The different ratio in which population and food increase – The necessary effects of these different ratios of increase – Oscillation produced by them in the condition of the lower classes of society – Reasons why this oscillation has not been so much observed as might be expected – Three propositions on which the general argument of the Essay depends – The different states in which mankind have been known to exist proposed to be examined with reference to these three propositions.

I SAID THAT POPULATION, WHEN UNCHECKED, increased in a geometrical ratio, and subsistence for man in an arithmetical ratio.

Let us examine whether this position be just. I think it will be allowed, that no state has hitherto existed (at least that we have any account of) where the manners were so pure and simple, and the means of subsistence so abundant, that no check whatever has existed to early marriages, among the lower classes, from a fear of not providing well for their families, or among the higher classes, from a fear of lowering their condition in life. Consequently in no state that we have yet known has the power of population been left to exert itself with perfect freedom.

Whether the law of marriage be instituted or not, the dictate of nature and virtue seems to be an early attachment to one woman. Supposing a liberty of changing in the case of an unfortunate choice, this liberty would not affect population till it arose to a height greatly vicious; and we are now supposing the existence of a society where vice is scarcely known.

In a state therefore of great equality and virtue, where pure and simple manners prevailed, and where the means of subsistence were so abundant that no part of the society could have any fears about providing amply for a family, the power of population being left to exert itself unchecked, the increase of the human species would evidently be much greater than any increase that has been hitherto known.

In the United States of America, where the means of subsistence have been more ample, the manners of the people more pure, and consequently the checks to early marriages fewer, than in any of the modern states of Europe, the population has been found to double itself in twenty-five years.

This ratio of increase, though short of the utmost power of population, yet as the result of actual experience, we will take as our rule, and say, that population, when unchecked, goes on doubling itself every twenty-five years or increases in a geometrical ratio.

Let us now take any spot of earth, this Island for instance, and see in what ratio the subsistence it affords can be supposed to increase. We will begin with it under its present state of cultivation.

If I allow that by the best possible policy, by breaking up more land and by great encouragements to agriculture, the produce of this Island may be doubled in the first twenty-five years, I think it will be allowing as much as any person can well demand.

In the next twenty-five years, it is impossible to suppose that the produce could be quadrupled. It would be contrary to all our knowledge of the qualities of land. The very utmost that we can conceive, is, that the increase in the second twenty-five years might equal the present produce. Let us then take this for our rule, though certainly far beyond the truth, and allow that, by great exertion, the whole produce of the Island might be increased every twenty-five years, by a quantity of subsistence equal to what it at present produces. The most enthusiastic speculator cannot suppose a greater increase than this. In a few centuries it would make every acre of land in the Island like a garden.

Yet this ratio of increase is evidently arithmetical.

It may be fairly said, therefore, that the means of subsistence increase in an arithmetical ratio. Let us now bring the effects of these two ratios together.

The population of the Island is computed to be about seven millions, and we will suppose the present produce equal to the support of such a number. In the first twenty-five years the population would be fourteen millions, and the food being also doubled, the means of subsistence would be equal to this increase. In the next twenty-five years the population would be twenty-eight millions, and the means of subsistence only equal to the support of twenty-one millions. In the next period, the population would be fifty-six millions, and the means of subsistence just sufficient for half that number. And at the conclusion of the first century the population would be one hundred and twelve millions and the means of subsistence only equal to the support of thirty-five millions, which would leave a population of seventy-seven millions totally unprovided for.

A great emigration necessarily implies unhappiness of some kind or other in the country that is deserted. For few persons will leave their families, connections, friends, and native land, to seek a settlement in untried foreign climes, without some strong subsisting causes of uneasiness where they are, or the hope of some great advantages in the place to which they are going.

But to make the argument more general and less interrupted by the partial views of emigration, let us take the whole earth, instead of one spot, and suppose that the restraints to population were universally removed. If the subsistence for man that the earth affords was to be increased every twenty-five years by a quantity equal to what the whole world at present produces, this would allow the power of production in the earth to be absolutely unlimited, and its ratio of increase much greater than we can conceive that any possible exertions of mankind could make it.

Taking the population of the world at any number, a thousand millions, for instance, the human species would increase in the ratio of – 1, 2, 4, 8, 16, 32, 64, 128, 256, 512, etc. and subsistence as – 1, 2, 3, 4, 5, 6, 7, 8, 9, 10, etc. In two centuries and a quarter, the population would be to the means of subsistence as 512 to 10: in three centuries as 4096 to 13, and in two thousand years the difference would be almost incalculable, though the produce in that time would have increased to an immense extent.

No limits whatever are placed to the productions of the earth; they may increase for ever and be greater than any assignable quantity. yet still the power of population being a power of a superior order, the increase of the human species can only be kept commensurate to the increase of the means of subsistence by the constant operation of the strong law of necessity acting as a check upon the greater power.

The effects of this check remain now to be considered.

Among plants and animals the view of the subject is simple. They are all impelled by a powerful instinct to the increase of their species, and this instinct is interrupted by no reasoning or doubts about providing for their offspring. Wherever therefore there is liberty, the power of increase is exerted, and the superabundant effects are repressed afterwards by want of room and nourishment, which is common to animals and plants, and among animals by becoming the prey of others.

The effects of this check on man are more complicated. Impelled to the increase of his species by an equally powerful instinct, reason interrupts his career and asks him whether he may not bring beings into the world for whom he cannot provide the means of subsistence. In a state of equality, this would be the simple question. In the present state of society, other considerations occur. Will he not lower his rank in life? Will he not subject himself to greater difficulties than he at present feels? Will he not be obliged to labour harder? and if he has a large family, will his utmost exertions enable him to support them? May he not see his offspring in rags and misery, and clamouring for bread that he cannot give them? And may he not be reduced to the grating necessity of forfeiting his independence, and of being obliged to the sparing hand of charity for support?

These considerations are calculated to prevent, and certainly do prevent, a very great number in all civilized nations from pursuing the dictate of nature in an early attachment to one woman. And this restraint almost necessarily, though not absolutely so, produces vice.

[...]

CHAPTER 9

Mr Condorcet's conjecture concerning the organic perfectibility of man, and the indefinite prolongation of human life – Fallacy of the argument, which infers an unlimited progress from a partial improvement, the limit of which cannot be ascertained, illustrated in the breeding of animals, and the cultivation of plants.

THE LAST QUESTION WHICH MR CONDORCET PROPOSES for examination is the organic perfectibility of man. He observes that if the proofs which have been already given and which, in their development will receive greater force in the work itself, are sufficient to establish the indefinite perfectibility of man upon the supposition of the same natural faculties and the same organization which he has at present, what will be the certainty, what the extent of our hope, if this organization, these natural faculties themselves, are susceptible of amelioration?

From the improvement of medicine, from the use of more wholesome food and habitations, from a manner of living which will improve the strength of the body by exercise without impairing it by excess, from the destruction of the two great causes of the degradation of man, misery, and too great riches, from the gradual removal of transmissible and contagious disorders by the improvement of physical knowledge, rendered more efficacious by the progress of reason and of social order, he infers that though man will not absolutely become immortal, yet that the duration between his birth and natural death will increase without ceasing, will have no assignable term, and may properly be expressed by the word 'indefinite'. He then defines this word to mean either a constant approach to an unlimited extent, without ever reaching it, or an increase. In the immensity of ages to an extent greater than any assignable quantity.

But surely the application of this term in either of these senses to the duration of human life is in the highest degree unphilosophical and totally unwarranted by any appearances in the laws of nature. Variations from different causes are essentially distinct from a regular and unretrograde increase. The average duration of human life will to a certain degree vary from healthy or unhealthy climates, from wholesome or unwholesome food, from virtuous or vicious manners, and other causes, but it may be fairly doubted whether there is really the smallest perceptible advance in the natural duration of human life since first we have had any authentic history of man. The prejudices of all ages have indeed been directly contrary to this supposition, and though I would not lay much stress upon these prejudices, they will in some measure tend to prove that there has been no marked advance in an opposite direction.

It may perhaps be said that the world is yet so young, so completely in its infancy, that it ought not to be expected that any difference should appear so soon.

If this be the case, there is at once an end of all human science. The whole train of reasonings from effects to causes will be destroyed. We may shut our eyes to the book of nature, as it will no longer be of any use to read it. The wildest and most improbable conjectures may be advanced with as much certainty as the most just and sublime theories, founded on careful and reiterated experiments. We may return again to the old mode of philosophising and make facts bend to systems, instead of establishing systems upon facts. The grand and consistent theory of Newton will be placed upon the same footing as the wild and eccentric hypotheses of Descartes. In short, if the laws of nature are thus fickle and inconstant, if it can be affirmed and be believed that they will change, when for ages and ages they have appeared immutable, the human mind will no longer have any incitements to inquiry, but must remain fixed in inactive torpor, or amuse itself only in bewildering dreams and extravagant fancies.

The constancy of the laws of nature and of effects and causes is the foundation of all human knowledge, though far be it from me to say that the same power which framed and executes the laws of nature may not change them all 'in a moment, in the twinkling of an eye.' Such a change may undoubtedly happen. All that I mean to say is that it is impossible to infer it from reasoning. If without any previous observable symptoms or indications of a change, we can infer that a change will take place, we may as well make any assertion whatever and think it as unreasonable to be contradicted in affirming that the moon will come in contact with the earth tomorrow, as in saying that the sun will rise at its usual time.

With regard to the duration of human life, there does not appear to have existed from the earliest ages of the world to the present moment the smallest permanent symptom or indication of increasing prolongation. The observable effects of climate, habit, diet, and other causes, on length of life have furnished the pretext for asserting its indefinite extension; and the sandy foundation on which the argument rests is that because the limit of human life is undefined; because you cannot mark its precise term, and say so far exactly shall it go and no further; that therefore its extent may increase for ever, and be properly termed indefinite or unlimited. But the fallacy and absurdity of this argument will sufficiently appear from a slight examination of what Mr Condorcet calls the organic perfectibility, or degeneration, of the race of plants and animals, which he says may be regarded as one of the general laws of nature.

I am told that it is a maxim among the improvers of cattle that you may breed to any degree of nicety you please, and they found this maxim upon another, which is that some of the offspring will possess the desirable qualities of the parents in a greater degree. In the famous Leicestershire breed of sheep, the object is to procure them with small heads and small legs. Proceeding upon these breeding maxims, it is evident that we might go on till the heads and legs were evanescent quantities, but this is so palpable an absurdity that we may be quite sure that the premises are not just and that there really is a limit, though we cannot see it or say exactly where it is. In this case, the point of the greatest degree of improvement, or the smallest size of the head and legs, may be said to be undefined, but this is very different from unlimited, or from indefinite, in Mr Condorcet's acceptation of the term. Though I may not be able in the present instance to mark the limit at which further improvement will stop, I can very easily mention a point at which it will not arrive. I should not scruple to assert that were the breeding to continue for ever, the head and legs of these sheep would never be so small as the head and legs of a rat.

It cannot be true, therefore, that among animals, some of the offspring will possess the desirable qualities of the parents in a greater degree, or that animals are indefinitely perfectible.

The progress of a wild plant to a beautiful garden flower is perhaps more marked and striking than anything that takes place among animals, yet even here it would be the height of absurdity to assert that the progress was unlimited or indefinite.

One of the most obvious features of the improvement is the increase of size. The flower has grown gradually larger by cultivation. If the progress were really unlimited it might be increased ad infinitum, but this is so gross an absurdity that we may be quite sure that among plants as well as among animals there is a limit to improvement, though we do not exactly know where it is. It is probable that the gardeners who contend for flower prizes have often applied stronger dressing without success. At the same time it would be highly presumptuous in any man to say that he had seen the finest carnation or anemone that could ever be made to grow. He might however assert without the smallest chance of being contradicted by a future fact, that no carnation or anemone could ever by cultivation be increased to the size of a large cabbage; and yet there are assignable quantities much greater than a cabbage. No man can say that he has seen the largest ear of wheat, or the largest oak that could ever grow; but he might easily, and with perfect certainty, name a point of magnitude at which they would not arrive. In all these cases therefore, a careful distinction should be made, between an unlimited progress, and a progress where the limit is merely undefined.

It will be said, perhaps, that the reason why plants and animals cannot increase indefinitely in size is, that they would fall by their own weight. I answer, how do we know this but from experience? – from experience of the degree of strength with which these bodies are formed. I know that a carnation, long before it reached the size of a cabbage, would not be supported by its stalk, but I only know this from my experience of the weakness and want of tenacity in the materials of a carnation stalk. There are many substances in nature of the same size that would support as large a head as a cabbage.

The reasons of the mortality of plants are at present perfectly unknown to us. No man can say why such a plant is annual, another biennial, and another endures for ages. The whole affair in all these cases, in plants, animals, and in the human race, is an affair of experience, and I only conclude that man is mortal because the invariable experience of all ages has proved the mortality of those materials of which his visible body is made:

What can we reason, but from what we know?

Sound philosophy will not authorize me to alter this opinion of the mortality of man on earth, till it can be clearly proved that the human race has made, and is making, a decided progress towards an illimitable extent of life. And the chief reason why I adduced the two particular instances from animals and plants was to expose and illustrate, if I could, the fallacy of that argument which infers an unlimited progress, merely because some partial improvement has taken place, and that the limit of this improvement cannot be precisely ascertained.

The capacity of improvement in plants and animals, to a certain degree, no person can possibly doubt. A clear and decided progress has already been made, and yet, I think, it appears that it would be highly absurd to say that this progress has no limits. In human life, though there are great variations from different causes, it may be doubted whether, since the world began, any organic improvement whatever in the human frame can be clearly ascertained. The foundations, therefore, on which the arguments for the organic perfectibility of man rest, are unusually weak, and can only be considered as mere conjectures. It does not, however, by any means seem impossible that by an attention to breed, a certain degree of improvement, similar to that among animals, might take place among men. Whether intellect could be communicated may be a matter of doubt: but size, strength, beauty, complexion, and perhaps even longevity are in a degree transmissible. The error does not seem to lie in supposing a small degree of improvement possible, but in not discriminating between a small improvement, the limit of which is undefined, and an improvement really unlimited. As the human race, however, could not be improved in this way, without condemning all the bad specimens to celibacy, it is not probable that an attention to breed should ever become general; indeed, I know of no well-directed attempts of this kind, except in the ancient family of the Bickerstaffs, who are said to have been very successful in whitening the skins and increasing the height of their race by prudent marriages, particularly by that very judicious cross with Maud, the milk-maid, by which some capital defects in the constitutions of the family were corrected.

It will not be necessary, I think, in order more completely to shew the improbability of any approach in man towards immortality on earth, to urge the very great additional weight that an increase in the duration of life would give to the argument of population.

Many, I doubt not, will think that the attempting gravely to controvert so absurd a paradox as the immortality of man on earth, or indeed, even the perfectibility of man and society, is a waste of time and words, and that such unfounded conjectures are best answered by neglect. I profess, however, to be of a different opinion. When paradoxes of this kind are advanced by ingenious and able men, neglect has no tendency to convince them of their mistakes. Priding themselves on what they conceive to be a mark of the reach and size of their own understandings, of the extent and comprehensiveness of their views, they will look upon this neglect merely as an indication of poverty, and narrowness, in the mental exertions of their contemporaries, and only think that the world is not yet prepared to receive their sublime truths.

On the contrary, a candid investigation of these subjects, accompanied with a perfect readiness to adopt any theory warranted by sound philosophy, may have a tendency to convince them that in forming improbable and unfounded hypotheses, so far from enlarging the bounds of human science, they are contracting it, so far from promoting the improvement of the human mind, they are obstructing it; they are throwing us back again almost into the infancy of knowledge and weakening the foundations of that mode of philosophising, under the auspices of which science has of late made such rapid advances. The present rage for wide and unrestrained speculation seems to be a kind of mental intoxication, arising, perhaps, from the great and unexpected discoveries which have been made of late years, in various branches of science. To men elate and giddy with such successes, every thing appeared to be within the grasp of human powers; and, under this illusion, they confounded subjects where no real progress could be proved with those where the progress had been marked, certain, and acknowledged. Could they be persuaded to sober themselves with a little severe and chastised thinking, they would see, that the cause of truth, and of sound philosophy, cannot but suffer by substituting wild flights and unsupported assertions for patient investigation, and well authenticated proofs.

Mr Condorcet's book may be considered not only as a sketch of the opinions of a celebrated individual, but of many of the literary men in France at the beginning of the Revolution. As such, though merely a sketch, it seems worthy of attention.

[...]

CHAPTER 18

The constant pressure of distress on man, from the principle of population, seems to direct our hopes to the future – State of trial inconsistent with our ideas of the foreknowledge of God – The world, probably, a mighty process for awakening matter into mind – Theory of the formation of mind – Excitements from the wants of the body – Excitements from the operation of general laws – Excitements from the difficulties of life arising from the principle of population.

THE VIEW OF HUMAN LIFE which results from the contemplation of the constant pressure of distress on man from the difficulty of subsistence, by shewing the little expectation that he can reasonably entertain of perfectibility on earth, seems strongly to point his hopes to the future. And the temptations to which he must necessarily be exposed, from the operation of those laws of nature which we have been examining, would seem to represent the world in the light in which it has been frequently considered, as a state of trial and school of virtue preparatory to a superior state of happiness. But I hope I shall be pardoned if I attempt to give a view in some degree different of the situation of man on earth, which appears to me to be more consistent with the various phenomena of nature which we observe around us and more consonant to our ideas of the power, goodness, and foreknowledge of the Deity.

[...]

A state of trial seems to imply a previously formed existence that does not agree with the appearance of man in infancy and indicates something like suspicion and want of foreknowledge, inconsistent with those ideas which we wish to cherish of the Supreme Being. I should be inclined, therefore, as I have hinted before, to consider the world and this life as the mighty process of God, not for the trial, but for the creation and formation of mind, a process necessary to awaken inert, chaotic matter into spirit, to sublimate the dust of the earth into soul, to elicit an ethereal spark from the clod of clay. And in this view of the subject, the various impressions and excitements which man receives through life may be considered as the forming hand of his Creator, acting by general laws, and awakening his sluggish existence, by the animating touches of the Divinity, into a capacity of superior enjoyment. The original sin of man is the torpor and corruption of the chaotic matter in which he may be said to be born.

It could answer no good purpose to enter into the question whether mind be a distinct substance from matter, or only a finer form of it. The question is, perhaps, after all, a question merely of words. Mind is as essentially mind, whether formed from matter or any other substance. We know from experience that soul and body are most intimately united, and every appearance seems to indicate that they grow from infancy together. It would be a supposition attended with very little probability to believe that a complete and full formed spirit existed in every infant, but that it was clogged and impeded in its operations during the first twenty years of life by the weakness, or hebetude, of the organs in which it was enclosed. As we shall all be disposed to agree that God is the creator of mind as well as of body, and as they both seem to be forming and unfolding themselves at the same time, it cannot appear inconsistent either with reason or revelation, if it appear to be consistent with phenomena of nature, to suppose that God is constantly occupied in forming mind out of matter and that the various impressions that man receives through life is the process for that purpose. The employment is surely worthy of the highest attributes of the Deity.

This view of the state of man on earth will not seem to be unattended with probability, if, judging from the little experience we have of the nature of mind, it shall appear upon investigation that the phenomena around us, and the various events of human life, seem peculiarly calculated to promote this great end, and especially if, upon this supposition, we can account, even to our own narrow understandings, for many of those roughnesses and inequalities in life which querulous man too frequently makes the subject of his complaint against the God of nature.

The first great awakeners of the mind seem to be the wants of the body. (It was my intention to have entered at some length into this subject as a kind of second part to the Essay. A long interruption, from particular business, has obliged me to lay aside this intention, at least for the present. I shall now, therefore, only give a sketch of a few of the leading circumstances that appear to me to favour the general supposition that I have advanced.) They are the first stimulants that rouse the brain of infant man into sentient activity, and such seems to be the sluggishness of original matter that unless by a peculiar course of excitements other wants, equally powerful, are generated, these stimulants seem, even afterwards, to be necessary to continue that activity which they first awakened. The savage would slumber for ever under his tree unless he were roused from his torpor by the cravings of hunger or the pinchings of cold, and the exertions that he makes to avoid these evils, by procuring food, and building himself a covering, are the exercises which form and keep in motion his faculties, which otherwise would sink into listless inactivity. From all that experience has taught us concerning the structure of the human mind, if those stimulants to exertion which arise from the wants of the body were removed from the mass of mankind, we have much more reason to think that they would be sunk to the level of brutes, from a deficiency of excitements, than that they would be raised to the rank of philosophers by the possession of leisure. In those countries where nature is the most redundant in spontaneous produce the inhabitants will not be found the most remarkable for acuteness of intellect. Necessity has been with great truth called the mother of invention. Some of the noblest exertions of the human mind have been set in motion by the necessity of satisfying the wants of the body. Want has not unfrequently given wings to the imagination of the poet, pointed the flowing periods of the historian, and added acuteness to the researches of the philosopher, and though there are undoubtedly many minds at present so far improved by the various excitements of knowledge, or of social sympathy, that they would not relapse into listlessness if their bodily stimulants were removed, yet it can scarcely be doubted that these stimulants could not be withdrawn from the mass of mankind without producing a general and fatal torpor, destructive of all the germs of future improvement.

Locke, if I recollect, says that the endeavour to avoid pain rather than the pursuit of pleasure is the great stimulus to action in life: and that in looking to any particular pleasure, we shall not be roused into action in order to obtain it, till the contemplation of it has continued so long as to amount to a sensation of pain or uneasiness under the absence of it. To avoid evil and to pursue good seem to be the great duty and business of man, and this world appears to be peculiarly calculated to afford opportunity of the most unremitted exertion of this kind, and it is by this exertion, by these stimulants, that mind is formed. If Locke's idea be just, and there is great reason to think that it is, evil seems to be necessary to create exertion, and exertion seems evidently necessary to create mind.

The necessity of food for the support of life gives rise, probably, to a greater quantity of exertion than any other want, bodily or mental. The Supreme Being has ordained that the earth shall not produce good in great quantities till much preparatory labour and ingenuity has been exercised upon its surface. There is no conceivable connection to our comprehensions, between the seed and the plant or tree that rises from it. The Supreme Creator might, undoubtedly, raise up plants of all kinds, for the use of his creatures, without the assistance of those little bits of matter, which we call seed, or even without the assisting labour and attention of man. The processes of ploughing and clearing the ground, of collecting and sowing seeds, are not surely for the assistance of God in his creation, but are made previously necessary to the enjoyment of the blessings of life, in order to rouse man into action, and form his mind to reason.

To furnish the most unremitted excitements of this kind, and to urge man to further the gracious designs of Providence by the full cultivation of the earth, it has been ordained that population should increase much faster than food. This general law (as it has appeared in the former parts of this Essay) undoubtedly produces much partial evil, but a little reflection may, perhaps, satisfy us, that it produces a great overbalance of good. Strong excitements seem necessary to create exertion, and to direct this exertion, and form the reasoning faculty, it seems absolutely necessary, that the Supreme Being should act always according to general laws. The constancy of the laws of nature, or the certainty with which we may expect the same effects from the same causes, is the foundation of the faculty of reason. If in the ordinary course of things, the finger of God were frequently visible, or to speak more correctly, if God were frequently to change his purpose (for the finger of God is, indeed, visible in every blade of grass that we see), a general and fatal torpor of the human faculties would probably ensue; even the bodily wants of mankind would cease to stimulate them to exertion, could they not reasonably expect that if their efforts were well directed they would be crowned with success. The constancy of the laws of nature is the foundation of the industry and foresight of the husbandman, the indefatigable ingenuity of the artificer, the skilful researches of the physician and anatomist, and the watchful observation and patient investigation of the natural philosopher. To this constancy we owe all the greatest and noblest efforts of intellect. To this constancy we owe the immortal mind of a Newton.

Editorial Headnotes

6. Erasmus Darwin, *Zoonomia; or the Laws of Organic Life*, (2 vols.) (Dublin: P. Byrne & W. Jones, 1794), 478, 498–9, 500–9, 519–21.

7. Erasmus Darwin, *The Temple of Nature. Or, The Origin of Society. A Poem, with Philosophical Notes* (London: J. Johnson, 1803), pp. 1, 19, 26, 33, 34, 35–7, 153–9.

The British naturalist, philosopher, and physician Erasmus Darwin (1731–1802) has long been noted as a pre-Darwinian evolutionist. However, the depth of his evolutionary thinking has often been overshadowed by that of his grandson, a view that was only advanced by the fact that the younger Darwin himself stated that he found his grandfather's works to be overly speculative and of no influence.[1] The fact that Erasmus Darwin wrote some of his scientific works in verse has added to the ease with which his ideas have been marginalized. In his own time, however, Erasmus Darwin was regarded across Europe as a notable man of science. He openly sympathized with the liberationist and progressive ambitions of the French Revolution, and across his works, he described a natural, developmentalist account of the origins and development of life on earth, from its simplest beginnings up to and including human life and society. In works such as *The Botanic Garden* (1789; 1791), *Zoonomia; or The Laws of Organic Life* (1794), and *The Temple of Nature* (1803), he developed and laid out his transmutationist ideas. In *Zoonomi*a, he anticipated many of Lamarck's transmutationist ideas by several years but also highlighted a number of themes that would remain central to evolutionary inquiry: movement in plants; cross-fertilization; the origin of the sense of beauty; adaptive and protective colouration; heredity; and the domestication of animals. He also described sexual selection, although not by that name, and clearly recognized that all of nature was subject to a struggle, conflict, and death:

> "From Hunger's arm the shafts of Death are hurl'd,
> And one great Slaughterhouse the warring world!"[2]

It is notable that it was in response to Darwin's works – and the radical Francophile ideas with which they were quite correctly associated – that the Reverend William Paley wrote his 1802 work *Natural Theology*, in which he reviewed the English tradition of natural theology, and to which Charles Darwin responded in his *On the Origin of Species*.[3]

Notes

1 In her commentary on the difference between grandfather and grandson, Nora Barlow stated that although both formulated an evolutionary theory, Erasmus Darwin "left no lasting mark on commonly held beliefs" and "Charles succeeded where Erasmus failed" (p. 149). Erasmus Darwin was guilty of "wild theorizing" (p. 150). For Darwin's views, see p. 49, Nora Barlow (ed), *The Autobiography of Charles Darwin 1809–1882* (London: Norton, 1958).
2 Erasmus Darwin, *The Temple of Nature*, p.48.
3 See other volumes in this series for a detailed analysis of Paley and natural theology, most notably those edited by Adam Shapiro.

6

ZOONOMIA; OR THE LAWS OF ORGANIC LIFE

Erasmus Darwin

Source: Erasmus Darwin, *Zoonomia; or the Laws of Organic Life*, (2 vols.) (Dublin: P. Byrne & W. Jones, 1794), 478, 498–9, 500–9, 519–21

SECT. XXXIX.
OF GENERATION.

006 "Portrait of Erasmus Darwin."

DOI: 10.4324/9781003490548-7

Felix, qui causas altâ caligine mersas
Pandit, et evolvit tenuissima vincula rerum. ANON.

[...]

6. From this account of reproduction it appears, that all animals have a similar origin, viz. from a single living filament; and that the difference of their forms and qualities has arisen only from the different irritabilities and sensibilities, or voluntarities, or associabilities, of this original living filament; and perhaps in some degree from the different forms of the particles of the fluids, by which it has been at first stimulated into activity. And that from hence, as Linnæus has conjectured in respect to the vegetable world, it is not impossible, but the great variety of species of animals, which now tenant the earth, may have had their origin from the mixture of a few natural orders. And that those animal and vegetable mules, which could continue their species, have done so, and constitute the numerous families of animals and vegetables which now exist; and that those mules, which were produced with imperfect organs of generation, perished without reproduction, according to the observation of Aristotle; and are the animals, which we now call mules. See Botanic Garden, Part II. Note on Dianthus.

Such a promiscuous intercourse of animals is said to exist at this day in New South Wales by Captain Hunter. And that not only amongst the quadrupeds and birds of different kinds, but even amongst the fish, and, as he believes, amongst the vegetables. He speaks of an animal between the opossum and the kangaroo, from the size of a sheep to that of a rat. Many fish seemed to partake of the shark; some with a skait's head and shoulders, and the hind part of a shark; others with a shark's head and the body of a mullet; and some with a shark's head and the flat body of a sting-ray. Many birds partake of the parrot; some have the head, neck, and bill of a parrot, with long straight feet and legs; others with legs and feet of a parrot, with head and neck of a sea-gull. Voyage to South Wales by Captain John Hunter, p. 68.

7. All animals therefore, I contend, have a similar cause of their organization, originating from a single living filament, endued indeed with different kinds of irritabilities and sensibilities, or of animal appetencies; which exist in every gland, and in every moving organ of the body, and are as essential to living organization as chemical affinities are to certain combinations of inanimate matter.

[...]

8. When we revolve in our minds, first, the great changes, which we see naturally produced in animals after their nativity, as in the production of the butterfly with painted wings from the crawling caterpillar; or of the respiring frog from the subnatant tadpole; from the feminine boy to the bearded man, and from the infant girl to the lactescent woman; both which changes may be prevented by certain mutilations of the glands necessary to reproduction.

Secondly, when we think over the great changes introduced into various animals by artificial or accidental cultivation, as in horses, which we have exercised for the different purposes of strength or swiftness, in carrying burthens or in running races; or in dogs, which have been cultivated for strength and courage, as the bull-dog; or for acuteness of his sense of smell, as the hound and spaniel; or for the swiftness of his foot, as the greyhound; or for his swimming in the water, or for drawing snow-sledges, as the rough-haired dogs of the north; or lastly, as a play-dog for children, as the lap-dog; with the changes of the forms of the cattle, which have been domesticated from the greatest antiquity, as camels, and sheep; which have undergone so total a transformation, that we are now ignorant from what species of wild animals they had their origin. Add to these the great changes of shape and colour, which we daily see produced in smaller animals from our domestication of them, as rabbits, or pidgeons; or from the difference of climates and even of seasons; thus the sheep of warm climates are covered with hair instead of wool; and the hares and partridges of the latitudes, which are long buried in snow, become white during the winter months; add to these the various changes produced in the forms of mankind, by their early modes of exertion; or by the diseases occasioned by their habits of life; both of which became hereditary, and that through many generations. Those who labour at the anvil, the oar, or the loom, as well as those who carry sedan-chairs, or who have been educated to dance upon the rope, are distinguishable by the shape of their limbs; and the diseases occasioned by intoxication deform the countenance with leprous eruptions, or the body with tumid viscera, or the joints with knots and distortions.

Thirdly, when we enumerate the great changes produced in the species of animals before their nativity; these are such as resemble the form or colour of their parents, which have been altered by the cultivation or accidents above related, and are thus continued to their posterity. Or they are changes produced by the mixture of species as in mules; or changes produced probably by the exuberance of nourishment supplied to the fetus, as in monstrous births with additional limbs; many of these enormities of shape are propagated, and continued as a variety at least, if not as a new species of animal. I have seen a breed of cats with an additional claw on every foot; of poultry also with an additional claw, and with wings to their feet; and of others without rumps. Mr. Buffon mentions a breed of dogs without tails, which are common at Rome and at Naples, which he supposes to have been produced by a custom long established of cutting their tails close off. There are many kinds of pidgeons, admired for their peculiarities, which are monsters thus produced and propagated. And to these must be added, the changes produced by the imagination of the male parent, as will be treated of more at large in No. VI. of this Section.

When we consider all these changes of animal form, and innumerable others, which may be collected from the books of natural history; we cannot but be convinced, that the fetus or embryon is formed by apposition of new parts, and not by the distention of a primordial nest of germs, included one within another, like the cups of a conjurer.

Fourthly, when we revolve in our minds the great similarity of structure, which obtains in all the warm-blooded animals, as well quadrupeds, birds, and amphibious animals, as in mankind; from the mouse and bat to the elephant and whale;

one is led to conclude, that they have alike been produced from a similar living filament. In some this filament in its advance to maturity has acquired hands and fingers, with a fine sense of touch, as in mankind. In others it has acquired claws or talons, as in tygers and eagles. In others, toes with an intervening web, or membrane, as in seals and geese. In others it has acquired cloven hoofs, as in cows and swine; and whole hoofs in others, as in the horse. While in the bird kind this original living filament has put forth wings instead of arms or legs, and feathers instead of hair. In some it has protruded horns on the forehead instead of teeth in the fore part of the upper jaw; in others tushes instead of horns; and in others beaks instead of either. And all this exactly as is daily seen in the transmutations of the tadpole, which acquires legs and lungs, when he wants them; and loses his tail, when it is no longer of service to him.

Fifthly, from their first rudiment, or primordium, to the termination of their lives, all animals undergo perpetual transformations; which are in part produced by their own exertions in consequence of their desires and aversions, of their pleasures and their pains, or of irritations, or of associations; and many of these acquired forms or propensities are transmitted to their posterity. See Sect. XXXI. 1.

As air and water are supplied to animals in sufficient profusion, the three great objects of desire, which have changed the forms of many animals by their exertions to gratify them, are those of lust, hunger, and security. A great want of one part of the animal world has consisted in the desire of the exclusive possession of the females; and these have acquired weapons to combat each other for this purpose, as the very thick, shield-like, horny skin on the shoulder of the boar is a defence only against animals of his own species, who strike obliquely upwards, nor are his tushes for other purposes, except to defend himself, as he is not naturally a carnivorous animal. So the horns of the stag are sharp to offend his adversary, but are branched for the purpose of parrying or receiving the thrusts of horns similar to his own, and have therefore been formed for the purpose of combating other stags for the exclusive possession of the females; who are observed, like the ladies in the times of chivalry, to attend the car of the victor.

The birds, which do not carry food to their young, and do not therefore marry, are armed with spurs for the purpose of fighting for the exclusive possession of the females, as cocks and quails. It is certain that these weapons are not provided for their defence against other adversaries, because the females of these species are without this armour. The final cause of this contest amongst the males seems to be, that the strongest and most active animal should propagate the species, which should thence become improved.

Another great want consists in the means of procuring food, which has diversified the forms of all species of animals. Thus the nose of the swine has become hard for the purpose of turning up the soil in search of insects and of roots. The trunk of the elephant is an elongation of the nose for the purpose of pulling down the branches of trees for his food, and for taking up water without bending his knees. Beasts of prey have acquired strong jaws or talons. Cattle have acquired a rough tongue and a rough palate to pull off the blades of grass, as cows and sheep. Some birds have acquired harder beaks to crack nuts, as the parrot. Others have acquired beaks adapted to break the harder seeds, as sparrows. Others for the softer seeds of flowers, or the buds of trees, as the finches. Other birds have acquired long beaks to penetrate the moister soils in search of insects or roots, as woodcocks; and others broad ones to filtrate the water of lakes, and to retain aquatic insects. All which seem to have been gradually produced during many generations by the perpetual endeavour of the creatures to supply the want of food, and to have been delivered to their posterity with constant improvement of them for the purposes required.

The third great want amongst animals is that of security, which seems much to have diversified the forms of their bodies and the colour of them; these consist in the means of escaping other animals more powerful than themselves. Hence some animals have acquired wings instead of legs, as the smaller birds, for the purpose of escape. Others great length of fin, or of membrane, as the flying fish, and the bat. Others great swiftness of foot, as the hare. Others have acquired hard or armed shells, as the tortoise and the echinus marinus.

The contrivances for the purposes of security extend even to vegetables, as is seen in the wonderful and various means of their concealing or defending their honey from insects, and their seeds from birds. On the other hand swiftness of wing has been acquired by hawks and swallows to pursue their prey; and a proboscis of admirable structure has been acquired by the bee, the moth, and the humming bird, for the purpose of plundering the nectaries of flowers. All which seem to have been formed by the original living filament, excited into action by the necessities of the creatures, which possess them, and on which their existence depends.

From thus meditating on the great similarity of the structure of the warm-blooded animals, and at the same time of the great changes they undergo both before and after their nativity; and by considering in how minute a portion of time many of the changes of animals above described have been produced; would it be too bold to imagine, that in the great length of time, since the earth began to exist, perhaps millions of ages before the commencement of the history of mankind, would it be too bold to imagine, that all warm-blooded animals have arisen from one living filament, which THE GREAT FIRST CAUSE endued with animality, with the power of acquiring new parts, attended with new propensities, directed by irritations, sensations, volitions, and associations; and thus possessing the faculty of continuing to improve by its own inherent activity, and of delivering down those improvements by generation to its posterity, world without end!

Sixthly, The cold-blooded animals, as the fish-tribes, which are furnished with but one ventricle of the heart, and with gills instead of lungs, and with fins instead of feet or wings, bear a great similarity to each other; but they differ, nevertheless, so much in their general structure from the warm-blooded animals, that it may not seem probable at first view, that the same living filament could have given origin to this kingdom of animals, as to the former. Yet are there some creatures, which unite or partake of both these orders of animation, as the whales and seals; and more particularly the frog, who changes from an aquatic animal furnished with gills to an aerial one furnished with lungs.

The numerous tribes of insects without wings, from the spider to the scorpion, from the flea to the lobster; or with wings, from the gnat and the ant to the wasp and the dragon-fly, differ so totally from each other, and from the red-blooded classes above described, both in the forms of their bodies, and their modes of life; besides the organ of sense, which they seem to possess in their antennæ or horns, to which it has been thought by some naturalists, that other creatures have nothing similar; that it can scarcely be supposed that this nation of animals could have been produced by the same kind of living filament, as the red-blooded classes above mentioned. And yet the changes which many of them undergo in their early state to that of their maturity, are as different, as one animal can be from another. As those of the gnat, which passes his early state in water, and then stretching out his new wings, and expanding his new lungs, rises in the air; as of the caterpillar, and bee-nymph, which feed on vegetable leaves or farina, and at length bursting from their self-formed graves, become beautiful winged inhabitants of the skies, journeying from flower to flower, and nourished by the ambrosial food of honey.

There is still another class of animals, which are termed vermes by Linnæus, which are without feet, or brain, and are hermaphrodites, as worms, leeches, snails, shell-fish, coralline insects, and sponges; which possess the simplest structure of all animals, and appear totally different from those already described. The simplicity of their structure, however, can afford no argument against their having been produced from a living filament as above contended.

Last of all the various tribes of vegetables are to be enumerated amongst the inferior orders of animals. Of these the anthers and stigmas have already been shewn to possess some organs of sense, to be nourished by honey, and to have the power of generation like insects, and have thence been announced amongst the animal kingdom in Sect. XIII. and to these must be added the buds and bulbs which constitute the viviparous offspring of vegetation. The former I suppose to be beholden to a single living filament for their seminal or amatorial procreation; and the latter to the same cause for their lateral or branching generation, which they possess in common with the polypus, tænia, and volvox; and the simplicity of which is an argument in favour of the similarity of its cause.

Linnæus supposes, in the Introduction to his Natural Orders, that very few vegetables were at first created, and that their numbers were increased by their intermarriages, and adds, suadent hæc Creatoris leges a simplicibus ad composita. Many other changes seem to have arisen in them by their perpetual contest for light and air above ground, and for food or moisture beneath the soil. As noted in Botanic Garden, Part II. Note on Cuscuta. Other changes of vegetables from climate, or other causes, are remarked in the Note on Curcuma in the same work. From these one might be led to imagine, that each plant at first consisted of a single bulb or flower to each root, as the gentianella and daisy; and that in the contest for air and light new buds grew on the old decaying flower stem, shooting down their elongated roots to the ground, and that in process of ages tall trees were thus formed, and an individual bulb became a swarm of vegetables. Other plants, which in this contest for light and air were too slender to rise by their own strength, learned by degrees to adhere to their neighbours, either by putting forth roots like the ivy, or by tendrils like the vine, or by spiral contortions like the honey-suckle; or by growing upon them like the misleto, and taking nourishment from their barks; or by only lodging or adhering on them, and deriving nourishment from the air, as tillandsia.

Shall we then say that the vegetable living filament was originally different from that of each tribe of animals above described? And that the productive living filament of each of those tribes was different originally from the other? Or, as the earth and ocean were probably peopled with vegetable productions long before the existence of animals; and many families of these animals long before other families of them, shall we conjecture, that one and the same kind of living filaments is and has been the cause of all organic life?

This idea of the gradual formation and improvement of the animal world accords with the observations of some modern philosophers, who have supposed that the continent of America has been raised out of the ocean at a later period of time than the other three quarters of the globe, which they deduce from the greater comparative heights of its mountains, and the consequent greater coldness of its respective climates, and from the less size and strength of its animals, as the tygers and allegators compared with those of Asia or Africa. And lastly, from the less progress in the improvements of the mind of its inhabitants in respect to voluntary exertions.

This idea of the gradual formation and improvement of the animal world seems not to have been unknown to the ancient philosophers. Plato having probably observed the reciprocal generation of inferior animals, as snails and worms, was of opinion, that mankind with all other animals were originally hermaphrodites during the infancy of the world, and were in process of time separated into male and female. The breasts and teats of all male quadrupeds, to which no use can be now assigned, adds perhaps some shadow of probability to this opinion. Linnæus excepts the horse from the male quadrupeds, who have teats; which might have shewn the earlier origin of his existence; but Mr. T. Hunter asserts, that he has discovered the vestiges of them on his sheath, and has at the same time enriched natural history with a very curious fact concerning the male pidgeon; at the time of hatching the eggs both the male and female pidgeon undergo a great change in their crops; which thicken and become corrugated, and secrete a kind of milky fluid, which coagulates, and with which alone they for a few days feed their young, and afterwards feed them with this coagulated fluid mixed with other food. How this resembles the breasts of female quadrupeds after the production of their young! and how extraordinary, that the male should at this time give milk as well as the female! See Botanic Garden, Part II. Note on Curcuma.

The late Mr. David Hume, in his posthumous works, places the powers of generation much above those of our boasted reason; and adds, that reason can only make a machine, as a clock or a ship, but the power of generation makes the maker of the machine; and probably from having observed, that the greatest part of the earth has been formed out of organic recrements; as the immense beds of limestone, chalk, marble, from the shells of fish; and the extensive provinces of clay, sandstone, ironstone, coals, from decomposed vegetables; all which have been first produced by generation, or by the secretions of organic life; he concludes, that the world itself might have been generated, rather than created; that is, it might have been gradually produced from very small beginnings, increasing by the activity of its inherent principles, rather than by a sudden evolution of the whole by the Almighty fiat. – What a magnificent idea of the infinite power of THE GREAT ARCHITECT! THE CAUSE OF CAUSES! PARENT OF PARENTS! ENS ENTIUM!

[...]

3. I conclude, that the imagination of the male at the time of copulation, or at the time of the secretion of the semen, may so affect this secretion by irritative or sensitive association, as described in No. 5. 1. of this section, as to cause the production of similarity of form and of features, with the distinction of sex; as the motions of the chissel of the turner imitate or correspond with those of the ideas of the artist. It is not here to be understood, that the first living fibre, which is to form an animal, is produced with any similarity of form to the future animal; but with propensities, or appetencies, which shall produce by accretion of parts the similarity of form, feature, or sex, corresponding to the imagination of the father.

Our ideas are movements of the nerves of sense, as of the optic nerve in recollecting visible ideas, suppose of a triangular piece of ivory. The fine moving fibres of the retina act in a manner to which I give the name of white; and this action is confined to a defined part of it; to which figure I give the name of triangle. And it is a preceding pleasurable sensation existing in my mind, which occasions me to produce this particular motion of the retina, when no triangle is present. Now it is probable, that the acting sibres of the ultimate terminations of the secreting apertures of the vessels of the testes, are as fine as those of the retina; and that they are liable to be thrown into that peculiar action, which marks the sex of the secreted embryon, by sympathy with the pleasurable motions of the nerves of vision, or of touch; that is, with certain ideas of imagination. From hence it would appear, that the world has long been mistaken in ascribing great power to the imagination of the female, whereas from this account of it, the real power of imagination, in the act of generation, belongs solely to the male. See Sect. XII. 3. 3.

It may be objected to this theory, that a man may be supposed to have in his mind, the idea of the form and features of the female, rather than his own, and therefore there should be a greater number of female births. On the contrary, the general idea of our own form occurs to every one almost perpetually, and is termed consciousness of our existence, and thus may effect, that the number of males surpasses that of females. See Sect. XV. 3. 4. and XVIII. 13. And what further confirms this idea is, that the male children most frequently resemble the father in form, or feature, as well as in sex; and the female most frequently resemble the mother, in feature, and form, as well as in sex.

It may again be objected, if a female child sometimes resembles the father, and a male child the mother, the ideas of the father, at the time of procreation, must suddenly change from himself to the mother, at the very instant, when the embryon is secreted or formed. This difficulty ceases when we consider, that it is as easy to form an idea of feminine features with male organs of reproduction, or of male features features with female ones, as the contrary; as we conceive the idea of a sphinx or mermaid, as easily and as distinctly as of a woman. Add to this, that at the time of procreation the idea of the male organs, and of the female features, are often both excited at the same time, by contact, or by vision.

7

THE TEMPLE OF NATURE. OR, THE ORIGIN OF SOCIETY. A POEM, WITH PHILOSOPHICAL NOTES

Erasmus Darwin

Source: Erasmus Darwin, *The Temple of Nature. Or, The Origin of Society. A Poem, with Philosophical Notes* (London: J. Johnson, 1803), pp. 1, 19, 26, 33, 34, 35–7, 153–9

007 "Frontispiece to Erasmus Darwin's *Temple of Nature* (London: J. Johnson, 1803)." *Image in the public domain.*

 DOI: 10.4324/9781003490548-8

THE TEMPLE OF NATURE.

CANTO I.
PRODUCTION OF LIFE.

I. By firm immutable immortal laws,
Impress'd on Nature by the Great First Cause,
Say, Muse! how rose from elemental strife
Organic forms, and kindled into life;
How Love and Sympathy, with potent charm,
Warm the cold heart, the lifted hand disarm;
Allure with pleasures, and alarm with pains,
And bind Society in golden chains.
[...]

Shrin'd in the midst majestic Nature stands,
Extends o'er earth and sea her hundred hands;
Tower upon tower her beamy forehead crests,
And births unnumber'd milk her hundred breasts;
Drawn round her brows a lucid veil depends,
O'er her fine waist the purfled woof descends;
Her stately limbs the gather'd folds surround,
And spread their golden selvage on the ground.
[...]
Whence the fine nerve to move and feel assign'd,
Contractile fibre, and ethereal mind:

"How Love and Sympathy the bosom warm,
Allure with pleasure, and with pain alarm,
With soft affections weave the social plan,
And charm the listening Savage into Man."
[...]

Next the long nerves unite their silver train,
And young Sensation permeates the brain;
Through each new sense the keen emotions dart,
Flush the young cheek, and swell the throbbing heart.
From pain and pleasure quick Volitions rise,
Lift the strong arm, or point the inquiring eyes;[a]
[...]

With Reason's light bewilder'd Man direct,
And right and wrong with balance nice detect.
Last, in thick swarms Associations spring,
Thoughts join to thoughts, to motions motions cling;

Whence in long trains of catenation flow
Imagined joy, and voluntary woe.
[...]

V. "ORGANIC LIFE beneath the shoreless waves
Was born, and nurs'd in Ocean's pearly caves;[b]

First forms minute, unseen by spheric glass,
Move on the mud, or pierce the watery mass;
These, as successive generations bloom,
New powers acquire, and larger limbs assume;
Whence countless groups of vegetation spring,
And breathing realms of fin, and feet, and wing.[c]

Thus the tall Oak, the giant of the wood,
Which bears Britannia's thunders on the flood;
The Whale, unmeasured monster of the main,
The lordly Lion, monarch of the plain,
The Eagle, soaring in the realms of air,
Whose eye undazzled drinks the solar glare; –
Imperious man, who rules the bestial crowd,
Of language, reason, and reflection proud,
With brow erect, who scorns this earthy sod,
And styles himself the image of his God;
Arose from rudiments of form and sense,
An embryon point, or microscopic ens![d]
[...]

CANTO IV.
OF GOOD AND EVIL.

I. "How few," the Muse in plaintive accents cries,
And mingles with her words pathetic sighs. –
"How few, alas! in Nature's wide domains,
The sacred charm of SYMPATHY restrains!
Uncheck'd desires from appetite commence,
And pure reflection yields to selfish sense!
– Blest is the Sage,[e] who learn'd in Nature's laws,
With nice distinction marks effect and cause;
Who views the insatiate Grave, with eye sedate,
Nor fears thy voice, inexorable Fate!

"When War, the Demon, lifts his banner high,
And loud artillery rends the affrighted sky;

Swords clash with swords, on horses horses rush,
Man tramples man, and nations nations crush;
Death, his vast scythe, with sweep enormous wields,
And shuddering Pity quits the sanguine fields.

"The wolf, escorted by his milk-drawn dam,
Unknown to mercy, tears the guiltless lamb;
The towering eagle, darting from above,
Unfeeling rends the inoffensive dove;
The lamb and dove on living nature feed,
Crop the young herb, or crush the embryon seed.[f]

Nor spares the loud owl, in her dusky flight,
Smit with sweet notes, the minstrel of the night;
Nor spares, enamour'd of his radiant form,
The hungry nightingale the glowing worm;
Who with bright lamp alarms the midnight hour,
Climbs the green stem, and slays the sleeping flower.

"Fell Oestrus buries in her rapid course
Her countless brood, in stag, or bull, or horse;
Whose hungry larva eats its living way,
Hatch'd by the warmth, and issues into day.[g]

The wing'd Ichneumon for her embryon young,
Gores, with sharp horn, the catterpiller throng.
The cruel larva mines its silky course,
And tears the vitals of its fostering nurse.[h]

While fierce Libellula, with jaws of steel,
Ingulfs an insect-province at a meal;
Contending bee-swarms rise on rustling wings,
And slay their thousands with envenom'd stings.

"Yes! smiling Flora, drives her armed car,
Through the thick ranks of vegetable war;
Herb, shrub, and tree, with strong emotions rise
For light and air, and battle in the skies;
Whose roots, diverging, with opposing toil,
Contend below for moisture and for soil;[i]

Round the tall Elm the flattering Ivies bend,
And strangle, as they clasp, their struggling friend;
Envenom'd dews from Mancinella flow,

And scald with caustic touch the tribes below;
Dense shadowy leaves, on stems aspiring borne,
With blight and mildew thin the realms of corn;
And insect hordes, with restless tooth, devour
The unfolded bud, and pierce the ravell'd flower.

"In ocean's pearly haunts, the waves beneath,
Sits the grim monarch of insatiate Death;
The shark rapacious, with descending blow,
Darts on the scaly brood that swims below;[j]

The crawling crocodiles, beneath that move,
Arrest, with rising jaw, the tribes above;
With monstrous gape, sepulchral whales devour
Shoals at a gulp, a million in an hour.
– Air, earth, and ocean, to astonish'd day,
One scene of blood, one mighty tomb display!
From Hunger's arm the shafts of Death are hurl'd,
And one great Slaughter-house the warring world![k]

Notes

a *And young Sensation*, l.270. Both sensation and volition consist in an affection of the central part of the sensorium, or of the whole of it; and hence cannot exist till the nerves are united in the brain. The motions of a limb of any animal cut from the body, are therefore owing to irritation, not to sensation or to volition. For the definitions of irritation, sensation, volition, and association, see additional Note II.

b *Beneath the shoreless waves*, l. 295. The earth was originally covered with water, as appears from some of its highest mountains, consisting of shells cemented together by a solution of part of them, as the limestone rocks of the Alps; Ferber's Travels. It must be therefore concluded, that animal life began beneath the sea.

Nor is this unanalogous to what still occurs, as all quadrupeds and mankind, in their embryon state, are aquatic animals; and thus may be said to resemble gnats and frogs. The fetus in the uterus has an organ called the placenta, the fine extremities of the vessels of which permeate the arteries of the uterus, and the blood of the fetus becomes thus oxygenated from the passing stream of the maternal arterial blood; exactly as is done by the gills of fish from the stream of water, which they occasion to pass through them.

c But the chicken in the egg possesses a kind of aerial respiration, since the extremities of its placental vessels terminate on a membranous bag, which contains air, at the broad end of the egg; and in this the chick in the egg differs from the fetus in the womb, as there is in the egg no circulating maternal blood for the insertion of the extremities of its respiratory vessels, and in this also I suspect that the eggs of birds differ from the spawn of fish; which latter is immersed in water, and which has probably the extremities of its respiratory organ inserted into the soft membrane which covers it, and is in contact with the water. *First forms minute*, l. 297. See Additional Note I. on Spontaneous Vitality.

d *An embryon point*, l. 314. The arguments, showing that all vegetables and animals arose from such a small beginning, as a living point or living fibre, are detailed in Zoonomia, Sect. 39, 4. 8. on generation.

e *Blest is the Sage*, l:7.

> Felix, qui potuit rerum cognoscere causas;
> Quique metus omnes, et inexorabile fatum,
> Subjecit pedibus, strepitumque Acherontis avari.
>
> VIRG. GEORG. II. 490.

f *The towering eagle*, l. 19.

> Torva leæna lupum sequitur, lupus ipse capellam,
> Florentem cytisum sequitur lasciva capella. VIRG.

g *Fell Oestrus buries*, l. 29. The gadfly, bot-fly, or sheep-fly: the larva lives in the bodies of cattle throughout the whole Winter; it is extracted from their backs by an African bird called Buphaga. Adhering to the anus, it artfully introduces itself into the intestines of horses, and becomes so numerous in their stomachs, as sometimes to destroy them; it climbs into the nostrils of sheep and calves, and producing a nest of young in a transparent hydatide in the frontal sinus, occasions the vertigo or turn of those animals. In Lapland it so attacks the rein deer, that the natives annually travel with the berds from the woods to the mountains. Lin. Syst. Nat.

h *The wing'd Ichneumon*, l. 33. Linneus describes seventy-seven species of the ichneumon fly, some of which have a sting as long, and some twice as long as their bodies. Many of them insert their eggs into various catterpillers, which when they are hatched, seem for a time to prey on the reservoir of silk in the backs of those animals, designed for their own use to spin a cord to support them, or a bag to contain them, while they change from their larva form to a butterfly; as I have seen in above fifty cabbage-catterpillers.

The ichneumon larva then makes its way out of the catterpillar, and spins itself a small cocoon like a silk-worm; these cocoons are about the size of a small pin's head, and I have seen about ten of them on each cabbage-catterpillar, which soon dies after their exclusion.

Other species of ichneumon insert their eggs into the aphis, and into the larva of the aphidivorous fly; others into the bedeguar of rose trees, and the gall-nuts of oaks; whence those excrescences seem to be produced, as well as the hydatides, in the frontal sinus of sheep and calves, by the stimulus of the larvæ deposited in them.

i *While fierce Libellula*, l. 37. The Libellula or Dragon-fly, is said to be a most voracious animal: Linneus says, in their perfect state, they are the hawks to naked winged flies; in their larva state, they run beneath the water, and are the cruel crocodiles of aquatic insects. Syst. Nat.

Contending bee-swarms, l. 39. Stronger bee-swarms frequently attack weak hives, and in two or three days destroy them, and carry away their honey; this I once prevented by removing the attacked hive, after the first day's battle, to a distinct part of the garden. See Phytologia, Sect. XIV. 3. 7.

j *The shark rapacious*, l. 57. The shark has three rows of sharp teeth within each other, which he can bend downwards internally to admit larger prey, and raise to prevent its return; his snout hangs so far over his mouth, that he is necessitated to turn upon his back, when he takes fish that swim over him, and hence seems peculiarly formed to catch those that swim under him.

k *The crawling crocodiles*, l. 59. As this animal lives chiefly at the bottom of the rivers, which he frequents, he has the power of opening the upper jaw as well as the under one, and thus with greater facility, catches the fish or water-fowl which swim over him.

One great slaughter-house, l. 66. As vegetables are an inferior order of animals fixed to the soil; and as the locomotive animals prey upon them, or upon each other; the world may indeed be said to be one great slaughter-house. As the digested food of vegetables consists principally of sugar, and from this is produced again their mucilage, starch, and oil, and since animals are sustained by these vegetable productions, it would seem that the sugar-making process, carried on in vegetable vessels, was the great source of life to all organized beings. And that if our improved chemistry should ever discover the art of making sugar from fossile or aerial matter without the assistance of vegetation, food for animals would then become as plentiful as water, and they might live upon the earth without preying on each other, as {135} thick as blades of grass, with no restraint to their numbers but the want of local room.

It would seem that roots fixed in the earth and leaves innumerable waving in the air were necessary for the decomposition of water and air, and the conversion of them into saccharine matter, which would have been not only cumberous but totally incompatible with the locomotion of animal bodies. For how could a man or quadruped have carried on his head or back a forest of leaves, or have had long branching lacteal or absorbent vessels terminating in the earth? Animals therefore subsist on vegetables; that is they take the matter so prepared, and have organs to prepare it further for the purposes of higher animation and greater sensibility.

Editorial Headnote

8. Johann Friedrich Blumenbach, *On the Natural Variety of Mankind (De generis humani)*, (3rd edn. 1795), reprinted in *The Anthropological Treatises of Johann Friedrich Blumenbach*, edited by Thomas Bendyshe, (London: Longman, Green, Longman, Roberts, & Green, 1865), pp. 188–206.

Johann Friedrich Blumenbach (1752–1840) was a German anatomist, anthropologist, physiologist, and physician. A part of the 'Göttingen school,' he was not only influential in the development of German biology[1] but also in France and among British anatomists and physiologists. A number of Blumenbach's books were translated into English and served as popular textbooks in comparative anatomy and physiology from the early 1800s onward. Notable translations by the surgeons William Lawrence and William Coulson were made in 1809 and 1827, respectively.[2] Blumenbach also explicitly applied his comparative anatomical work to his understanding of human natural history. Indeed, his 1775 doctoral work *On the Natural Variety of Mankind* was among the first works in this period to advance the idea of a natural history of humanity. It was in this work that Blumenbach developed methodologies that were to shape the subsequent development of physical anthropology and craniometry, and despite Blumenbach's insistence that there was no racial hierarchy, many of those who applied his methods reached other conclusions.[3] In a later chapter of this work, Blumenbach identified five human racial types, each of which had their origin in the primordial Caucasian race but had 'degenerated' to fit their environment. Like Buffon, Blumenbach used this term to describe adaptative change and thus did not intend the pejorative or hierarchical association that this word might imply today.[4] Here, though, we consider the extract in which he articulated his understanding of species, the mechanisms of variation, and heredity.

In 1781, Blumenbach advanced the idea that there was a natural, vital, formative force that drove the self-development of every living organism. This he termed the *nisus formativus*, or *Bildungstrieb* (developmental or formative driving force). He first published this idea in a short book, *Über den Bildungstrieb,* in which he described it as a power of generation, development, and regeneration. It was in the 2nd and subsequent editions of his *Natural History of Mankind* (*De generis humani*), published in the same year, though, that he explicitly extended his claims for the *Bildungstrieb* as responsible for the formation of varieties and subspecies.[5]

Blumenbach's works were widely used in Britain as basic introductory texts in physiology, anatomy, natural history, and anthropology. Indeed, such was Blumenbach's influence on Lawrence that he dedicated his 1822 book *Lectures on Physiology, Zoology, and The Natural History of Man* to Blumenbach.[6] Blumenbach's essay on the *Bildungstrieb* was also translated and published in Britain, and his *Natural Variety of Mankind* was later translated and reprinted by Thomas Bendyshe (1827–1886), the vice-president of the Anthropological Society of London, in 1865. Bendyshe published the first and third editions of Blumenbach's work alongside an essay on the varieties of man by the anatomist John Hunter (1728–1793). Bendyshe's editorial introduction remains controversial, as he made efforts to frame Blumenbach's conception of human races in a way that fit the prevailing opinion of racial hierarchy within the Anthropological Society at that time.

Notes

1 Tim Lenoir, *The Strategy of Life. Teleology and Mechanics in Nineteenth Century German Biology* (Chicago: University of Chicago Press, 1982); John H. Zammito, *The Gestation of German Biology. Philosophy and Physiology from Stahl to Schelling* (Chicago: University of Chicago Press, 2018).

2 Johann Friedrich Blumenbach, *A Manual of Comparative Anatomy Translated from the German of J.F. Blumenbach with Additional Notes by William Lawrence Esq. FRS. Second Edition, Revised and Augmented, by William Coulson, Demonstrator of Anatomy at the Medical School, Aldersgate Street, and Member of the Zoological Society* (London: W. Simpkin & R. Marshall, 1827).

3 See other collections and texts on scientific racism and the science of race in this set of volumes and in other volumes in this series.

4 Thomas Junker, "Blumenbach's theory of human races and the natural unity of mankind," in Nicolaus Rupke and Gerhard Lauer, (eds.), *Johann Friedrich Blumenbach: Race and Natural History, 1750–1850*, (London: Routledge 2019), pp. 96–112, on p. 104.

5 Robert J. Richards, "Kant and Blumenbach on the Bildungstrieb: A historical misunderstanding," *Studies in the History and Philosophy of the Biological and Biomedical Sciences*, (2000), Vol.31, No.1, 11–32, on p. 20.

6 William Lawrence, *Lectures on Physiology, Zoology, and the Natural History of Man*, (London: Smith, 1822).

8

ON THE NATURAL VARIETY OF MANKIND (DE GENERIS HUMANI)

Johann Friedrich Blumenbach

Source: Johann Friedrich Blumenbach, *On the Natural Variety of Mankind* (*De generis humani*), (3rd edn. 1795), in The Anthropological Treatises of Johann Friedrich Blumenbach, edited by Thomas Bendyshe, (London: Longman, Green, Longman, Roberts, & Green, 1865), pp. 188–206

OF THE CAUSES AND WAYS BY WHICH THE SPECIES OF ANIMALS DEGENERATE IN GENERAL.

008 "Johann Friedrich Blumenbach, 1823 etching by Ludwig Emil Grimm."

Image in the public domain.

DOI: 10.4324/9781003490548-9

22. *Subject proposed.* Hitherto we have investigated those things in which man differs from the rest of the animals. Now we come nearer to the primary object of the whole treatise, for we are to inquire of what kind and how great is the natural diversity which separates the races and the multifarious nations of men; and to consider whether the origin of this diversity can be traced to degeneration, or whether it is not so great as to compel us rather to conclude that there is more than one original species of man. Before this can be done, there are two questions which must be considered: First, what is *species* in zoology? Secondly, how in general a primordial species may degenerate into varieties? and now of each separately.

23. *What is species?* We say that animals belong to one and the same species, if they agree so well in form and constitution, that those things in which they do differ may have arisen from degeneration. We say that those, on the other hand, are of different species, whose essential difference is such as cannot be explained by the known sources of degeneration, if I may be allowed to use such a word. So far well in the abstract, as they say. Now we come to the real difficulty, which is to set forth the characters by which, *in the natural world*, we may distinguish mere varieties from genuine species.

The immortal Ray, in the last century, long before Buffon, thought those animals should be referred to the same species, which copulate together, and have a fertile progeny. But, as in the domestic animals which man has subdued, this character seemed ambiguous and uncertain, on account of the enslaved life they lead; in the beginning of this century, the sagacious Frisch restricted it to wild animals alone, and declared that those were of the same species, who copulate in a natural state[a].

But it must be confessed that, even with this limitation, we make but little progress. For, in the first place, what very little chance is there of bringing so many wild animals, especially the exotic ones, about which it is of the greatest possible interest for us to know whether they are to be considered as mere varieties, or as different species, to that test of copulation? especially if their native countries are widely apart; as is the case with the *Satyrus Angolensis* (Chimpanzee) and that of the island of Borneo (Orang-utan).

Then it is universally the case that the obscurity and doubt is much smaller, and of much less importance, in the case of wild animals on the point in question, than of those very animals which are excluded by this argument, that is, the domestic. Here, in truth, is the great difficulty. Hence the wonderful differences of opinion about, for example, the common dog, whose races you see are by some referred to many primitive species; by others are considered as mere degenerated varieties from that stock which is called the domestic dog (*Chien de berger*); again, there are others who think that all these varieties are derived from the jackal; and, finally, others contend that the latter, together with all the domestic dogs and their varieties, are descended from the wolf, and so forth.

As then the principle sought to be deduced from copulation is not sufficient to define the idea of species and its difference from variety, so neither are the other things which are adduced with this object, for example, the constancy of any character. Thus the snowy colour and the red pupils of the white *variety* of rabbit are as constant as any specific character could possibly be. So that I almost despair of being able to deduce any notion of species in the study of zoology, except from *analogy* and resemblance. I see, for example, that the molar teeth of the African elephant differ most wonderfully in their conformation from those of the Asiatic. I do not know whether these elephants, which come from such different parts of the world, have ever copulated together; nor do I know any more how constant this conformation of the teeth may be in each. But since, so far in all the specimens which I have seen, I have observed the same difference; and since I have never known any example of molar teeth so changed by mere *degeneration*, I conjecture from analogy that those elephants are not to be considered as mere varieties, but must be held to be different species.

The ferret, on the contrary, does not seem to me a separate species, but must be considered as a mere variety of the polecat, not so much because I have known them copulate together, as because the former has red pupils, and from all analogy I consider that those mammals in whom the internal eye is destitute of the dark pigment, must be held to be mere varieties which have degenerated from their original stocks.

24. *Application of what has been said to the question whether we should divide mankind into varieties or species.* It is easily manifest whither what we have hitherto said has been tending. We have no other way, but that of analogy, by which we are likely to arrive at a solution of the problem above proposed. But as we enter upon this path, we ought always to have before our eyes the two golden rules which the great Newton has laid down for philosophizing. First, *That the same causes should be assigned to account for natural effects of the same kind.* We must therefore assign the same causes for the bodily diversity of the races of mankind to which we assign a similar diversity of body in the other *domestic* animals which are widely scattered over the world. Secondly, *That we ought not to admit more causes of natural things than what are sufficient to explain the phenomena.* If therefore it shall appear that the causes of degeneration are sufficient to explain the phenomena of the corporeal diversity of mankind, we ought not to admit anything else deduced from the idea of the plurality of human species.

25. *How does the primitive species degenerate into varieties?* As we are now about to treat of the modes of degeneration, I hope best to consult perspicuity in dealing with the subject if I arrange it again under two heads; of which the first will briefly relate the principal phenomena of the degeneration of brute animals; and the second will inquire into the causes of this degeneration. This being done, it will be easier in the following section to compare the phenomena of variety in mankind as well with those phenomena of degeneration in brute animals as with the causes of them.

26. *Principal phenomena of the degeneration of brute animals.* A few instances, and those taken from the warm-blooded animals alone, and also as far as possible from the mammals which are most like man in their corporeal economy, will

be enough to show that there is no native variety in mankind which may not be observed to arise amongst other animals as a mere variety and by degeneration. But it is better to go over these things in separate chapters.

27. *Colour*. Thus in the way of colour, the pigs in Normandy are all white; in Savoy, black; in Bavaria[b], chesnut. The *Pecus bubulum* in Hungary generally varies from white to grey; in Franconia they are red, &c. In Corsica the dogs and horses are beautifully spotted. In Normandy, the peacocks are black; ours, on the other hand, are generally white. On the Guinea coast, the birds, especially of the hen tribe[c], and the dogs, are black like the aborigines; and, what is particularly remarkable, the Guinea dog (which Linnæus calls *C. Ægyptius*, I do not know why) is, like the men of that climate, distinguished for the velvety softness of his smooth skin, and the great and nearly specific cutaneous perspiration[d].

28. *Texture of the hair*. As to the texture of hair, what a difference is there not, I ask, in the wool alone of the sheep of different climates, from the tender Tibetan up to the thick and almost stiff Ethiopian? Or in the bristles of the sow, which are so soft in those of Normandy, that they are not fit for scouring-brushes? And what a difference there is, in this respect, between the boar and the domestic sow, especially as to the short wool which grows between the bristles[e]. How remarkable too is the effect of every region of the globe upon the hair of more than one kind of the domestic mammals, as the effect of the climate of Galatia on the bearded cattle of Angora, and on the rabbits and cats, who are so conspicuous for their woolly softness and the extraordinary length and generally snowy whiteness of their coats.

29. *Stature*. As to stature the difference between the Patagonian and the Laplander is much smaller than what is observed everywhere in other domestic animals of different parts of the world. Thus pigs, when transported to Cuba from Europe, grow to double their natural size[f]. So also do cows when transported to Paraguay[g].

30. *Figure and proportion of parts*. As to the proportion of parts, what a great difference there is between the horses of Arabia or Syria and of northern Germany; between the thick-footed cows of the Cape of Good Hope and the thin-footed ones of England! The hinder legs of the sows of Normandy are much higher than the front legs, &c. The cows in some parts of England and Ireland have no horns at all[h]; in Sicily, on the other hand, they have very large ones; but I must not say anything of the vast horns of the Abyssinian oxen, which Sir Joseph Banks showed me, for they, if we are to trust Bruce, ought rather to be referred to some morbific disposition. We may however mention here the *Ovis polycerata;* and as to the variety of hoofs, there are whole races of sows with solid and with three-cloven hoofs[i]. As to some other parts, we have sheep with broad tails; the fringes of the crested canary (what our people call *kapp. vögel*) and other things of this kind.

31. *Above all, the shape of the skull*. The shape has been observed to differ everywhere in the varieties of mankind; but all this difference is not a whit greater, if indeed it can be compared to that which may be observed amongst the different races of other domestic animals. The skull of the Ethiopian does not differ more

from that of the European than that of the domestic sow from the osseous head of the boar; or than the head of the Neapolitan horse, which is called from its shape ram-headed, from that of the Hungarian horse, which the learned know well is conspicuous for its singular lowness and the size of its inferior jaw. In the urus, the progenitor of our domestic race of bulls, according to the observations of Camper, very large foveæ lacrymales are visible; which, on the contrary, are entirely obsolete in our country cattle. I say nothing of that manifestly monstrous degeneration of skull in the variety of hen they call the Paduan[j].

32. *Causes of degeneration.* Animal life supposes two faculties, depending upon the vital forces as primary conditions and principles of all and singular its functions; the one, namely, of so receiving the force of the stimuli which act upon the body that the parts are affected by it; the other of so reacting from this affection that the living motions of the body are in this way set in action and perfected. So there is no motion in the animal machine without a preliminary stimulus and a consequent reaction. These are the hinges on which all the physiology of the animal economy turns. And these are the fountains from which, just as the business itself of generation, so also the *causes of degeneration* flow; but in order to make this clear to those even who know but little of physiology, it will be as well to premise with a few words from that science.

33. *Formative force.* I have in another place professedly, and in a separate book devoted to this subject, endeavoured to show that the vulgar system of evolution, as it is called (according to which it is taught that no animal or plant is generated, but that all individual organic bodies were at the very earliest dawn of creation already formed in the shape of undeveloped germs and are now being only successively evolved), answers neither to the phenomena themselves of nature, nor to sound philosophic reasoning. But on the contrary, by properly joining together the two principles which explain the nature of organic bodies, that is the physico-mechanical with the teleological, we are conducted both by the phenomena of generation, and by sound reasoning, to lay down this proposition: That the genital liquid is only the shapeless material of organic bodies, composed of the innate matter of the inorganic kingdom, but differing in the force it shows, according to the phenomena; by which its first business is under certain circumstances of maturation, mixture, place, &c. to put on the form destined and determined by them; and afterwards through the perpetual function of nutrition to preserve it, and if by chance it should be mutilated, as far as lies in its power to restore it by reproduction.

Let me be allowed to distinguish this energy, so as to prevent its being confused with the other kinds of vital force, or with the vague and undefined words of the ancients, the plastic force, &c. by the name of the formative force (*nisus formativus*); by which name I wish to designate not so much the cause as some kind of perpetual and invariably consistent effect, deduced *à posteriori*, as they say, from the very constancy and universality of phenomena. Just in the same way as we use the name of attraction or gravity to denote certain forces, the *causes* of which however still remain hid, as they say, in Cimmerian darkness.

As then other vital forces, when they are excited by their appointed and proper stimuli, become active and ready for reaction, so also the formative force is excited by the stimuli which belong to it, that is, by the kindling of heat in the egg during the process of incubation. But as other vital forces, as contractility, irritability, &c. put themselves out only by the mode of motion, this, on the other hand, of which we are talking, manifests itself by increase, and by giving a determinate form to matter; by which it happens that every plant and every animal propagates its species in its offspring (either immediately, or gradually by the successive access and change of other stimuli, through metamorphosis).

Now the way in which the formative force may sometimes turn aside from its determined direction and plan is principally in three forms. First, by the production of monsters; then by hybrid generation through the mixture of the genital liquid of different species; finally, by degeneration into varieties, properly so called. The production of monsters, by which, whether through some disturbance and as it were mistake of the formative force, or even through accidental or adventitious circumstances, as by external pressure, &c. a structure manifestly faulty and unnaturally deformed is intruded upon organic bodies, has nothing to do with our present purpose. Nor is this the place to consider hybrids sprung from the commingling of the generation of different species, since by a most wise law of nature (by which the infinite confusion of specific forms is guarded against) hybrids of this kind, especially in the animal kingdom, scarcely ever occur except through the interference of man: and then they are almost invariably sterile, so as to be unable to propagate any further their new ambiguous shape sprung from anomalous venery.

Still, meanwhile, this subject we are now discussing may be illustrated by the history of hybrids sprung from different species; partly on account of their analogy with those hybrids which spring from different *varieties*, of which we shall speak by and by; partly, because, like everything else, they go as proofs to refute that theory about the evolution of pre-formed germs, and to display clearly the power and efficacy of the formative force; a consideration, which will escape no one who rightly appreciates those well-known and very remarkable experiments, in which, in the very rare instances of *prolific* hybrids, when their fecundation has been frequently repeated for many generations by the aid of the male seed of the same species, that new appearance of hybrid posterity has so sensibly deflected from the maternal form as more and more to pass into the paternal form of the other species, and so, finally, the former seems to become quite transmuted into the latter, by a sort of arbitrary metamorphosis[k].

But the mixture of specifically different generation, although it cannot overturn, or as it were suffocate, all the excitability of the formative force, still can impart to it a singular and anomalous direction. And so it happens that the continuous action, carried on for several series of generations of some peculiar stimuli in organic bodies, again has great influence in sensibly diverting the formative force from its accustomed path, which deflection is the most bountiful source of degeneration, and the mother of varieties properly so called. So now let us go to work and examine one by one the chief of these stimuli.

34. *Climate*. That the power of climate must be almost infinite, as on all organic bodies, so especially on warm-blooded animals, will quickly appear to any one who considers first, by how intimate and how constant a bond these animals are bound while alive to the action of the atmospheric air in which they dwell. Besides, how wonderfully this air (which was once held to be a simple element of itself) is made up of what they call multifarious elements, such as gasiform constituents, the accessories of light, heat, electricity, &c. Then of what different proportions of these matters does it not consist, and in consequence of this variety how different must be the atmospheric action on those we call animals! Especially when we throw in the consideration of so many other things, by whose accession climates differ so much, as the position of countries in respect of the zones of the globe, the elevation of the soil, mountains, the vicinity of the sea or lakes and rivers, the customary winds, and innumerable other things of this kind.

This air, then, which those we call animals suck in by breathing from the time of birth, modified so greatly by the variety of climates, is decomposed in their lungs as it were in a living laboratory. Part of what they inhale is distributed with the arterial blood over the whole body; but as a balance to another portion of this point, elements are liberated, which are partly deposited on the peripheral integuments of the body, and partly are carried back by the flow of venous blood to the respiratory organs; hence arise the various modifications of the blood itself, and the remarkable influxes of these humours, especially of fat, bile, &c. into the secretions. Hence finally the action of all these things as so many stimuli on a living solid, and hence the resulting reaction as well of this thus affected solid, as what especially belongs to our discussion, the direction and determination of the formative force. This great and perpetual influence of climate on the animal economy and the habit and conformation of the body, although there has been no time when it has not attracted the attention of good observers, has in our own time above all been illustrated and confirmed by the great advance that has been made in chemistry, and by a deeper study of physiology. Still it is always a difficult and arduous thing, in the discussion of these varieties, to settle what is to be attributed exclusively to climate, what rather to other causes of degeneration, and finally to the joint action of both. Meanwhile, I will bring forward one or two instances of degeneration which seem most clearly to be derived from the effects of climate. For example, the white colour of many animals in northern regions, which have other colours in the temperate zones. Instances are, those of wolves, hares, cattle, falcons, crows, blackbirds, thrushes, chaffinches, &c. That this whiteness must be attributed to cold, we learn from the analogy of animals of the same kind who, under the same climate during winter, change their summer colour into white or grey; as weasels and ermines, hares, squirrels, reindeer, the ptarmigan, snow-bunting, and others[1]. So also I am more inclined to attribute to climate that snowy fleece so conspicuous for its silky softness of some of the animals of Angora than to the kind of diet, because that is shared by those who feed on all sorts of different things, by the carnivorous, as the cat for example, equally with the herbivorous ruminants, as goats, &c.

Such too seems to be the explanation of the coally blackness which under some districts of the torrid zone, as on the coasts of Guinea, animals of different orders, mammalia as well as birds, are seen to put on with the colour of the Ethiopians (s. 27). And it is above all worthy of remark that this Ethiopic blackness, just like that Syrian whiteness, although the animals may be transported into regions of a very different climate, is still preserved permanently for many series of generations. Nor is the power and influence of climate on the stature of organic bodies at all inferior; since cold obstructs their increase, which, on the contrary is manifestly augmented and promoted by heat. Thus the horses of Scotland, or cold North Wales, are small; in Scandinavia the horses and the cattle, like the indigenous races, are of tall and stalwart stature; in Smaland they are sensibly smaller, and in the north of East Gothland are in proportion smallest of all.

35. *Diet*. It seems extremely probable, what has been demonstrated principally by the sagacity of G. Fordyce, that the primary elements, as they are called, of every kind of alimentary substance, whether it be taken from the animal or the vegetable kingdom, are the same. Hence the same sort of chyle, and universally the same kind of blood, is elaborated by all the multifarious warm-blooded animals, carnivorous as well as herbivorous, out of the most different kinds of nourishment, if only it has been properly submitted to the organs of digestion. Still, however much this may appear to be true, it cannot be denied that the innumerable adventitious qualities of different matters of food, have had great power in changing the natures and properties of animals; to prove which a few instances will be enough.

Singing birds show that there is some specific power in some kinds of food to change the colour of animals; especially some sorts of larks and finches, which it has been proved, if they are fed upon hemp seed alone, sensibly grow black. The African sheep when transported to England is a proof how wonderfully, when the diet is changed, the texture of the hairs will change also; for its wool which is common by nature, and stiff like the hair of a camel, after it has been fed one year upon English pastures becomes of a most magnificent delicacy[m]. The influence food has towards changing the stature and the proportions, is plain from the comparison of domestic animals. Horses which in marshy countries (called in the vernacular *Maschländer*) live upon rich food, as the Frisian especially, grow large; whereas, on the contrary, in rocky and stony countries, such as those of Œland, or on dry heathy soils, they remain stunted. Thus it is surprising how fat and bellied horses become on a fat soil, though their legs become shorter in proportion. But when they are fed upon drier grass, as, for example, the Cape grass, they secrete less fat, but are remarkable for their strong and fleshy legs; to say nothing of the multifarious diversities of the taste and weight of flesh, which again depend upon the variety of diet.

36. *Mode of life*. When I speak of the kind of life as a cause of degeneration, I include under that head all those points besides climate and diet which so far have to do with the natural economy of animals, that when they act long and continuously upon the same condition of body they are at length strong enough to change

it to some extent. The principal of these are cultivation and the force of custom, whose power and influence are again so manifestly conspicuous in our domestic animals.

Consider, for instance, the vast difference which separates the conformation and the proportions of the parts of the generous horse trained in the school, and the wild horse, which they call a wild beast. The latter, when it fights with others bites rather than kicks; the former, on the other hand, when bridled and armed with iron feet, prefers to attack his enemy with them, and almost unlearns to bite. Many kinds of mammals when subdued by man show by the hanging of their tails and the lapping of their ears a spirit tamed and subdued by slavery. In many the very corporeal functions of secretion, generation, &c. are changed in a wonderful way. In the domestic pig, for example, the adipose membrane appears in a vast mass, which is quite wanting in the boar, whose tender and as it were woolly hairs, on the contrary, inserted between the bristles, sensibly disappear in that domestic variety. These domestic animals are much more liable to monstrous births than their wild aborigines; and also to troops of new diseases, and especially to new kinds of worms of which no vestige is to be found in their wild and original variety; the truth of which assertion, though paradoxical, is not to be invalidated, as may be proved by the instance of the *Hydas intercutis*, called, in the vernacular, *Finnen*, Ital. *Lazaroli*[n]. I place under this head also stunted stature from premature and unseasonable venery, and everything of that kind.

37. *Hybrid generation.* So much for the triple sources of degeneration which only by long and daily action, continued through many series of generations, are sufficiently strong, slowly, and by little and little, to change the primeval character of animals and produce varieties. But the case is different, and a new character is imparted to the immediate offspring, when different varieties of this kind, sprung at length from those causes, come to copulate together, for thus they give rise to a hybrid offspring, like neither parent altogether, but participating in the form of each, and being as it were a mean between the two. Hybrid is the name commonly given to the offspring of parents of manifestly different species, as mules sprung from the horse and ass, or birds from the union of the crested canary with the linnet. But this is not the place for us to speak of these, for there is no account to be taken of them in varieties of the human race. Not indeed that horrid stories are wanting of the union of men with brutes, when either men have had to do with the females of beasts (whether carried away by unbridled lust[o], or from some mad idea of continence[p], or because they expected some medicinal aid from this sort of crime[q]), or when we are told that women have been made use of by male brutes (whether that has happened through any violent rape[r]. or because women have solicited them in the madness of lust[s], or have prostituted themselves from religious superstition[t]), still we have never known any instance related on good authority of any such connexion being fruitful, or that any hybrid has ever been produced from the horrid union of beast and man. But we have only to do with those hybrids which spring from the intercourse of different varieties of one and the same species, as when, for example, the green canary bird is paired with the

white variety, &c., which connexion has a wonderful effect in changing the colour and conformation of the new progeny which results therefrom; so that this is often applied with the greatest advantage in the impregnation of domestic animals for the purpose of improving and ennobling the offspring, especially in the case of horses and sheep.

38. *Hereditary peculiarities of animals from diseased temperament.* An hereditary disposition to disease would seem at first sight rather to belong to the pathology than to the natural history of animals. But when the matter is more carefully looked into, it is plain that in more ways than one it has something to do with those causes of degeneration we are concerned with. For, in the first place, some external qualities of animals, although according to common ideas they are never referred to a truly diseased constitution, still seem to come very nearly to that, since they are for the most part found in conjunction with an unnaturally weak affection. I include among these, for example, that peculiar whiteness of some animals, which the wise Verulam long ago called the colour of defect. We learn by the example of the Hungarian oxen, whose woolly skin only comes after castration, that we may frequently recognize as a cause the vicious constitution and defect of the corporeal economy. On the other hand, it is proved by the instances of the Angora cats and dogs, that morbid symptoms follow extraordinary whiteness of that kind, for it is a common observation that those animals are almost always hard of hearing.

It is also the case that some genuine diseases when the animal nature has been as it were used to them for a long series of generations seem to get sensibly milder and milder and less inconvenient, so that at last they can scarcely be considered more than a diseased affection. An example is afforded by that vicious species of whiteness which, when united to a deficiency of the black pigment which lines the internal eye of hot-blooded animals, is known by the name of leucæthiopia. This when it seizes sporadically one or other of a family (for it is always a congenital affection) exhibits plainly the symptoms of cachexia, which everywhere comes very near to a leprous constitution. But in other cases when it has been established by a sort of hereditary right for many generations, it becomes a second nature, so that in the white variety of rabbits not a vestige remains of the original morbific affection, the existence of which however is determined by the analogy of other animals which have anomalously white pupils and red eyes. The ferret has been considered by some zoologists as a peculiar species of the genus *Mustela*, whereas, unless I am altogether deceived, it is as I have said above (s. 23) a mere variety of the pole-cat, and that of diseased origin through leucæthiopia.

39. *Problem proposed. Can mutilations and other artifices give a commencement to native varieties of animals?* It is disputed whether deformities or mutilations, effected upon animals either by accident or advisedly, especially in those cases where they have been repeated for many series of generations, can at length in progress of time terminate in a sort of second nature, so that what before was done by art now degenerates into a congenital conformation. Some[u] have asserted this, whilst others[v] on the contrary have denied it. Those who are for the

affirmative point to the examples of the young of different kinds of animals, dogs and cats for example, which are born without tails or ears after those parts have been cut off from their parents, as is proved by credible witnesses. And of boys among circumcised nations who are frequently born naturally *apellæ*[w]; and of scars which parents bear from wounds, whose marks afterwards are congenital in the infants. Buffon, indeed, went so far as to derive from the same source the peculiar characters of some animals, as the callosities on the breast and legs of camels, or the bald scurfy forehead of the rook (*Corvus frugilegus*). Those who do not allow these last instances will not unwisely reject this opinion of Buffon, as what is called a *petitio principii;* but the other instances we spoke of they will think should rather be attributed to chance.

I have not at present adopted as my own either the affirmative or the negative of these opinions; I would willingly give my suffrage with those on the negative side, if they could explain why peculiarities of the same sort of conformation, which are first made intentionally or accidentally, cannot in any way be handed down to descendants, when we see that other marks of race which have come into existence from other causes which up to the present time are unknown, especially in the face, as noses, lips, and eye-brows are universally propagated in families for few or many generations with less or greater constancy, just in the same way as *organic*[x] disorders, as deficiencies of speech and pronunciation, and such like; unless perhaps they prefer saying that all these occur also by chance.

40. *Some considerations to be observed in the examination of the causes of degeneration.* Many of the causes of degeneration we have already spoken of are so very clear, and so placed beyond all possibility of doubt, that most phenomena of degeneration above enumerated may by an easy process be undoubtedly referred to them, as effects to their causes. But on the other hand even in that very way there is frequently such a concurrence or such a conflicting opposition of many of them; such a diverse and multifarious proneness of organic bodies to degeneration, or reaction from it; and besides, these causes have such effects upon these bodies according as they act immediately (so to speak) or otherwise; and finally, such is the difference of these effects by which they are preserved unimpaired by a sort of tenacious constancy through long series of generations, or by some power of change withdraw themselves again in a short space of time, that in consequence of this diversified and various relation there is need of the greatest caution in the examination of varieties.

Let me then, if only for the benefit of the student, at the end of this discourse, before we pass to the varieties of men themselves, lay down some maxims of caution at least, as corollaries to be carefully borne in mind in the discussion we are entering upon:

1 The more causes of degeneration which act in conjunction, and the longer they act upon the same species of animals, the more palpably that species may fall off from its primeval conformation. Now no animal can be compared to man in this respect, for he is omnivorous, and dwells in every climate, and

is far more domesticated and far more advanced from his first beginnings than any other animal; and so on him the united force of climate, diet, and mode of life must have acted for a very long time.

2 On the other hand an otherwise sufficiently powerful cause of degeneration may be changed and debilitated by the accession of other conditions, especially if they are as it were opposed to it. Hence everywhere in various regions of the terraqueous globe, even those which lie in the same geographical latitude, still a very different temperature of the air and an equally different and generally a contrary effect on the condition of animals may be observed, according as they differ in the circumstances of a higher or lower position, proximity to the sea, or marshes, or mountains, or woods, or of a cloudy or serene sky, or some peculiar character of soil, or other circumstances of that kind.

3 Sometimes a remarkable phenomenon of degeneration ought to be referred not so much to the immediate, as to the mediate, more remote, and at the first glance concealed influence of some cause. Hence the darker colour of peoples is not to be derived solely from the direct action of the sun upon the skin, but also from its more remote, as its powerful influence upon the functions of the liver.

4 Mutations which spring from the mediate influence of causes of this sort seem to strike root all the deeper, and so to be all the more tenaciously propagated to following generations. Hence, if I mistake not, we are to look for the reason why the brown colour of skin contracted in the torrid zone will last longer in another climate than the white colour of northern animals if they are transported towards the south.

5 Finally, the mediate influences of those sort of causes may lie hid and be at such a distance, that it may be impossible even to conjecture what they are, and hence we shall have to refer the enigmatical phenomena of degeneration to them, as to their fountains. Thus, without doubt, we must refer to mediate causes of this kind, which still escape our observation, the racial and constant forms of skulls, the racial colour of eyes, &c.

Notes

a "When beasts by nature copulate with each other, it is an unfailing sign that they are of the same species." Berthout van Berchem fil. has lately adopted the same test of species, "if animals mix when in a natural state." But he makes no mention of Frisch, or even of Ray, nay, he says, "M. de Buffon, who was the first to abandon the little-to-be-depended-upon distinctions of the nomenclators, was also the first to make it understood that copulation was the best criterion for ascertaining species." See *Mem. de la Sociéte des Sciences Physiques de Lausanne,* T. II. p. 49.

b Comp. Voigt, *Magazin.* T. VI. P. I. p. 10.

c See Dan. Beeckman's *Voyage to and from Borneo,* Lond. 1718, 8vo. p. 14.

d Pechlin, *De Habitu et Colore Æthiopum,* Kilon. 1677, 8vo. p. 56.

e [Pechlin, *De Habitu et Colore Æthiopum*, Kilon. 1677, 8vo. p. 56.]

f Voigt, *Magazin. l. c.*

g F. Saver. Clavigero, *Storia Antica del Messico,* T. IV. p. 142.

h Comp. also Hippocrates, *De aeribus, aquis, et locis,* s. 44.
i Voigt, *Magazin, l. c.*
j Pallas, *spicileg. zoologic.* fasc. IV. p. 22, and Sandifort, *Muséum Anatom. Acad. Lugd. Batav*. T. I. p. 306.
k Kölreuter. Third account of the news of some experiments relating to the sex of plants, &c., p. 51, s. 24, with the title, "An entirely complete change of one kind of plant into another."
l Comp. besides others, Linnæus, in *Flora Lapponica,* p. 55, 352, ed. Smith.
m Comp. Jam. Bates *On the Literal Doctrine of Original Sin,* Lond. 1766, 8vo. p. 224.
n Malpighii *Opera Posthuma,* p. 84, ed. Lond. 1697, fol. – so J. A. E. Goeze, *Discovery; that the hydatids in swine's flesh are no glander disease, but true bladderworms*. 8vo. Hal. 1784.
o Comp. Th. Warton on Theocriti *Idyll.* I. 88, p. 19. "I have been told by a certain learned friend, that when he was travelling in Sicily and investigating closely not only the ancient monuments but also the manners of the people, that even their own priests used to ask the shepherds, who spend a solitary life in the Sicilian mountains, as a matter of course among the articles of confession, whether they had had anything to do with the she-goats."
p See Mart. à Baumgarten Equ. Germ. *Travels in Egypt, Arabia,* &c. p. 73. "As we went out of Alkan, in Egypt, we came to a village called Belbes, where we joined a caravan going to Damascus. There we saw a Saracenic saint, sitting on the heaps of sand, as naked as he came out of his mother's womb. We heard this saint whom we saw in that place publicly praised above all things; that he was a holy man, divine and perfect beyond all measure, because he never had any connexion with women or boys, but only with asses and mules."
q With this object, Pallas says that when the Persians suffer from hip-gout they copulate with the onagra. *Neue Nordische beyträge,* P. II. P. 38.
r Baboons. Comp. Ph. Phillips's *Travels in Guinea* in Churchill's *Collection of Voyages,* T. VI. P. 211. "Here are a vast number of overgrown large baboons, some as big as a large mastiff dog, which go in droves of 50 and 100 together, and are very dangerous to be met with, especially by women, who, I have been credibly assured, they have often seized upon, ravished, and in that kind abused one after another, till they have killed them."
s Thus, Steller says that the women of Kamtschatka formerly copulated with dogs. *Beschreibung Von Kamtschatka,* p. 289.
t As the women of Mendes with the sacred goat; on which singular custom, see a copious dissertation by D'Hancarville, *Recherches sur l'origine des Arts de la Grèce,* T. I. p. 320.
u Hippocrates and Aristotle. And very recently Klügel, in Tom. I. of the Encyclopedia, p. 541, ed. 2nd.
v See Kant, in *Berliner Monatsschrift,* 1785, T. vi. p. 400.
w Voigt, *Magazin,* T. VI. P. I. p. 22, and P. IV. p. 40.
x A remarkable instance is related by Hacquet in the *Magazin* of Voigt just cited, T. VI. P. IV. p. 34.

Editorial Headnote

9. Jean-Baptiste Lamarck, *Zoological Philosophy* (tr. Hugh Elliot; London: Macmillan and Co., 1914), pp. 35–40, 44–6, 106–13, 128–9, 169–73.

Jean-Baptiste Lamarck (1744–1829) was a French botanist and invertebrate zoologist who is now remembered primarily for publishing one of the first systematic theories of evolution, or *transformisme*. Born to a semi-impoverished noble family in northern France, Lamarck enlisted in the military at age seventeen before settling in Paris. He spent the early 1770s studying medicine before turning to botany, an interest he had developed in his military days. Lamarck's non-Linnaean approach to botanical classification impressed the famous naturalist Georges Louis Leclerc, the Comte de Buffon (1707–1788), who arranged to have Lamarck's *Flore française* (1778) published and then secured him a place in the botanical section of the Académie des Sciences.[1] In 1781, Buffon also created the position of *correspondant* to the Jardin du Roi for Lamarck, and in 1789, following Buffon's death, he was appointed to a salaried affiliation to the Jardin as botanist to the king and keeper of the herbaria.

Following the 1789 French Revolution, the Jardin du Roi was eventually converted into the Muséum Nationale d'Histoire Naturelle, which was intended to be a more republican institution and one that would better serve national and public interests.[2] Lamarck, who was named professor of "insects and worms" (he had not previously specialized in invertebrates but was a collector of shells), published works in the 1790s criticizing Lavoisierian chemistry; he also attempted to synthesize physics, chemistry, and physiology into a single unified theory of nature. It was around the turn of the nineteenth century that Lamarck came to believe in *transformisme*, or the evolution of species, an idea that numerous other naturalists were already receptive to[3] and that was consistent with his observations of 'transitional forms' of both plant and invertebrate specimens.

Lamarck first published his theory of *transformisme* in 1802, although the extracts below are quoted from his *Philosophie zoologique* of 1809 (trans. 1914), a more comprehensive and systematic presentation of the theory. In these extracts, Lamarck discusses the argument that species could change over time and that organisms exist along a continuum of individual variation rather than as perfectly distinct species – that is, he denies that species are a real natural category, arguing rather that they are an artifice created by and for the convenience of naturalists and systematists. He then introduces his argument for how an individual's habitual behaviours could change its physiology and how those changes could be passed on to its offspring. Finally, the fifth extract introduces a hypothetical scenario in which Lamarck suggests that the orangutan could have developed into a human-like creature with intelligence, language, and an upright posture.

Lamarck put forward these transformist ideas in response to ongoing debates in France about social reform and the progress of society. Species transformation meant that all of an organism's biological characteristics could change, including not only its external appearance but its internal physiology and intelligence. In light of this, Lamarck believed that naturalists had important contributions to make to conversations about social reform and good governance. Altering the social environment could produce physiological changes among the populace, and such biological alterations could, in turn, produce better citizens and a more virtuous society.

Notes

1 For further consideration of Buffon and his work, see Caden C. Testa and Piers J. Hale, *Evolution Before Darwin* (London: Routledge, 2025).

2 See Emma Spary, *Utopia's Garden: French Natural History from Old Regime to Revolution* (Chicago: University of Chicago Press, 2000).

3 See Richard Burkhardt, *The Spirit of System: Lamarck and Evolutionary Biology* (Cambridge, MA: Harvard University Press, 1997), pp. 201–209. Also the consideration of other evolutionary thinkers in Caden C. Testa and Piers J. Hale, *Evolution Before Darwin* (London: Routledge, 2025).

9

ZOOLOGICAL PHILOSOPHY

Jean-Baptiste Lamarck

Source: Jean-Baptiste Lamarck, *Zoological Philosophy* (tr. Hugh Elliot; London: Macmillan and Co., 1914), pp. 35–40, 44–6, 106–13, 128–9, 169–73

CHAPTER III.

OF SPECIES AMONG LIVING BODIES AND THE IDEA THAT WE SHOULD ATTACH TO THAT WORD.

009 "Jean-Baptiste Lamarck"

Source: Biodiversity Heritage Library

Image in the Public Domain.

DOI: 10.4324/9781003490548-10

It is not a futile purpose to decide definitely what we mean by the so-called *species* among living bodies, and to enquire if it is true that species are of absolute constancy, as old as nature, and have all existed from the beginning just as we see them to-day; or if, as a result of changes in their environment, albeit extremely slow, they have not in course of time changed their characters and shape.

The solution of this question is of importance not only for our knowledge of zoology and botany, but also for the history of the world.

I shall show in one of the following chapters that every species has derived from the action of the environment in which it has long been placed the *habits* which we find in it. These habits have themselves influenced the parts of every individual in the species, to the extent of modifying those parts and bringing them into relation with the acquired habits. Let us first see what is meant by the name of species.

Any collection of like individuals which were produced by others similar to themselves is called a species.

This definition is exact; for every individual possessing life always resembles very closely those from which it sprang; but to this definition is added the allegation that the individuals composing a species never vary in their specific characters, and consequently that species have an absolute constancy in nature.

It is just this allegation that I propose to attack, since clear proofs drawn from observation show that it is ill-founded.

The almost universally received belief is that living bodies constitute species distinguished from one another by unchangeable characteristics, and that the existence of these species is as old as nature herself. This belief became established at a time when no sufficient observations had been taken, and when natural science was still almost negligible. It is continually being discredited for those who have seen much, who have long watched nature, and who have consulted with profit the rich collections of our museums.

Moreover, all those who are much occupied with the study of natural history, know that naturalists now find it extremely difficult to decide what objects should be regarded as species.

They are in fact not aware that species have really only a constancy relative to the duration of the conditions in which are placed the individuals composing it; nor that some of these individuals have varied, and constitute races which shade gradually into some other neighbouring species. Hence, naturalists come to arbitrary decisions about individuals observed in various countries and diverse conditions, sometimes calling them varieties and sometimes species. The work connected with the determination of species therefore becomes daily more defective, that is to say, more complicated and confused.

It has indeed long been observed that collections of individuals exist which resemble one another in their organisation and in the sum total of their parts, and which have kept in the same condition from generation to generation, ever since they have been known. So much so that there seemed a justification for regarding any collection of like individuals as constituting so many invariable species. Now attention was not paid to the fact that the individuals of the species perpetuate themselves without variation only so long as the conditions of their existence do not vary in essential particulars. Since existing prejudices harmonise well with these successive regenerations of like individuals, it has been imagined that every species is invariable and as old as nature, and that it was specially created by the Supreme Author of all existing things.

Doubtless, nothing exists but by the will of the Sublime Author of all things, but can we set rules for him in the execution of his will, or fix the routine for him to observe? Could not his infinite power create an *order of things* which gave existence successively to all that we see as well as to all that exists but that we do not see?

Assuredly, whatever his will may have been, the immensity of his power is always the same, and in whatever manner that supreme will may have asserted itself, nothing can diminish its grandeur.

I shall then respect the decrees of that infinite wisdom and confine myself to the sphere of a pure observer of nature. If I succeed in unravelling anything in her methods, I shall say without fear of error that it has pleased the Author of nature to endow her with that faculty and power.

The idea formed of species among living bodies was quite simple, easy to understand, and seemed confirmed by the constancy in the shapes of individuals, perpetuated by reproduction or generation. Such are a great number of these alleged species that we see every day.

Meanwhile, the farther we advance in our knowledge of the various organised bodies which cover almost every part of the earth's surface, the greater becomes our difficulty in determining what should be regarded as a species, and still more in finding the boundaries and distinctions of genera.

According as the productions of nature are collected and our museums grow richer, we see nearly all the gaps filled up and the lines of demarcation effaced. We find ourselves reduced to an arbitrary decision which sometimes leads us to take the smallest differences of varieties and erect them into what we call species, and sometimes leads us to describe as a variety of some species slightly differing individuals which others regard as constituting a separate species.

Let me repeat that the richer our collections grow, the more proofs do we find that everything is more or less merged into everything else, that noticeable differences disappear, and that nature usually leaves us nothing but minute, nay puerile, details on which to found our distinctions.

How many genera there are both among animals and plants, among which the number of species referred to them is so great that the study and determination of these species are well nigh impracticable! The species of these genera, arranged in series according to their natural affinities, exhibit such slight differences from those next them as to coalesce with them. These species merge more or less into one another, so that there is no means of stating the small differences that distinguish them.

It is only those who have long and diligently studied the question of species, and who have examined rich collections, that are in a position to know to what extent species among living bodies merge into one another. And no one else can know that species only appear to be isolated, because others are lacking which are close to them but have not yet been collected.

I do not mean that existing animals form a very simple series, regularly graded throughout; but I do mean that they form a branching series, irregularly graded and free from discontinuity, or at least once free from it. For it is alleged that there is now occasional discontinuity, owing to some species having been lost. It follows that the species terminating each branch of the general series are connected on one side at least with other neighbouring species which merge into them. This I am now able to prove by means of well-known facts.

I require no hypothesis or supposition; I call all observing naturalists to witness.

Not only many genera but entire orders, and sometimes even classes, furnish instances of almost complete portions of the series which I have just indicated.

When in these cases the species have been arranged in series, and are all properly placed according to their natural affinities, if you choose one, and then, jumping over several others, take another a little way off, these two species when compared will exhibit great differences. It is thus in the first instance that we began to see such of nature's productions as lay nearest to us. Generic and specific distinctions were then quite easy to establish; but now that our collections are very rich, if you follow the above-mentioned series from the first species chosen to the second, which is very different from it, you reach it by slow gradations without having observed any noticeable distinctions.

I ask, where is the experienced zoologist or botanist who is not convinced of the truth of what I state?

How great the difficulty now is of studying and satisfactorily deciding on species among that multitude of every kind of polyps, radiarians, worms, and especially insects, such as butterflies, *Phalaena, Noctua, Tinea*, flies, *Ichneumon, Curculio, Cerambix*, chafers, rose-chafers, etc.! These genera alone possess so many species which merge indefinably into one another.

What a swarm of mollusc shells are furnished by every country and every sea, eluding our means of distinction and draining our resources.

Consider again, fishes, reptiles, birds and even mammals; you will see that except for gaps still to be filled, neighbouring species and even genera are separated by the finest differences, so that we have scarcely any foothold for setting up sound distinctions.

Is there not an exactly similar state of affairs in the case of botany, which deals with the other series, consisting of plants?

How great indeed are the difficulties of the study and determination of species in the genera *Lichen, Fucus, Carex, Poa, Piper, Euphorbia, Erica, Hieracium, Solanum, Geranium, Mimosa*, etc., etc.

When these genera were constituted only a small number of species belonging to them were known, and it was then easy to distinguish them; but now that nearly all the gaps are filled, our specific differences are necessarily minute and usually inadequate.

Let us see what are the causes which have given rise to this undoubted state of affairs; let us see if nature affords any explanation, and whether observation can help us.

We learn from a number of facts that, according as the individuals of one of our species change their abode, climate, habits, or manner of life, they become subject to influences which little by little alter the consistency and proportions of their parts, their shape, properties and even their organisation; so that in course of time everything in them shares in these mutations.

In the same climate, very different habitats and conditions at first merely cause variations in the individuals exposed to them; but in course of time the continued change of habitat in the individuals of which I speak, living and reproducing in these new conditions, induces alterations in them which become more or less essential to their being; thus, after a long succession of generations these individuals, originally belonging to one species, become at length transformed into a new species distinct from the first.

Suppose, for example, that the seeds of a grass or any other plant that grows normally in a damp meadow, are somehow conveyed first to the slope of a neighbouring hill where the ground although higher is still rich enough to allow the plant to maintain its existence. Suppose that then, after living there and reproducing itself many times, it reaches little by little the dry and almost barren ground of a mountain side. If the plant succeeds in living there and perpetuating itself for a number of generations, it will have become so altered that botanists who come across it will erect it into a separate species.

The same thing happens in the case of animals that are forced by circumstances to change their climate, habits, and manner of life: but in their case more time is required to work any noticeable change than in the case of plants.

The idea of bringing together under the name of species a collection of like individuals, which perpetuate themselves unchanged by reproduction and are as old as nature, involved the assumption that the individuals of one species could not unite in reproductive acts with individuals of a different species.

Unfortunately, observation has proved and continues every day to prove that this assumption is unwarranted; for the hybrids so common among plants, and the copulations so often noticed between animals of very different species, disclose the fact that the boundaries between these alleged constant species are not so impassable as had been imagined.

It is true that often nothing results from these strange copulations, especially when the animals are very disparate; and when anything does happen the resulting individuals are usually infertile; but we also know that when there is less disparity these defects do not occur. Now this cause is by itself sufficient gradually to create varieties, which then become races, and in course of time constitute what we call species.

To assist us to a judgment as to whether the idea of species has any real foundation, let us revert to the principles already set forth; they show:

(1) That all the organised bodies of our earth are true productions of nature, wrought successively throughout long periods of time.
(2) That in her procedure, nature began and still begins by fashioning the simplest of organised bodies, and that it is these alone which she fashions immediately, that is to say, only the rudiments of organisation indicated in the term *spontaneous generation*.
(3) That, since the rudiments of the animal and plant were fashioned in suitable places and conditions, the properties of a commencing life and established organic movement necessarily caused a gradual development of the organs, and in course of time produced diversity in them as in the limbs.
(4) That the property of growth is inherent in every part of the organised body, from the earliest manifestations of life; and then gave rise to different kinds of multiplication and reproduction, so that the increase of complexity of organisation, and of the shape and variety of the parts, has been preserved.
(5) That with the help of time, of conditions that necessarily were favourable, of the changes successively undergone by every part of the earth's surface, and, finally, of the power of new conditions and habits to modify the organs of living bodies, all those which now exist have imperceptibly been fashioned such as we see them.
(6) That, finally, in this state of affairs every living body underwent greater or smaller changes in its organisation and its parts; so that what we call species were imperceptibly fashioned among them one after another and have only a relative constancy, and are not as old as nature.

But objections may be raised to the allegation that nature has little by little fashioned the various animals known to us by the aid of much time and an infinite variation of environment. It may be asked whether this allegation is not refuted by the single fact of the wonderful variety observed in the *instinct* of various animals, and in the marvellous *skill* of all kinds which they exhibit.

Will anyone, it may be asked, venture to carry his love of system so far as to say that nature has created single-handed that astonishing diversity of powers, artifice, cunning, foresight, patience and skill, of which we find so many examples among animals? Is not what we see in the single class of insects far more than enough to convince us that nature cannot herself produce so many wonders; and to compel the most obstinate philosopher to recognise that the will of the Supreme Author of all things must be here invoked, and could alone suffice for bringing into existence so many wonderful things?

[...]

Of the Species Alleged to be Lost.

I am still doubtful whether the means adopted by nature to ensure the preservation of species or races have been so inadequate that entire races are now extinct or lost.

Yet the fossil remains that we find buried in the soil in so many different places show us the remains of a multitude of different animals which have existed, and among which are found only a very small number of which we now know any living analogues exactly alike.

Does this fact really furnish any grounds for inferring that the species which we find in the fossil state, and of which no living individual completely similar is known to us, no longer exist in nature? There are many parts of the earth's surface to which we have never penetrated, many others that men capable of observing have merely passed through, and many others again, like the various parts of the sea-bottom, in which we have few means of discovering the animals living there. The species that we do not know might well remain hidden in these various places.

If there really are lost species, it can doubtless only be among the large animals which live on the dry parts of the earth; where man exercises absolute sway, and has compassed the destruction of all the individuals of some species which he has not wished to preserve or domesticate. Hence arises the possibility that animals of the genera *Palaeotherium, Anoplotherium, Megalonix, Megatherium, Mastodon,* of M. Cuvier, and some other species of genera previously known, are no longer extant in nature: this however is nothing more than a possibility.

But animals living in the waters, especially the sea waters, and in addition all the races of small sizes living on the surface of the earth and breathing air, are protected from the destruction of their species by man. Their multiplication is so rapid and their means of evading pursuit or traps are so great, that there is no likelihood of his being able to destroy the entire species of any of these animals.

It is then only the large terrestrial animals that are liable to extermination by man. This extermination may actually have occurred; but its existence is not yet completely proved.

Nevertheless, among the fossil remains found of animals which existed in the past, there are a very large number belonging to animals of which no living and exactly similar analogue is known; and among these the majority belong to molluscs with shells, since it is only the shells of these animals which remain to us.

Now, if a quantity of these fossil shells exhibit differences which prevent us, in accordance with prevailing opinion, from regarding them as the representatives of similar species that we know, does it necessarily follow that these shells belong to species actually lost? Why, moreover, should they be lost, since man cannot have compassed their destruction? May it not be possible on the other hand, that the fossils in question belonged to species still existing, but which have changed since that time and become converted into the similar species that we now actually find. The following consideration, and our observations throughout this work, will give much probability to such an assumption.

Every qualified observer knows that nothing on the surface of the earth remains permanently in the same state. Everything in time undergoes various mutations, more or less rapid according to the nature of the objects and the conditions; elevated ground is constantly being denuded by the combined action of the sun, rainwaters and yet other causes; everything detached from it is carried to lower ground; the beds of streams, of rivers, even of seas change in shape and depth, and shift imperceptibly; in short, everything on the surface of the earth changes its situation, shape, nature and appearance, and even climates are not more stable.

Now I shall endeavour to show that variations in the environment induce changes in the needs, habits and mode of life of living beings, and especially of animals; and that these changes give rise to modifications or developments in their organs and the shape of their parts. If this is so, it is difficult to deny that the shape or external characters of every living body whatever must vary imperceptibly, although that variation only becomes perceptible after a considerable time.

Let us then no longer be astonished that among the numerous fossils found in all the dry parts of the world, and constituting the remains of so many animals which formerly existed, there are so few of which we recognise the living representatives.

What we should wonder at, on the contrary, is finding amongst these numerous fossil remains of once living bodies, any of which the still existing analogues are known to us. This fact, proved by our collections of fossils, suggests that the fossil remains of animals whose living analogues we know are the least ancient fossils. The species to which each of them belongs doubtless has not had time to undergo variation.

Naturalists who did not perceive the changes undergone by most animals in course of time tried to explain the facts connected with fossils, as well as the commotions known to have occurred in different parts of the earth's surface, by the supposition of a universal catastrophe which took place on our globe. They imagined that everything had been displaced by it, and that a great number of the species then existing had been destroyed.

Unfortunately this facile method of explaining the operations of nature, when we cannot see their causes, has no basis beyond the imagination which created it, and cannot be supported by proof.

Local catastrophes, it is true, such as those produced by earthquakes, volcanoes and other special causes are well known, and we can observe the disorder ensuing from them.

But why are we to assume without proof a universal catastrophe, when the better known procedure of nature suffices to account for all the facts which we can observe?

Consider on the one hand that in all nature's works nothing is done abruptly, but that she acts everywhere slowly and by successive stages; and on the other hand that the special or local causes of disorders, commotions, displacements, etc., can account for everything that we observe on the surface of the earth, while still remaining subject to nature's laws and general procedure. It will then be recognised that there is no necessity whatever to imagine that a universal catastrophe came to overthrow everything, and destroy a great part of nature's own works.

I have said enough on a subject which presents no difficulty. Let us now consider the general principles and essential characters of animals.

[...]

CHAPTER VII.

OF THE INFLUENCE OF THE ENVIRONMENT ON THE ACTIVITIES AND HABITS OF ANIMALS, AND THE INFLUENCE OF THE ACTIVITIES AND HABITS OF THESE LIVING BODIES IN MODIFYING THEIR ORGANISATION AND STRUCTURE.

WE are not here concerned with an argument, but with the examination of a positive fact – a fact which is of more general application than is supposed, and which has not received the attention that it deserves, no doubt because it is usually very difficult to recognise. This fact consists in the influence that is exerted by the environment on the various living bodies exposed to it.

It is indeed long since the influence of the various states of our organisation on our character, inclinations, activities and even ideas has been recognised; but I do not think that anyone has yet drawn attention to the influence of our activities and habits even on our organisation. Now since these activities and habits depend entirely on the environment in which we are habitually placed, I shall endeavour to show how great is the influence exerted by that environment on the general shape, state of the parts and even organisation of living bodies. It is, then, with this very positive fact that we have to do in the present chapter.

If we had not had many opportunities of clearly recognising the result of this influence on certain living bodies that we have transported into an environment altogether new and very different from that in which they were previously placed, and if we had not seen the resulting effects and alterations take place almost under our very eyes, the important fact in question would have remained for ever unknown to us.

The influence of the environment as a matter of fact is in all times and places operative on living bodies; but what makes this influence difficult to perceive is that its effects only become perceptible or recognisable (especially in animals) after a long period of time.

Before setting forth to examine the proofs of this fact, which deserves our attention and is so important for zoological philosophy, let us sum up the thread of the discussions that we have already begun.

In the preceding chapter we saw that it is now an unquestionable fact that on passing along the animal scale in the opposite direction from that of nature, we discover the existence, in the groups composing this scale, of a continuous but irregular degradation in the organisation of animals, an increasing simplification in their organisation, and, lastly, a corresponding diminution in the number of their faculties.

This well-ascertained fact may throw the strongest light over the actual order followed by nature in the production of all the animals that she has brought into existence, but it does not show us why the increasing complexity of the organisation of animals from the most imperfect to the most perfect exhibits only an *irregular gradation*, in the course of which there occur numerous anomalies or deviations with a variety in which no order is apparent.

Now on seeking the reason of this strange irregularity in the increasing complexity of animal organisation, if we consider the influence that is exerted by the infinitely varied environments of all parts of the world on the general shape, structure and even organisation of these animals, all will then be clearly explained.

It will in fact become clear that the state in which we find any animal, is, on the one hand, the result of the increasing complexity of organisation tending to form a regular gradation; and, on the other hand, of the influence of a multitude of very various conditions ever tending to destroy the regularity in the gradation of the increasing complexity of organisation.

I must now explain what I mean by this statement: *the environment affects the shape and organisation of animals*, that is to say that when the environment becomes very different, it produces in course of time corresponding modifications in the shape and organisation of animals.

It is true if this statement were to be taken literally, I should be convicted of an error; for, whatever the environment may do, it does not work any direct modification whatever in the shape and organisation of animals.

But great alterations in the environment of animals lead to great alterations in their needs, and these alterations in their needs necessarily lead to others in their activities. Now if the new needs become permanent, the animals then adopt new habits which last as long as the needs that evoked them. This is easy to demonstrate, and indeed requires no amplification.

It is then obvious that a great and permanent alteration in the environment of any race of animals induces new habits in these animals.

Now, if a new environment, which has become permanent for some race of animals, induces new habits in these animals, that is to say, leads them to new activities which become habitual, the result will be the use of some one part in preference to some other part, and in some cases the total disuse of some part no longer necessary.

Nothing of all this can be considered as hypothesis or private opinion; on the contrary, they are truths which, in order to be made clear, only require attention and the observation of facts.

We shall shortly see by the citation of known facts in evidence, in the first place, that new needs which establish a necessity for some part really bring about the existence of that part, as a result of efforts; and that subsequently its continued use gradually strengthens, develops and finally greatly enlarges it; in the second place, we shall see that in some cases, when the new environment and the new needs have altogether destroyed the utility of some part, the total disuse of that part has resulted in its gradually ceasing to share in the development of the other parts of the animal; it shrinks and wastes little by little, and ultimately, when there has been total disuse for a long period, the part in question ends by disappearing. All this is positive; I propose to furnish the most convincing proofs of it.

In plants, where there are no activities and consequently no habits, properly so-called, great changes of environment none the less lead to great differences in the development of their parts; so that these differences cause the origin and development of some, and the shrinkage and disappearance of others. But all this is here brought about by the changes sustained in the nutrition of the plant, in its absorption and transpiration, in the quantity of caloric, light, air and moisture that it habitually receives; lastly, in the dominance that some of the various vital movements acquire over others.

Among individuals of the same species, some of which are continually well fed and in an environment favourable to their development, while others are in an opposite environment, there arises a difference in the state of the individuals which gradually becomes very remarkable. How many examples I might cite both in animals and plants which bear out the truth of this principle! Now if the environment remains constant, so that the condition of the ill-fed, suffering or sickly individuals becomes permanent, their internal organisation is ultimately modified, and these acquired modifications are preserved by reproduction among the individuals in question, and finally give rise to a race quite distinct from that in which the individuals have been continuously in an environment favourable to their development.

A very dry spring causes the grasses of a meadow to grow very little, and remain lean and puny; so that they flower and fruit after accomplishing very little growth.

A spring intermingled with warm and rainy days causes a strong growth in this same grass, and the crop is then excellent.

But if anything causes a continuance of the unfavourable environment, a corresponding variation takes place in the plants: first in their general appearance and condition, and then in some of their special characters.

Suppose, for instance, that a seed of one of the meadow grasses in question is transported to an elevated place on a dry, barren and stony plot much exposed to the winds, and is there left to germinate; if the plant can live in such a place, it will always be badly nourished, and if the individuals reproduced from it continue to exist in this bad environment, there will result a race fundamentally different from that which lives in the meadows and from which it originated. The individuals of this new race will have small and meagre parts; some of their organs will have developed more than others, and will then be of unusual proportions.

Those who have observed much and studied large collections, have acquired the conviction that according as changes occur in environment, situation, climate, food, habits of life, etc., corresponding changes in the animals likewise occur in size, shape, proportions of the parts, colour, consistency, swiftness and skill.

What nature does in the course of long periods we do every day when we suddenly change the environment in which some species of living plant is situated.

Every botanist knows that plants which are transported from their native places to gardens for purposes of cultivation, gradually undergo changes which ultimately make them unrecognisable. Many plants, by nature hairy, become glabrous or nearly so; a number of those which used to lie and creep on the ground, become erect; others lose their thorns or excrescences; others again whose stem was perennial and woody in their native hot climates, become herbaceous in our own climates and some of them become annuals; lastly, the size of their parts itself undergoes very considerable changes. These effects of alterations of environment are so widely recognised, that botanists do not like to describe garden plants unless they have been recently brought into cultivation.

Is it not the case that cultivated wheat (*Triticum sativum*) is a plant which man has brought to the state in which we now see it? I should like to know in what country such a plant lives in nature, otherwise than as the result of cultivation.

Where in nature do we find our cabbages, lettuces, etc., in the same state as in our kitchen gardens? and is not the case the same with regard to many animals which have been altered or greatly modified by domestication?

How many different races of our domestic fowls and pigeons have we obtained by rearing them in various environments and different countries; birds which we should now vainly seek in nature?

Those which have changed the least, doubtless because their domestication is of shorter standing and because they do not live in a foreign climate, none the less display great differences in some of their parts, as a result of the habits which we have made them contract. Thus our domestic ducks and geese are of the same type as wild ducks and geese; but ours have lost the power of rising into high regions of the air and flying across large tracts of country; moreover, a real change has come about in the state of their parts, as compared with those of the animals of the race from which they come.

Who does not know that if we rear some bird of our own climate in a cage and it lives there for five or six years, and if we then return it to nature by setting it at liberty, it is no longer able to fly like its fellows, which have always been free? The slight change of environment for this individual has indeed only diminished its power of flight, and doubtless has worked no change in its structure; but if a long succession of generations of individuals of the same race had been kept in captivity for a considerable period, there is no doubt that even the structure of these individuals would gradually have undergone notable changes. Still more, if instead of a mere continuous captivity, this environmental factor had been further accompanied by a change to a very different climate; and if these individuals had by degrees been habituated to other kinds of food and other activities for seizing it, these factors when combined together and become permanent would have unquestionably given rise imperceptibly to a new race with quite special characters.

Where in natural conditions do we find that multitude of races of dogs which now actually exist, owing to the domestication to which we have reduced them? Where do we find those bull-dogs, grey-hounds, water-spaniels, spaniels, lap-dogs, etc., etc.; races which show wider differences than those which we call specific when they occur among animals of one genus living in natural freedom?

No doubt a single, original race, closely resembling the wolf, if indeed it was not actually the wolf, was at some period reduced by man to domestication. That race, of which all the individuals were then alike, was gradually scattered with man into different countries and climates; and after they had been subjected for some time to the influences of their environment and of the various habits which had been forced upon them in each country, they underwent remarkable alterations and formed various special races. Now man travels about to very great distances, either for trade or any other purpose; and thus brings into thickly populated places, such as a great capital, various races of dogs formed in very distant countries. The crossing of these races by reproduction then gave rise in turn to all those that we now know.

The following fact proves in the case of plants how the change of some important factor leads to alteration in the parts of these living bodies.

So long as *Ranunculus aquatilis* is submerged in the water, all its leaves are finely divided into minute segments; but when the stem of this plant reaches the surface of the water, the leaves which develop in the air are large, round and simply lobed. If several feet of the same plant succeed in growing in a soil that is merely damp without any immersion, their stems are then short, and none of their leaves are broken up into minute divisions, so that we get *Ranunculus hederaceus*, which botanists regard as a separate species.

There is no doubt that in the case of animals, extensive alterations in their customary environment produce corresponding alterations in their parts; but here the transformations take place much more slowly than in the case of plants; and for us therefore they are less perceptible and their cause less readily identified.

As to the conditions which have so much power in modifying the organs of living bodies, the most potent doubtless consist in the diversity of the places where they live, but there are many others as well which exercise considerable influence in producing the effects in question.

It is known that localities differ as to their character and quality, by reason of their position, construction and climate: as is readily perceived on passing through various localities distinguished by special qualities; this is one cause of variation for animals and plants living in these various places. But what is not known so well and indeed what is not generally believed, is that every locality itself changes in time as to exposure, climate, character and quality, although with such extreme slowness, according to our notions, that we ascribe to it complete stability.

Now in both cases these altered localities involve a corresponding alteration in the environment of the living bodies that dwell there, and this again brings a new influence to bear on these same bodies.

Hence it follows that if there are extremes in these alterations, there are also finer differences: that is to say, intermediate stages which fill up the interval. Consequently there are also fine distinctions between what we call species.

It is obvious then that as regards the character and situation of the substances which occupy the various parts of the earth's surface, there exists a variety of environmental factors which induces a corresponding variety in the shapes and structure of animals, independent of that special variety which necessarily results from the progress of the complexity of organisation in each animal.

In every locality where animals can live, the conditions constituting any one order of things remain the same for long periods: indeed they alter so slowly that man cannot directly observe it. It is only by an inspection of ancient monuments that he becomes convinced that in each of these localities the order of things which he now finds has not always been existent; he may thence infer that it will go on changing.

Races of animals living in any of these localities must then retain their habits equally long: hence the apparent constancy of the races that we call species, – a constancy which has raised in us the belief that these races are as old as nature.

But in the various habitable parts of the earth's surface, the character and situation of places and climates constitute both for animals and plants environmental influences of extreme variability. The animals living in these various localities must therefore differ among themselves, not only by reason of the state of complexity of organisation attained in each race, but also by reason of the habits which each race is forced to acquire; thus when the observing naturalist travels over large portions of the earth's surface and sees conspicuous changes occurring in the environment, he invariably finds that the characters of species undergo a corresponding change.

Now the true principle to be noted in all this is as follows:

1. Every fairly considerable and permanent alteration in the environment of any race of animals works a real alteration in the needs of that race.
2. Every change in the needs of animals necessitates new activities on their part for the satisfaction of those needs, and hence new habits.
3. Every new need, necessitating new activities for its satisfaction, requires the animal, either to make more frequent use of some of its parts which it previously used less, and thus greatly to develop and enlarge them; or else to make use of entirely new parts, to which the needs have imperceptibly given birth by efforts of its inner feeling; this I shall shortly prove by means of known facts.

Thus to obtain a knowledge of the true causes of that great diversity of shapes and habits found in the various known animals, we must reflect that the infinitely diversified but slowly changing environment in which the animals of each race have successively been placed, has involved each of them in new needs and corresponding alterations in their habits. This is a truth which, once recognised, cannot be disputed. Now we shall easily discern how the new needs may have been satisfied, and the new habits acquired, if we pay attention to the two following laws of nature, which are always verified by observation.

First Law.

In every animal which has not passed the limit of its development, a more frequent and continuous use of any organ gradually strengthens, develops and enlarges that organ, and gives it a power proportional to the length of time it has been so used; while the permanent disuse of any organ imperceptibly weakens and deteriorates it, and progressively diminishes its functional capacity, until it finally disappears.

Second Law.

All the acquisitions or losses wrought by nature on individuals, through the influence of the environment in which their race has long been placed, and hence through the influence of the predominant use or permanent disuse of any organ; all these are preserved by reproduction to the new individuals which arise, provided that the acquired modifications are common to both sexes, or at least to the individuals which produce the young.

Here we have two permanent truths, which can only be doubted by those who have never observed or followed the operations of nature, or by those who have allowed themselves to be drawn into the error which I shall now proceed to combat.

Naturalists have remarked that the structure of animals is always in perfect adaptation to their functions, and have inferred that the shape and condition of their parts have determined the use of them. Now this is a mistake: for it may be easily proved by observation that it is on the contrary the needs and uses of the parts which have caused the development of these same parts, which have even given birth to them when they did not exist, and which consequently have given rise to the condition that we find in each animal.

If this were not so, nature would have had to create as many different kinds of structure in animals, as there are different kinds of environment in which they have to live; and neither structure nor environment would ever have varied.

[...]

CHAPTER VIII.

OF THE NATURAL ORDER OF ANIMALS, AND THE WAY IN WHICH THEIR CLASSIFICATION SHOULD BE DRAWN UP SO AS TO BE IN CONFORMITY WITH THE ACTUAL ORDER OF NATURE.

I HAVE already observed that the true aim of a classification of animals should not be merely the possession of a list of classes, genera and species, but also the provision of the greatest facilities for the study of nature and for obtaining a knowledge of her procedure, methods and laws.

I do not hesitate to say, however, that our general classifications of animals up to the present have been in the inverse order from that followed by nature when bringing her living productions successively into existence; thus, when we proceed from the most complex to the simplest in the usual way, we increase the difficulty of acquiring a knowledge of the progress in complexity of organisation; and we also find it less easy to grasp both the causes of that progress and of the interruptions in it.

When once we have recognised that a thing is useful and indeed indispensable for the end in view and that it is free from drawbacks, we should hasten to carry it into execution although it is contrary to custom.

This is the case with regard to the way in which a general classification of animals should be drawn up.

We shall see that it is not a matter of indifference from which end we begin this general classification of animals, and that the beginning of the order is not a mere matter of choice.

The existing custom of placing at the head of the animal kingdom the most perfect animals, and of terminating this kingdom with the most imperfect and simplest in organisation, is due, on the one hand, to that natural prejudice towards giving the preference to the objects which strike us most or in which we are most pleased or interested; and, on the other hand, to the preference for passing from the better known to what is less known.

When the study of natural history began to occupy attention, these reasons were no doubt very plausible; but they must now yield to the needs of science and especially to those of facilitating the progress of natural knowledge.

With regard to the numerous and varied animals which nature has produced, if we cannot flatter ourselves that we possess an exact knowledge of the real order which she followed in bringing them successively into existence, it is nevertheless true that the order which I am about to set forth is probably very near it: reason and all our acquired knowledge testify in favour of this probability.

If indeed it is true that all living bodies are productions of nature, we are driven to the belief that she can only have produced them one after another and not all in a moment. Now if she shaped them one after another, there are grounds for thinking that she began exclusively with the simplest, and only produced at the very end the most complex organisations both of the animal and vegetable kingdoms.

[...]

Chiroptera.

Galeopithecus [Insectivore. H. E.].	Noctilio.
Rhinolophus.	Bat.
Phyllostome.	Flying Fox.

Quadrumana.

Galago, Tarsius, Loris, Makis, Indris } [Lemurs. H. E.].	Baboon.
	Sapajou.
	Cebus.
	African Baboon.
	Pongo.
Guenon (Old-world Monkey).	Orang.

Remark. According to the order which I have adopted, the quadrumanous family comprises the most perfect of known animals, and especially the later genera of this family; and as a matter of fact the genus orang (*Pithecus*) is at the end of the entire order, just as the monas is at the beginning of it. How great is the difference in organisation and faculties between these two genera!

Naturalists who have considered man exclusively according to the affinities of his organisation, have formed a special genus for him with six known varieties, thus making him a separate family which they have described in the following manner.

Bimana.

Mammals with differentiated unguiculate limbs; with three kinds of teeth and opposable thumbs on the hands only.

Man.

Varieties.
- Caucasian.
- Hyperborean.
- Mongolian.
- American.
- Malayan.
- Ethiopian or Negro.

This family has received the name of Bimana, because in man it is only the hands that have a separate thumb opposite to the fingers while in the Quadrumana the hands and feet have the same character as regards the thumb.

Some Observations with regard to Man.

If man was only distinguished from the animals by his organisation, it could easily be shown that his special characters are all due to longstanding changes in his activities and in the habits which he has adopted and which have become peculiar to the individuals of his species.

As a matter of fact, if some race of quadrumanous animals, especially one of the most perfect of them, were to lose, by force of circumstances or some other cause, the habit of climbing trees and grasping the branches with its feet in the same way as with its hands, in order to hold on to them; and if the individuals of this race were forced for a series of generations to use their feet only for walking, and to give up using their hands like feet; there is no doubt, according to the observations detailed in the preceding chapter, that these quadrumanous animals would at length be transformed into bimanous, and that the thumbs on their feet would cease to be separated from the other digits, when they only used their feet for walking.

Furthermore, if the individuals of which I speak were impelled by the desire to command a large and distant view, and hence endeavoured to stand upright, and continually adopted that habit from generation to generation, there is again no doubt that their feet would gradually acquire a shape suitable for supporting them in an erect attitude; that their legs would acquire calves, and that these animals would then not be able to walk on their hands and feet together, except with difficulty.

Lastly, if these same individuals were to give up using their jaws as weapons for biting, tearing or grasping, or as nippers for cutting grass and feeding on it, and if they were to use them only for mastication; there is again no doubt that their facial angle would become larger, that their snout would shorten more and more, and that finally it would be entirely effaced so that their incisor teeth became vertical.

Let us now suppose that a quadrumanous race, say the most perfect, acquired through constant habit among all its individuals the conformation just described, and the faculty of standing and walking upright, and that ultimately it gained the supremacy over the other races of animals, we can then easily conceive:

1. That this race having obtained the mastery over others through the higher perfection of its faculties will take possession of all parts of the earth's surface, that are suitable to it;
2. That it will drive out the other higher races, which might dispute with it the fruits of the earth, and that it would compel them to take refuge in localities which it does not occupy itself;
3. That it will have a bad effect on the multiplication of allied races, and will keep them exiled in woods or other deserted localities, that it will thus arrest the progress of their faculties towards perfection; whereas being able itself to spread everywhere, to multiply without obstacle from other races and to live in large troops, it will create successively new wants, which will stimulate its skill and gradually perfect its powers and faculties;
4. Finally, that this predominant race, having acquired an absolute supremacy over all the rest, will ultimately establish a difference between itself and the most perfect animals, and indeed will leave them far behind.

The most perfect of the quadrumanous races might thus have become dominant; have changed its habits as a result of the absolute sway exercised over the others, and of its new wants; have progressively acquired modifications in its organisation, and many new faculties; have kept back the most perfect of the other races to the condition that they had reached; and have wrought very striking distinctions between these last and themselves.

The orang of Angola (*Simia troglodytes*, Lin.) is the most perfect of animals: it is much more perfect than the orang of the Indies (*Simia satyrus*, Lin.), called the orang-outang; yet they are both very inferior to man in bodily faculties and intelligence.[a] These animals often stand upright; but as that attitude is not a confirmed habit, their organisation has not been sufficiently modified by it, so that the standing position is very uncomfortable for them.

We know from the stories of travellers, especially as regards the orang of the Indies, that when it has to fly from some pressing danger it immediately falls on to its four feet. Thus, it is said, the true origin of this animal is disclosed, since it is obliged to abandon a deceptive attitude that is alien to it.

No doubt this attitude is alien to it, since it adopts it less when moving about, and its organisation is hence less adapted to it; but does it follow that, because the erect position is easy to man, it is therefore natural to him?

Although a long series of generations has confirmed the habit of moving about in an upright position, yet this attitude is none the less a tiring condition in which man can only remain for a limited period, by means of the contraction of some of his muscles.

If the vertebral column were the axis of the human body, and kept the head and other parts in equilibrium, man would be in a position of rest when standing upright. Now we all know that this is not the case; that the head is out of relation with the centre of gravity; that the weight of the chest and belly, with their contained viscera, falls almost entirely in front of the vertebral column; that the latter has a slanting base, etc. Hence it is necessary as M. Richerand observes, to keep a constant watch when standing, in order to avoid the falls to which the body is rendered liable by the weight and arrangement of its parts.

After discussing the questions with regard to the erect position of man, this observer expresses himself as follows: "The relative weight of the head, and of the thoracic and abdominal viscera, gives a forward inclination to the axial line of the body, as regards the plane on which it rests; a line which should be exactly perpendicular to this plane, if standing is to be perfect. The following fact may be cited in support of this assertion: I have observed that children, among whom the head is bulky, the belly protruding and the viscera burdened with fat, find it difficult to get accustomed to standing upright; it is only at the end of their second year that they venture to trust their own strength; they continue liable to frequent falls and have a natural tendency to adopt the position of a quadruped" (*Physiologie*, vol. ii., p. 268).

This arrangement of parts, as a result of which the erect position is a tiring one for man, instead of being a state of rest, would disclose further in him an origin analogous to that of the other mammals, if his organisation alone were taken into consideration.

In order to follow out the hypothesis suggested at the beginning of these observations, some further considerations must now be added.

The individuals of the dominant race in question, having seized all the places of habitation which were suitable to them and having largely increased their needs according as the societies which they formed became larger, had to multiply their ideas to an equivalent extent, and thus felt the need for communicating them to their fellows. We may imagine that this will have compelled them to increase and vary in the same degree the signs which they used for communicating these ideas; hence it is clear that the individuals of this race must have made constant efforts, and turned all their resources towards the creation, multiplication and adequate variation of the signs made necessary by their ideas and numerous wants.

This is not the case with other animals; for although the most perfect of them such as the Quadrumana mostly live in troops, they have made no further progress in the perfection of their faculties subsequent to the high supremacy of the race named; for they have been chased away and banished to wild and desert places where they had little room, and lived a wretched, anxious life, incessantly compelled to take refuge in flight and concealment. In this situation these animals contract no new needs and acquire no new ideas; their ideas are but few and unvaried; and among them there are very few which they need to communicate to others of their species. Very few different signs therefore are sufficient to make themselves understood by their fellows; all they require are a few movements of the body or parts of it, a few hissings and cries, varied by simple vocal inflections.

Individuals of the dominant race already mentioned, on the other hand, stood in need of making many signs, in order rapidly to communicate their ideas, which were always becoming more numerous and could no longer be satisfied either with pantomimic signs or with the various possible vocal inflections. For supplying the large quantity of signs which had become necessary, they will by various efforts have achieved the formation of articulate sounds. At first they will only have used a small number, in conjunction with inflexions of the voice; gradually they will have increased, varied and perfected them, in correspondence with the growth in their needs and their gain of practice. In fact, habitual exercise of their throat, tongue and lips in the articulation of sounds will have highly developed that faculty in them.

Hence would arise for this special race the marvellous faculty of speaking; and seeing that the remote localities to which the individuals of the race would have become distributed, would favour the corruption of the signs agreed upon for the transmission of each idea, languages would arise and everywhere become diversified.

In this respect, therefore, all will have been achieved by needs alone: they will have given rise to efforts, and the organs adapted to the articulation of sounds will have become developed by habitual use.

Such are the reflections which might be aroused, if man were distinguished from animals only by his organisation, and if his origin were not different from theirs.

Note

a See in my *Recherches sur les corps vivants,* p. 136, some observations on the orang of Angola.

Editorial Headnote

10. Jean-Baptiste Lamarck, "Gall's System," in reprinted in Wheeler and Barbour's *The Lamarck Manuscripts at Harvard* (Cambridge, Mass.: Harvard University Press, 1933), pp. 133, 134–5.

Jean-Baptiste Lamarck (1744–1829) was a French botanist and invertebrate zoologist who is now remembered primarily for publishing one of the first systematic theories of evolution, or *transformisme*. Although he did not make his living as a physician, in this unpublished lecture, he clearly discusses treating several suicidal patients, indicating that he did practice medicine at some point. William Morton Wheeler and Thomas Barbour have dated this manuscript to sometime between 1798 and 1806.[1] Lamarck studied medicine during the early 1770s before becoming embedded in Parisian botanical circles, and the contents of his personal library at the time of his death indicate that he kept up with developments in medical science throughout his life.

Lamarck's ongoing interest in medicine is apparent from his physiological and evolutionary arguments, which were built on several different medical theories that were in vogue at the time, including the cerebral localization of Franz Joseph Gall (1758–1828).[2] Lamarck disagreed with Gall's claim to be able to discern a person's moral qualities by measuring their skull; however, he was more sympathetic to the general argument that the intellectual faculties depended on the physical brain and even that specific mental functions could be associated with particular anatomical structures or regions.[3]

Lamarck and Gall also shared a common interest in the heritability of intellectual faculties, neurological anatomy, and associated diseases. In this extract, Lamarck argued that insanity, and particularly suicidality, was one such disease of the brain, resulting from a disruption of its normal anatomy or functioning. Such a pathological state could be brought on or worsened by environmental factors, including seasonal and atmospheric conditions; however, Lamarck also argued that it could be hereditary and could run in families. This was a consequence of his general belief in the heritability of physiological variation, including changes an individual had undergone during their lifetime that were the result of their habits and way of living. Thus, part of what was at stake in Lamarck's theory of species transformation was the possibility that people could inherit not just favourable traits but also harmful and even pathological ones. In light of this, Lamarck believed that naturalists had important contributions to make to conversations about social reform and good governance. Altering the social environment could produce physiological changes among the populace, and such biological alterations could, in turn, produce better citizens and a more virtuous society.[4]

Notes

1 William Morton Wheeler and Thomas Barbour, *The Lamarck Manuscripts at Harvard* (Cambridge, MA: Harvard University Press, 1933), p. xiv.

2 Gall's work was later popularized as "phrenology." For more on this, see Caden C. Testa and Piers J. Hale, *Evolution Before Darwin* (London: Routledge, 2025); John van Wyhe, *Phrenology and the Origins of Scientific Naturalism* (London: Routledge, 2004); Roger Cooter, *The Cultural Meaning of Popular Science. Phrenology and the Organization of Consent in Nineteenth-Century Britain* (Cambridge: Cambridge University Press, 1985).

3 For more of Lamarck's comments on Gall, see *Zoological Philosophy* (tr. Hugh Elliot; London: Macmillan and Co., 1914), pp. 368–371.

4 See also extracts from Lamarck's work and commentary in Caden C. Testa and Piers J. Hale, Evolution Before Darwin (London: Routledge, 2025); also Caden C. Testa, "Species Transformation and Social Reform. The Role of the Will in Jean-Baptiste Lamarck's Transformist Theory," *Journal of the History of Biology*, (2023), 56 (1):125–51.

10

"GALL'S SYSTEM"

Jean-Baptiste Lamarck

Source: Jean-Baptiste Lamarck, "Gall's System" in Wheeler and Barbour's *The Lamarck Manuscripts at Harvard* (Cambridge, Mass.: Harvard University Press, 1933), pp. 133, 134–5

IS THE DISPOSITION TO MENTAL ALIENATION HEREDITARY? IT IS.

There is for that reason an organ which is more developed. Men of genius are nearer to madness than ordinary individuals, because they occupy themselves strenuously and without relaxation with the same object for a long time. Eventually the particular organ becomes fatigued and insanity is the result.

[...]

ON SUICIDE

This is another cerebral malady, of a moral nature. I believe that the man who commits suicide was prepared for suicide in advance by a peculiar disposition of the cerebral organ and that the immediate cause merely serves to determine the occurrence.

HISTORY OF SUICIDE

(1) This malady may be endemic or peculiar to certain climates. It is very rare in Vienna, very common in London, rather frequent in Paris.
(2) Sometimes suicide is epidemic or peculiar to certain seasons or conditions of the atmosphere. It is such in England when the northeast wind, which brings the fog, prevails.
(3) Suicide may also be hereditary, and then there are always in each family stock certain individuals who commit it. I knew a family in which suicide was hereditary and in which in my time two brothers took their lives and the women always had the desire to do so when they were pregnant. Legislators, moralists and physicians should have a knowledge of such facts.

It is commonly believed that great events are produced by insignificant causes. The same maxim is applied to suicide, but this is not true, because in individuals

DOI: 10.4324/9781003490548-11

with suicidal tendency there is an organic and very remarkable predisposition and the last, or decisive, cause may be very insignificant.

The malady which leads to suicide is called melancholia, disgust with existence, aversion from all pleasures, and despair. Those afflicted complain of pain in the forehead, at the base of the nose, and conceal their discomfort from their whole environment. Sometimes the malady may return even after recovery.

As the disease progresses, the ideas become fixed, and the patients fear that they will lack sufficient means of support for themselves or their families. Others suffer from religious notions, or regard life as a means of sin and in order to avoid it contemplate self-destruction.

Editorial Headnote

11. [Lockhart Muirhead], Review of Lamarck's *Philosophie Zoologique*, *Monthly Review*, 55 (August 1811), pp. 473–84, 55 (April 1813), pp. 481–90.

In this two-part review of Jean-Baptiste Lamarck's 1809 work *Philosophie Zoologique*, which was published in the *Monthly Review* in 1811, Lockhart Muirhead (1765–1829), professor of zoology at Glasgow University, introduced many Anglophone readers to Lamarck and his evolutionary arguments. Although Muirhead expressed agreement with many of Lamarck's general arguments about organic plasticity and the artificial nature of biological taxonomies, he ultimately concluded that *Philosophie* was "an ingenious but excentric publication"[1] and that Lamarck lacked sufficient evidence for some of his major claims, such as the assertion that nature had successively formed all organisms, and that some rudimentary organisms were produced through spontaneous generation.

The review was, if not the first, then certainly one of the first instances in which the English concept of *evolution* was associated with Lamarck's theory of species transformation. In the eighteenth century, evolution referred more often to a theory of embryological development; however, Muirhead referred multiple times throughout the 1811 review to Lamarck's process of increasing biological complexity over time as being evolutionary. Although Muirhead disagreed that Lamarck actually had sufficient empirical evidence to support this idea, the review was still significant for the dissemination of Lamarck's ideas. This was particularly the case within the Edinburgh (and Edinburgh-trained) naturalist and medical circles, the members of which tended to be more receptive to transmutationist ideas than was the case in the more orthodox established English scientific community. This publication shows that Muirhead considered Lamarck's evolutionary theory to be worthy of serious consideration, even if his conclusions were still unproven.

Muirhead also took issue with Lamarck's habit of talking about nature as having been constrained in its operations. Because Lamarck considered nature to be equivalent, rather than subordinate, to God, Muirhead argued that Lamarck ought to have made it more clear that nature could in no way be limited or impelled in its actions. Muirhead did not suggest that evolutionary theory was inherently at odds with the Christian religion or an omnipotent God; however, it is clear that, in his mind, some of Lamarck's more materialist ideas risked treading on delicate theological grounds.

Note

1 Lockhart Muirhead, Review of Lamarck's *Philosophie Zoologique*, *Monthly Review*, 55 (August 1811), pp. 473–484, 55 (April 1813), pp. 481–490.

11

REVIEW OF LAMARCK'S *PHILOSOPHIE ZOOLOGIQUE*

[Lockhart Muirhead]

Source: [Lockhart Muirhead], Review of Lamarck's *Philosophie Zoologique*, *Monthly Review*, 55 (August 1811), pp. 473–84, 5 (April 1813), pp. 481–90

> ART. IV. *Philosophie Zoologique*, &c.; *i. e.* Zoological Philosophy, or an Exposition of those Considerations which relate to the Natural History of Animals; to the Diversity of their Organization, and of the Faculties derived from it; to the physical Causes which maintain Life within them, and give rise to the Movements which they perform; to those, in short, which produce either the Feeling or the Intellect with which they may be indued. By J. B. P. A. LAMARCK, Professor of Zoology in the Museum of Natural History, &c. &c. 2 Vol. 8vo. Paris. 1809.

FOR the accommodation of his pupils, Professor LAMARCK was desirous of exhibiting a written sketch of those general principles which refer to the study of animals. With the view, however, of arriving at legitimate results, he perceived the necessity of surveying the various organization of living creatures, and of noting the most important differences which obtain among those of each family, order, and class, until he found himself insensibly drawn into inquiries of more extended range, and into the examination of some of the most difficult zoological problems. That his speculations on the various modes and forms of animal existence should have led him to the discovery of a scale of being, which may be reckoned either in the ascending or the descending series, needs not to excite our surprize: but it does not follow, as an infallible consequence, that nature first produced the most simple and then the more complex organized structures, regularly proceeding in the order of their complication. How, indeed, can our limited faculties ever be able to ascertain in what degrees of succession the different races of animals assumed their stations on our globe, or whether all did not receive their origin, simultaneously, by the fiat of the Almighty? While, however, the existence of a gradation of beings does not necessarily imply its primary evolution in the order of that gradation, it will be readily conceded that the elementary constituents of vitality

DOI: 10.4324/9781003490548-12

will be best understood by contemplating the phænomena of life, in its lowest and most imperfect stages; and that, accordingly, the laws of its economy begin to be unfolded with more satisfactory evidence, since the history of the least perfect animals with which we are acquainted has formed an object of serious study.

As we may safely presume that the conditions necessary to vitality are not only complete in the least compounded state of organization, but are reduced, moreover, to their most simple term, a question of some consequence naturally occurs; namely, are modes of organization, destined to acquire gradually greater degrees of complexity, deducible from these simple forms? One great object of the present treatise is to prove the affirmative of this inquiry, from the two following considerations:

> 'First, it is evinced by many known facts that the continued employment of an organ tends to its developement, strengthens it, and even enlarges it; whereas defect of employment, become habitual with respect to the organ, is prejudicial to its developement, injures it, reduces it by slow degrees, and if permitted to subsist in all the individuals which have sprung from one another, during a long series of generations, will terminate in its extinction. Hence, we may conceive that a change of circumstances constraining the individuals of a race of animals to change their habits, the organs which are least employed gradually die away, while those which are more frequently exercised are more completely unfolded, and acquire vigour and dimensions proportioned to the habitual employment to which these individuals have subjected them.
>
> 'Secondly, when I reflected on the moving power of the fluids, in the very yielding parts which contain them, I was soon convinced that, in proportion as the fluids of an organized body have their motion accelerated, these fluids modify the cellular tissue in which they move, opening passages, and forming various canals; – creating, in short, different organs, according to the state of organization in which they occur.
>
> 'From these two considerations, I deemed it certain that the motion of fluids in the interior of animals, (a motion which is progressively accelerated with the increased composition of the organization,) and that the influence of new circumstances, in proportion as animals were exposed to them in the course of their diffusion over the habitable parts of the earth, were the two general causes which have placed the various animals in the respective states in which we find them.'

We have expressed these positions, as nearly as we could, in M. LAMARCK's own terms, because they form the basis of his theory, and because he repeatedly refers to them in the progress of his work. His inquiries, however, have led him a step farther, and induced him to assign, as the physical cause of the phænomena of feeling, a more compound state of the nervous system than that which takes place in animals of the lowest order, and, as the cause of the phænomena of intellect, a still more compounded state of the same system. Feeling he conceives to be essentially distinct from irritability, and derived from a separate source; the former resulting from a faculty peculiar to restrain animals, and requiring a specific apparatus of organs for its exercise, and the latter being an inherent property of all animal organization. To internal feeling, in the more perfect animals, he ascribes the producing power of their movements; which, again, in the more imperfect tribes, originate in external excitements. Having remarked that the motions of animals are never communicated, but always excited, he observed that nature borrowed from surrounding media the exciting power of the vital motions and actions of imperfect animals, and was enabled, by compounding their organization, first to convey that power into the animals themselves, and secondly to place it at the disposal of the individuals:

> 'Thus,' says he, 'this *Zoological Philosophy* exhibits the results of my studies on animals, their general and particular characters, their organization, the causes of its developement and diversity, and the faculties which they derive from it. I have here brought together the principal materials which I had collected for a projected work on living bodies, under the title of *Biology;* a work which, on my part, will remain unexecuted.'

The learned Professor offers his speculations to the public merely as his opinions and reasonings; to which, he admits, the opinions and reasonings of others may be opposed. In meditating on the ideas which he now submits to the judgment of the world, he acknowleges the pleasure which he received from their likeness to truth, and the compensation which they afforded him for a life of protracted and fatiguing study.

> 'I shall attain,' he adds, 'the end which I proposed, if lovers of the natural sciences shall find in this work some views and principles of which they can avail themselves; if the observations which I have advanced, and which are my own, are confirmed or approved by those who have had occasion to be occupied with the same pursuits; and if the ideas which they are calculated to suggest can in any degree forward our knowlege, or put us in the way of arriving at unknown truths.'

This is the becoming language of modesty: but, like other theorists, the author, in the course of his argumentation, seems to regard his favourite notions as so many propositions established by fact, and susceptible of complete demonstration.

Our attention is next called to the *Preliminary Discourse*, which is very unnecessarily separated from the *Advertisement*, the materials of both being obviously reducible to the form of a single preface or introduction. Following, however, the author's train of thinking, in the order in which he has consigned it to writing, we cordially assent to the truth of the observation, that the most important phænomena of animal life have chiefly attracted the attention of physiologists, from the period in which the more limited degrees of organization were assumed as fixed and referable points, in the course of their investigations. In our attempts to explain the origin and nature of the vital principle, and of the physical constitution of the various races of animals, we shall most probably be baffled on the very threshold of our inquiries, if we commence with man, the system of whose organization is by far the most refined and complicated with which we are acquainted: but the most simple forms, in which animal motion has been perceived, may be contemplated with better hopes of success; and, from a review of the first ascertained rudiments of life, we might gradually proceed to that of its more complex and intricate modifications. That this is the most eligible and philosophical mode of conducting our inquiries into the origin and operations of the laws of organic life will scarcely admit of dispute: but that the results of inquiries, so conducted, and prosecuted with ability and discernment, would entirely coincide with the author's deductions, is far less apparent.

The presence or the absence of a vertebral column in animals gives occasion to an obvious and important distinction in the multiplied diversities of their structure. The *invertebral* tribes, either from their minuteness or their fancied insignificance, were long overlooked, or very superficially examined. Yet the study of them ought particularly to excite the attention and interest of the naturalist, both because they are more numerous and varied than the other classes of animals, and because the differences of their structure are more singular and more distinctly defined.

Fully aware of these truths, and of the importance of investigating the productions of nature first in general masses or groupes, and afterward in their most minute details, – while the rapid progress of comparative anatomy, the recent discoveries of zoologists, and his own additional observations, loudly called for a new edition of his system of invertebral animals, – M. LAMARCK was induced to concentrate, in a separate work, those general principles which have a reference to the study of the animal kingdom, the essential facts which demand our attention in the prosecution of that study, the grounds of their natural distribution, and the most important consequences which flow from the facts and observations thus collected, so as to exhibit the foundation of the true philosophy of zoological science. The present publication, he informs us, is only a remoulded edition corrected, and greatly enlarged, of his Researches into the nature of Living Bodies, divided into three principal parts, each of which is subdivided into chapters:

> 'Thus, in the first part, of which the object is to exhibit the essential facts that have been observed, and the general principles of the natural sciences, I shall first consider what I term the *artificial parts* of the sciences in question, the importance of attending to the doctrine of *relations*, and the idea which we should form of what is denominated *species* among living bodies. Then, after having unfolded the general principles relative to animals, I shall state, on the one hand, the proofs of that series of

> organization which extends from one extremity of the animal scale to the other, the most perfect animals being placed at the anterior extremity of this scale; and I shall shew, on the other hand, the influence of *circumstances and habits* on the organs of animals, as producing those causes which advance or arrest their developement. I shall close this part with a view of the *natural order* of animals, and an exposition of the most eligible mode of *distributing* and *classifying* them.
>
> 'In the second part, I shall offer my ideas on the order and state of things which constitute the essence of animal life, and shall indicate the conditions essential to the existence of that admirable phænomenon of nature. I shall next endeavour to determine the exciting cause of organic motions, that of tension and irritability, the properties of the cellular texture, the only circumstance in which *spontaneous generations* can take place, the evident consequences of the acts of life, &c.
>
> 'Lastly, the third part will present my opinion on the physical causes of feeling, the power of action, and the acts of intellect performed by certain animals.
>
> 'Under this division of the subject, I shall treat, 1. Of the origin and formation of the nervous system: 2. Of the nervous fluid, which can be known only indirectly, but of which the existence is attested by the phænomena which it alone can produce: 3. Of the physical sensibility and mechanism of sensations: 4. Of the producing force of the motions and actions of animals: 5. Of the source of volition, or of the faculty of the will: 6. Of ideas, and of their different orders: 7. Finally, Of certain particular acts of the understanding, as attention, reflection, imagination, memory, &c.
>
> 'The views unfolded in the second and third parts, doubtless, involve subjects of very difficult examination, and even questions apparently incapable of solution: but they present, at the same time, so much interest, that our attempts may prove of advantage, either by bringing to light truths which had escaped observation, or by preparing the way which may conduct us to the attainment of them.'

In pursuance of this plan, the author commences with the consideration of what he aukwardly terms the *parts of art* in natural science; by which we are to understand the artificial steps to which men have had recourse, in their attempts to arrange the multiplied productions of nature. Under these he includes *Systematic Distributions, Classes, Orders, Families, Genera*, and *Nomenclature;* none of which, he maintains, are founded in nature, though he admits their practical utility and expediency. The productions of nature, he assures us, are merely individuals, which appear after one another in continued succession, which resemble those that have produced them, and which belong to races infinitely diversified; shading into all forms and degrees of organization, and each remaining unchanged as long as no cause of change happens to act on it. The order of nature, which, in proportion as it is studied and known, ought to supersede all conventional and arbitrary arrangements, is, in her organic domain, that of organization; proceeding from the most simple to the most complex structures; and presenting, at its two extreme points, living bodies, which are the most remote from one another, in respect of relations, and offer in their constitution and form the greatest possible differences. In the animal kingdom, the outlines of the grand departments are already sketched agreeably to this principle: but the limits of their subordinate classes, orders, families, and genera, are still exposed to the unsteadiness of arbitrary regulations. That some of the classes of animal productions have their boundaries visibly defined by nature herself is a mere illusion, the offspring of our circumscribed knowlege of the animal forms which exist, and which have existed; and the farther our observation extends, the more we may be convinced that distinctions, apparently the most marked and insulated, are effaced by new discoveries. Thus, in Australasia, we find a class of animals intermediate between the Mammalia and birds. The idea of particular and subordinate *classes*, within those of more general import, is justly condemned as unphilosophical. The multitude and diversity of objects included under one general class may, it is true, completely justify the institution of inferior sections and groupes: but these latter ought not to be denominated by the more general title. The interests of science require that the artificial disposition of its materials should be regulated by the utmost possible simplicity, which is not incompatible with all the divisions and subdivisions which may be found necessary, provided that each has a special denomination; otherwise, nomenclature itself, by the licence of growing abuse, may become a subject of more difficult comprehension than the objects which it is destined to denote.

It would be needless to shew that the secondary divisions of our artificial systems are, according to M. LAMARCK's principles, still more liable to exception than the primary divisions of classes; and that the frequent alterations, which take place in the institution of orders and genera, are attended with very serious trouble and inconvenience. Yet, until we arrive at the knowlege of all living beings, similar changes, we apprehend, must be occasionally introduced in the progress of natural science, whatever be the basis of our distributions. If that basis, however, be assumed from the relations which animals really bear to one another, and if such relations can be easily and accurately defined, a very important step will, no doubt, be gained in the construction of a natural method. From the aid of comparative anatomy, therefore, the author would derive the sources of his arrangement; selecting three special organs in the order of their importance, as the fittest from which to deduce the requisite relations. These are, 1. the organ of feeling; 2. the organ of respiration; and, 3. the organ of circulation. The first two are the most generally employed by nature, and, consequently, more important than the third, which disappears after the crustaceous animals; whereas the first two extend to the animals of two classes subordinate to the crustaceous. Again, of the first two, the organ of feeling is the most important with respect to relations, because it gives birth to the most eminent of the animal faculties, and because, without this organ, muscular action could not exist. Such ideas are strictly philosophical: yet it will, at the same time, be obvious, that every student of nature cannot conveniently avail himself of nice anatomical operations, when roaming on land or water in quest of *prey*.

M. Lamarck next labours, and, we are inclined to think, not without success, to prove that the commonly received notion of the permanency of specific characters is erroneous:

> 'I repeat it,' he says, 'the more our collections are enriched, the more are we convinced that every thing is more or less shaded; that the remarkable differences vanish; and that, for the most part, nature leaves at our disposal, for the establishment of distinctions, only minute and, if we may say so, childish particularities.
>
> 'How many genera, among animals and vegetables, have acquired such an extent by the multitude of species referred to them, that the study and determination of these species have become almost impracticable! The *species* of these genera, ranged in regular succession, and brought together according to their natural relations, present such slight differences from those which are conterminous, that they graduate into one another, and are mutually blended, so as scarcely to permit us to express in language any trifling points of distinction.
>
> 'They only who have long and assiduously applied to the determination of *species*, and who have consulted such collections, know how much the kinds of living bodies run into one another, and are convinced that, when we see insulated species, we consider them as such merely because we are unacquainted with others more nearly related to them, and which are wanting to complete ourassortment.
>
> 'I would not, however, be understood to say that the animals which exist do not form a very simple series, and equally shaded throughout, but that *that* series is ramified, and graduated in various directions; and that it has not, or at least that it has not always had, any discontinuity of parts, if it be true that, in consequence of the loss of some species, a few breaks are now perceptible. It follows that the *species*, which terminate each branch of the general series, are connected, at least on one side, with other adjoining species which harmonize with them. This much the well known state of things now enables me to demonstrate.
>
> 'To accomplish my purpose, I need not have recourse to any hypothesis or supposition, but merely appeal to all naturalists who are accustomed to observe.
>
> 'Not only many genera, but whole orders, and sometimes even classes, already present us with portions almost complete, of the state of things which I have just indicated.
>
> 'Now, in such cases, when the *species* are arranged in a series, and all stationed according to their natural relations, if we select any one of them, and, passing over several others, fix on a second, somewhat removed from the first, these two *species*, when compared, will present

us with striking dissimilarities. It is thus that we began to view the productions of nature which were most within our reach, and whose generic and specific distinctions were easily settled: but, now that our collections are so richly stored, were we to follow the series to which I have just alluded, from the *species* selected in the first instance to that which is singled out in the second, and which is so different from the first, we should reach it by gradual shadings, without remarking any distinctions worthy of being noted.

'I ask, where is the zoologist or botanist, who is not perfectly convinced of the justness of this view of the subject?

'How can we now study or determine the *species*, in a satisfactory manner, amid that multitude of polypes, of all orders, of radiarii, vermes, and especially of insects, among which the genera of *papilio, phalæna, noctuella, tinia, musca, ichneumon, weevil, capricorn, scarabæus, cetonia*, &c. &c. already, of themselves, furnish so many species which closely approximate, graduate into one another, and are almost confounded together?

'What a crowd of shells do not the molluscous tribes offer to our contemplation, from all countries and seas, which elude our means of distinction, and exhaust our discriminative resources?

'If we ascend to the fishes, the reptiles, birds, and even the mammalia, we shall see, with the exception of the vacancies which still remain to be filled up, connecting links every where uniting the neighbouring species, – nay, even the genera, – and leaving to our industry scarcely any ground of good distinctions.

'Does not botany, which considers the other series which the vegetable tribes compose, exhibit, in its different parts, a state of things perfectly similar?

'In fact, what difficulties do we not now experience in the study and determination of species in the genera, *lichen, fucus, carex, poa, piper, euphorbia, erica, hieracium, solanum, geranium, mimosa?* &c. &c.

'When these genera were first instituted, a few species only were known; and it was then an easy matter to distinguish them: but now, when nearly all the intervals between them are closed, our specific differences are necessarily minute, and generally insufficient.'

That facts correspond with these representations will not be readily disputed, and we have more than once hinted our suspicion that the system of nature is a system of insensible gradations, that even the boundaries of animal and vegetable life are obscurely defined, and that the two extremes of organic and inorganic matter may meet at a common point. To the will of the sovereign Creator we must ultimately ascribe this astonishing yet graduating variety in the productions of nature, which the philosopher contemplates with admiration and complacency, and which involves the mere framers of modern nomenclature in a labyrinth of perplexity. We are permitted, however, to trace some of the secondary agency which is employed for the production of the effect in question. A change of climate, for example, or domestication, or intercourse between animals nearly allied in physical constitution, are observed to alter the forms and habits of the progeny; and to induce and multiply such changes on the original diversities, that we feel ourselves lost among breeds, species, and varieties. M. Lamarck does not stop here, but, assuming that the most simple forms of animation were first produced, and are still directly generated by nature, he contends that, from these rudiments, or germs of organization, the finer and more complex forms have been slowly evolved, in the lapse of ages. This doctrine he deliberately expounds in the following canons, or maxims; and with the same degree of confidence, as if he had watched the progress of nature in different ages of the world:

'1. All the organized bodies of our globe are genuine productions of nature, which she has successively formed, during long periods of time.

'2. In her mode of procedure, nature commenced, and still every day recommences, by forming the most simple organized bodies, and directly forms them only, namely, the first rudiments of organization, or, as they have been termed, *spontaneous generations*.

'3. When the first outlines of the animal and vegetable were formed in suitable places and circumstances, the faculties of incipient life and of organic motion being established, have necessarily, and by slow degrees, unfolded the organs, and, in progress of time, diversified them, as well as their parts.

'4. The faculty of increment in each portion of the organized body, being inherent in the first effects of life, has given rise to different modes of the multiplication and regeneration of individuals, and thus the progress acquired in the composition of organization, and in the form and diversity of the parts, has been preserved.

'5. By the aid of an adequate length of time, of circumstances necessarily favourable, of changes which all the points of the globe's surface have successively undergone in their condition, – in short, from the powerful influence which new situations and new habits exercise in modifying the organs of bodies indued with life, – all those which now exist have been insensibly formed such as we see them.

'6. Lastly, according to a similar order of things, all living bodies having experienced more or less important changes in the state of their organization and parts, what is termed *species* among them has been thus insensibly and successively formed, has but a relative permanence in its condition, and cannot be so antient as nature.'

Here the author pauses, to meet two objections which may be urged against this theory of gradual evolution and change. 1. Shall we ascribe to nature, it may be said, and not to the sovereign author of all things, the wonderful instincts and various modes of industry that are exercised by animals? – The answer is obvious, that the supreme Creator may educe his works in the way and manner which he deems most fit; that he may bid all tribes of beings spring into existence at once; or may so constitute and indue the more rude and simple forms, that they shall prove the proximate sources of gradually unfolding varieties: – but M. LAMARCK seems to talk of nature as if it were really a *being* or *agent* subordinate to the Almighty power, instead of the mere expression of an abstract idea; which indeed we often personify, but which, when strictly analyzed, can only denote that order and constitution of things which the great first cause has established and maintains. – 2. It has been long received as an admitted truth, that the constancy of species is inalterable; and the fact seems to have been decided by the precise conformity of individuals of living species in Egypt with such of the same species as were embalmed two or three thousand years ago. In vain, therefore, is it alleged that the same species of animals undergoes any material alteration in the progress of ages. – This celebrated zoologist pretends not to dispute the accuracy of *Geoffroi*, and others of his countrymen, who have ascertained the striking identity between preserved and recent specimens: but he attempts to destroy the force of the argument grounded on this identity, by maintaining that the climate and physical constitution of Egypt have undergone no material change since the date of our most antient records; and that, in course, none could take place on its native animals. This opinion, however, is more easily asserted than proved. Will M. LAMARCK convince his readers that the soil, climate, and produce of Egypt are at this day precisely what they were in the reign of Sesorists? or that animals deemed *sacred* do not, from that very circumstance, receive a certain degree of indulgent and domestic treatment, which, according to his own reasoning, ought to influence not only their habits but their organization?

We cannot, in short, subscribe to the full extent of this author's doctrine; because the facts and analogies, with which we are acquainted, would not bear us out in the support of it. That a few thousand years seem to be a period of long duration to man, while, in regard to the system of the universe, they may be estimated as an instant, we will not dispute: but, beyond the verge of these few thousand years, the data of human reasoning disappear; and though we may roam with more or less felicity in the regions of fancy and conjecture, in these dark regions we must be contented to abide: for where is the *proof* that nature has *successively formed all her productions after long periods of time?* If we are to have recourse to the once exploded doctrine of *spontaneous generation* in some instances, why not in others? Are we certain that the *monas*, for example, is of a less complex structure that the *hydatid?* or that the latter, the *fasciola*, and the solitary *tænia*, are produced by creatures like themselves? If so, where are the parent animals to be found? or how do they penetrate into the inmost substance of the human viscera? Again, why should it follow, as a *necessary consequence*, that the establishment of the first germs of vegetable and animal life should imply the gradual evolution of organs, and the diversities of their forms and parts? or, if this must *necessarily* be, is there any point at which the unfolding of the scene is destined to stop? or are new organs, and new diversities of parts, to be formed and acquired to all eternity? – Lastly, we are at a loss to conceive how the faculty of *growth*, inherent in organized bodies, should give rise to the various modes of *multiplication* and *regeneration*.

When obvious illogical deductions pervade the leading principles of any system, it is surely unnecessary to pursue its examination in detail. In the present instance, particularly, we beg leave to waive the consideration of the author's *generalities on animals*, and his *scale of gradation*: not only because the chapters so intitled frequently recall us to the ideas which have been already stated, but because they are principally occupied with his classification of animals; a subject with which our zoological readers (it is to be presumed) are sufficiently familiar.

We are unwilling, however, to close our report of the first part of this work, without adverting to an improper mode of expression to which the writer has more than once had recourse, and which implies the idea that nature, as if circumscribed in her means, had been *obliged* or *constrained* to act as she has done. Now, on the strange supposition that Nature is a *real personage*, inferior to the Deity, and operating merely in virtue of a limited and delegated power, such language might be tolerated: but if by Nature we understand that order of things in the physical world which has been ordained by the Supreme Being, (and we can attach no other idea to the term,) the slightest insinuation of compulsion or incompetency must be regarded as highly irreverent. In justice, however, to M. LAMARCK, it behoves us to observe that in one passage he interprets the *power of nature* by the *order of existing things*; that in another he strongly reprobates the idea of assigning limits to the Supreme power; and that, in a third, he emphatically asks whether that power be less an object of admiration, because it has willed that things should be as they are.

In a subsequent article, we purpose to notice the two remaining parts of this ingenious but excentric publication.

[*To be continued.*]

[...]

ART IV. *Philosophie Zoologique*, &c.; *i. e.* Zoological Philosophy, &c.
By J. B. P. A. LAMARCK.
[*Art. concluded from p.* 484. *of our lxvth Volume.*]

SINCE we first invited the attention of our readers to this extraordinary publication, circumstances over which we had no controul have occurred, to retard our notice of its two remaining parts. Yet, if our estimate of their merits be as correct as we conceive it to be, the public, while they give us credit for the present redemption of our pledge to continue our report, will not perhaps be much disposed to quarrel with us for the lateness of its accomplishment.

The Second Part commences with a more comfortable analytical definition of *nature* than various passages, interspersed through the volumes, would lead us to apprehend; since the author limits the term to the physical bodies which exist, to the general and special laws which regulate the changes of their state and situation, and to their infinitely diversified motions, whence results the admirable order of things in the grand whole that is exhibited to our view. He even condescends to regard the *eternity of nature* as vague, abstract, and unsatisfactory; to contemplate the universe as an effect; and to ascribe the whole of nature to a First Cause, or Supreme Power, which has created every thing. Without examining too critically how far his subsequent attempts to explain the origin of life, sensation, and intellect, are reconcileable with these professions, the latter will sufficiently warrant the inference that the physical causes to which he refers are to be regarded as of merely a secondary description; so that, even on the supposition that his reasoning were in all respects unassailable, he has only pointed out those laws or processes which the Deity has ordained for the formation of organized Beings, of various ranks and conditions. When he asserts, however, that by studying the bodies which we know, or which have been observed, with their properties, qualities, relations, and motions, we can really ascertain even the proximate causes of the vast multitude of natural phænomena, and especially those of the existence of the living principle, he must pardon the tardiness of our conceptions, or the inaptitude of our judgment, if we do not fully comprehend the meaning nor feel the force of his arguments. His decided propensity to deduce intellectual and moral appearances from circumstances of physical organization bespeaks no novelty of doctrine: but his mode of illustrating his primary positions, however extravagant, frequently implies both originality and ingenuity of thought; and, if we cannot acquiesce in his conclusions, it is merely for want of that proof which probably can never be attained, but which the author, in the plentitude of his imagination, flatters himself that he has unveiled. – Some idea of the scope and purport of this second part may be formed from his own sketch of its outlines:

> 'If we survey with continued attention the different phænomena of organization, more especially those which are connected with the developement of that organization, and chiefly in the more imperfect animals, we shall be convinced,
>
> '1. That every operation to which nature has recourse, in forming her direct creations, consists in organizing into *cellular texture* the small masses of gelatinous or mucilaginous matter which she finds at her disposal, and under favourable circumstances, in filling these small cellular masses with containable fluids, put in motion by means of subtle exciting fluids, which constantly flow on them from the surrounding media.
>
> '2. That the *cellular texture* is the matrix in which all organization has been formed, and in the heart of which the different organs have been successively developed, by means of the motion of the containable fluids which have gradually modified this cellular texture.
>
> '3. That, in fact, the inherent property of the motion of the fluids, in the yielding parts of the living bodies which contain them, is to form in them passages for themselves, places of deposit, and issues; to create in them canals, and, in process of time, different organs; to vary

these canals and organs, according to the diversities either of the motions or of the nature of the fluids which give rise to them, and assume their modifications in the parts; to enlarge, elongate, divide, and gradually to impart solidity to these canals and organs, from the matters which are incessantly forming and separated from the essential moving fluids: matters, of which one portion is assimilated and united to the organs, while the other is discharged outwardly.

'4. Lastly, That organic motion is endued with the properties not only of unfolding organization, but of multiplying the organs and their appropriate functions.

'After having expounded these grand considerations, which appear to me to present incontrovertible truths, though hitherto unperceived, I shall inquire what are the faculties that are common to all animals, and shall then pass in review the most remarkable of those which are necessarily peculiar to certain animals, and incompatible with the condition of others.

'I may confidently assert that the progress of our physiological knowlege is very materially impeded by entertaining the inconsiderate supposition that all animals possess the same organs, and enjoy the same faculties, as if nature were constrained, on all occasions, to employ the same means to attain its ends.

'An object which I ought not to overlook, in this second part of my work, is the consideration of the immediate results of life in any body. Now I can demonstrate that these results give rise to combinations among principles which, independently of this circumstance, could never coalesce. These combinations even teem and multiply, in proportion as the vital energy increases; insomuch that, in the more perfect animals, they present much intricacy, and a considerable accumulation of their combined principles. Thus, living bodies constitute, in virtue of the power of life which they possess, the principal means which nature employs in giving existence to a multitude of different compounds that would never have taken place without this remarkable cause.

'It is in vain to allege that living bodies find, in the alimentary substances on which they feed, matter ready formed to enter into the composition of their frame, their solids, and their fluids of every description; for, in these alimentary substances, they meet only with materials fitted to form the combinations which I have just mentioned, and not the combinations themselves.

'Such are the subjects which compose the second part of this work. Their importance would doubtless require the most ample illustration: but I have limited my design to a succinct exposition of the points which are essential to the comprehension of my observations.'

With a view to the elucidation of these positions, M. LAMARCK states what he conceives to be the leading features of distinction between organic and inorganic or brute matter, and between the animal and vegetable departments of the former; as well as the points in which the two last-mentioned divisions of beings may be supposed to agree. As *Duméril, Richerand*, and others, have drawn similar parallels, we shall here only take leave to remark, in passing, that one or two of the reputed grounds of distinction will be questioned by those who ascribe an imperfect sort of organization to the symmetrical portions of matter; and that the doctrine of an abrupt and wide chasm between the elective attractions of mineral particles, and vegetable assimilation, will not be readily conceded by the advocates of a system of imperceptible gradations in the works of nature: a system for the existence of which, in other parts of his work, the writer eagerly contends. Indeed, we are not prepared to maintain the absolute negation of motion in mineral substances; and chemical changes and combinations, in many cases at least, proceed within them, which seem to make no contemptible approaches to the first obscure indications of the elements of life.

Let us now grant to M. LAMARCK his four conditions for the residence of life in any body; namely, the existence of cellular texture, of contained fluid, of the favourable state of these parts for the excitement of organic motion in the fluid, and an exciting cause: the origin of the very first requisite remains in the same degree of darkness as ever; since we are still completely ignorant of the circumstances which dispose the particles of matter to assume the cellular in preference to the crystalline or any other arrangement: then, the formation of fluids susceptible of organic motions is not more easily apprehended; and we are equally at a loss to perceive what causes the favourable state of the parts to admit of vitality. Hence, on the supposition that these conditions are observed in every living being, we can consider them only as physical *facts*, and not as an *explanation* of the principle of life. With regard to the exciting cause, the author presumes that he has detected it in the form of caloric and the electric fluid; the first producing and maintaining the *orgasm*, or tension of the supple part of living bodies, and the latter causing the organic motions and actions of animals. That a certain temperature is requisite to the existence of every animal, we do not deny; but how can we prove that this temperature is the *cause* of life, any more than that atmospheric air is the cause of respiration? With respect to the electrical fluid, again, we know hardly any thing of its nature and properties, except when it is manifest, for a moment, in a state of accumulation; and when, like the accumulation of heat, it is destructive of the principle of life. M. LAMARCK admits, however, that other elements may enter into the exciting cause of life; and that, in the more perfect and complex animals, even caloric and electricity are so modified as to be maintained and renewed within the machine. The hybernation of certain animals in the northern latitudes, the perishing of others by excessive cold, and the luxuriant pullulation of animal and vegetable life in the tropical climates, sufficiently establish the fact that a due degree of warmth is indispensable to the diffusion and maintenance of vitality: but they no more imply the cause of that wonderful phænomenon, than the residence of fishes in water implies that the liquid element is the cause of their existence.

By the term *orgasm*, is here denoted that state of active tone, or tension of the flexible parts of animals and vegetables, which others have denominated *latent sensibility*, and which retains the parts in question in a proper state to be affected by the exciting cause of vital movement. If its intensity be increased to such a degree as to overcome the cohesion of the particles of matter that are affected by its influence, it is expressed by the term *erethism*, as its maximum is by that of *inflammation*, and its minimum by that of *atony. Irritability*, he maintains, has its source in orgasm, and cannot properly be predicated of plants; the orgasm which they manifest (and which others have, without hesitation, termed *irritability*) being too feeble to occasion those sudden contractions and distensions to which he thinks the name of irritability should be restricted.

An entire chapter is somewhat needlessly devoted to illustrate a position which few physiologists, we presume, are disposed to question; namely, that a *cellular texture* forms a more or less considerable portion of all those substances which are included under the title of *organic:* but this is followed by the assertion of a more problematical doctrine, namely, that of *direct* or *spontaneous generation.* The antients, it is observed, had the sagacity to perceive that heat tended to generate certain animals, without the intervention of sexual union: but they overlooked the important consideration of its combination with moisture as essential to the eduction of animality; and, from their ignorance of the history of the smaller animals, they ascribed the existence of many of them to heat, or fermentation alone, though it has since been demonstrated that they proceed from parents like themselves. From this latter circumstance, again, the moderns (it is insinuated) have run into the opposite extreme, and cancelled from their philosophy the very idea of equivocal generation. We are, at all events, compelled to admit that the origin of various living beings, as of some of the byssi and mucors, infusorial animalcules, hydatids, intestinal worms, &c., is veiled in the most profound mystery, and seems to be scarcely reconcileable to our common notions of vegetable and animal reproduction. The present writer supposes that he has solved the difficulty by referring to nature the direct production of the first traces of animation, indued with propensities and dispositions to compound organization, to create particular organs, to insulate those organs and their functions, – to divide, in short, and to multiply their different centres of activity. A subtile and penetrating fluid, he alleges, perhaps analogous to that which constitutes fecundating vapour; is diffused over the globe, and constantly maintains the stimulus that imparts tension and motion to the gelatinous or mucilaginous bodies fitted to receive them. 'Nature, in a word,' to adopt the language of his own theorem, 'by the aid of heat, light, electricity, and humidity, forms *spontaneous* or direct generations, at the extremity of each kingdom of living bodies, where the most simple of these bodies have their stations.'

As to the more imperfect animals and vegetables which are not propagated by sexual intercourse, he maintains that the direct interference of nature becomes even necessary as often as these frail sketches of vitality are obliterated by the rigour of the season. From the simple lineaments of the vital structure, are gradually deduced (though we are left to trace the steps) greater specialty of organization, more appropriate habits, and more complex fabrics; which are afterward continued by the pairing of the respective races. Aware, perhaps, that the direct mode of production may also be predicated of the tæniæ, and other intestinal worms, M. LAMARCK is willing to allow that nature may exercise her immediate influence in other parts of the scale of being than at its lower extremities. Now, all this, even if perfectly consonant with fact, does not very satisfactorily explain the principle of life, although it may indicate under what circumstances it is found to exist. The growth and multiplication of lichens are, somewhat inconsiderately, ascribed to the combined influence of moisture and *warmth;* whereas, like many of the mosses, they preferably flourish in winter, and in the colder regions of the globe.

In opposition to the common sentiment, that living bodies are endued with peculiar properties by which they are enabled to resist the influence of physical attraction and chemical affinities, M. LAMARCK contends that they are incessantly liable to the operations of mechanical and chemical laws: but that the tendency of these laws to decompose their structure is, during the period of their existence, counteracted by the repairing energies of the vital force. This is, at least, different language from that which is commonly held on the subject: but, if we rightly understand its import, its amount is nearly the same. Nature, we are told, is, on the one hand, every where busied in destroying all existing compounds, in disengaging their principles from the state of combination, and restoring them to that freedom which reinstates them in the use of their peculiar faculties: while, on the other, a particularly powerful and continually active cause *likewise exists in nature*, which possesses the property of forming combinations, of multiplying and diversifying them, and which uniformly tends to charge them with principles; having its residence in the organic actions of living bodies, in which it continually forms combinations that otherwise would never have existed. Our readers will easily excuse us from dragging them farther into this labyrinth of metaphysical speculation; of which the windings have conducted us to this notable conclusion, that 'all compound mineral matters, such as earths, stones, metallic, sulphureous, bituminous, and saline substances, &c., proceed from the residues of living beings; residues which have undergone successive alterations in their composition at the surface and in the bosom of the earth and the waters; and that living bodies are the primary source and origin of all known compound matters.' How, then, we may ask, did vegetables and animals exist before their recrements had formed the solid parts of the globe?

The faculties which are common to all living bodies are, according to the author, those of nourishing, – of producing their bodies, (for he is convinced that living creatures form their own substance and organs,) – of unfolding their system, or growing to a limited extent, – and of regenerating themselves, or producing bodies in all respects like themselves: – phænomena which he deduces from his own principles of orgasm and vitality, and from the more obvious and *tangible* doctrine of assimilation and excess of nourishment. He ascribes the progressive decay and final dissolution of the body to the gradual stiffening of the originally flexible parts, which at length become incapable of performing the organic motions. The increasing rigidity of the organs, and the contraction of the vessels, are represented as inevitable consequences of the laws of animal nourishment, by which more matter is assimilated to the system than is lost by the natural discharges.

Among the numerous special or particular faculties, which appertain only to certain races or descriptions of living beings, the author reckons those of digestion, respiration by means of an appropriate organ, the performance of actions and locomotion by muscular organs, feeling, or the power of experiencing sensation, multiplication by sexual congress, circulation of the essential fluids, some degree of intelligence, &c. The original formation, however, of the special organs on which appropriate functions depend, remains to be explained; and the creation of a particularity of structure, destined to the performance of an exclusive office, is a problem scarcely less profound than that which is exhibited in the creation of a very imperfect vegetable, or animal, endued with the mere rudiments of vitality. Sensation, it may be alleged, is a general apanage of life: but, if it results from a system of nerves, concentrating in a focus or sensorium, we should then concur with the present zoologist in withholding it from the molluscous and infusorial tribes, and assign its first indications to the class of insects.

The special organ of *intellect*, in the theory before us, is a portion super-added to the brain, constituting the two-fold hemispheres which invest it.

> 'Let it here suffice to remark that, among animals provided with a nervous system, none but the more perfect have their brain furnished with the two hemispheres which I have just mentioned; and that, probably, all the invertebral kinds, except perhaps some *mollusca*, of the last order, are generally destitute of them; though many of them have a brain, to which the nerves, from one or more particular senses, proceed directly, although the brain is usually separated into two lobes, or divided by a furrow.
>
> 'Agreeably to these views, the faculty of performing acts of *intellect* scarcely begins sooner than with the fishes, or, at most, with the *cephalopode mollusca*. In these stages, it exists in its greatest imperfection: it is somewhat gradually unfolded in the *reptiles*, especially in those of the last orders: it has made great advances in the *birds;* and it presents in the *mammiferous* families of the last orders the utmost limits to which it can attain in the animal creation.'

Surely he is a sturdy physiologist who presumes to ascertain the boundaries of intelligence, which Supreme Power and Goodness may prescribe to the higher ranks of created beings!

The main object of the Third Part of the treatise under review is to illustrate the nature and cause of animal feeling, or sensation; and the origin of ideas through the intervention of the *hypocephalus*, for so M. LAMARCK has technically denominated the supplementary organ of the brain to which we have just alluded. From the contents of two elaborate chapters, in which he details his notions of the nervous system and fluid, he attempts to prove, 1. that all animals cannot possess this system of organs; 2. that in its origin, and consequently in the state of its greatest simplicity, it imparts to the animals which possess it only the faculty of muscular motion; 3. that, in the progress of its composition, it moreover communicates the power of feeling; and, 4. that, when complete in all its parts, it supplies the animals to which it belongs, not only with muscular motion and sensation, but 'with the capacity of forming and comparing ideas, – in other words, of exerting judgment, or understanding.' The ingenuity of the reasoning which is bestowed on each of these particulars, and on the various modifications of the nervous system, bespeaks an intimate acquaintance with the structure and habits of various races of animals: but we shall not stay to analyze it, both because it contains a recurrence to some of those arguments to which we have already adverted, and because we are fully persuaded that it must ultimately fail of producing conviction on the mind of the cool and candid inquirer. Its general tendency is to refer not only animal motions and actions but mental energies, to particular modifications of organic matter, and to real or fancied medullary centres of relation, with which the nervous filaments are immediately connected. In the scale of intellect, too, fishes are (somewhat preposterously) exalted above insects; though their discernment and affections are, to all appearance, much less acute and prepossessing. It would be idle to refute the existence or non-existence of a nervous fluid, since even M. LAMARCK admits that it is not cognizable by any of the senses; and motion may be referred either to the influence of a fluid, or to some other unknown agency: but, if we are incapable of demonstrating the presence of any such subtile fluid in the nerves, much less can we prove that it is of an electrical or galvanic nature, modified by its residence in some part of the animal economy. However this may be, we are here taught that it is the vehicle of physical sensation; transmitting an impression from any part of the body, through the nerve, to the common centre or centres of relation, on which it produces a shock that affects the whole nervous system, with the solitary exception of the transmitting nerve. This general commotion occasions an instant re-action; which, terminating in the extremity of the nerve first affected, not only constitutes what we term a sensation, but gives rise to our fallacious reference of the feeling to the part affected. Transmission, commotion, re-action, and re-transmission, are all accomplished without our consciousness, (how, then, do we know that they take place?) and with the rapidity of lightning. No particular portion of the material system, we are repeatedly admonished, is capable of producing sensation, which is merely a general effect, or result of the operation of the sensitive system: but this system, we are told, is identified with that of the nerves; and hence we are driven to the alternative of either rejecting the theory, or admitting that certain delicate arrangements of matter, pervaded by a subtile and invisible fluid, are capable, in certain circumstances, of producing sensation: a conclusion certainly not very remote from the theory of vibrations, which M. Lamarck himself regards as visionary and absurd. To meet the charge of materializing all animal actions and emotions, he occasionally reverts to the principle, that no particular organ is inherently capable of producing them; and that they are only the consequences of the whole machine being affected by a stimulating cause. Yet this cause we must presume to be material, since no species of excitement is suggested but caloric, and the electric, galvanic, or some analogous fluid. Besides, the actions and emotions themselves are distinctly represented as flowing from certain arrangements and dispositions of the corporeal organs, and the mention of any spiritual agency does not once occur. The multiplication of organs is invariably represented as the contrivance employed by nature to increase the powers and capacities of animals. The presence of a nervous system is asserted to superadd to irritability, sensation, an intimate feeling or consciousness of existence; and the property of acting by an internal power, or by the will. In all cases, the nervous fluid is extremely accommodating; since, notwithstanding its exquisite subtlety, and the continuousness of its particles, one portion of it is at the disposal of the animal, for the performance of actions, and another is reserved for muscular excitement.

It would be truly tiresome to pursue these excentric lucubrations to a greater length, or to dwell on the deduction of the mental processes of attention, memory, and reasoning, from the compartments and agitations of the wonder-working hypocephalus. Enough, we conceive, has been said to prove that the learned and celebrated Professor is more ingenious than convincing; that he more frequently resorts to doubtful principles than to well attested facts; and that, like many of his countrymen, he is more enamoured of bold and paradoxical discussion, than of sober observation and cool induction. Nothing short of direct proof will fully satisfy our scientific readers, that certain orders of animals are incapable of volition; that others are unconscious of existence; that all have originally inhabited the waters; and that life and thought have been evolved precisely in the manner which is set forth in the Zoological Philosophy of this redoubtable member of the Legion of Honour. Muir.

Part 2

EDINBURGH TRANSMUTATIONISTS

Editorial Headnotes

12. [Albrecht Rengger], "Observations on the Nature and Importance of Geology," *Edinburgh New Philosophical Journal,* (Oct. 1826), pp. 293–302.

13. [André Étienne d'Audebert de Férussac], "Of the Changes which Life has Experienced on the Globe", *Edinburgh New Philosophical Journal*, (1827), pp. 298–301.

14. [Étienne Geoffroy St-Hilaire (trans. Alexandre Bertand], "For the Continuity of the Animal Kingdom by Means of Generation from the First Ages of the World to the Present Times," *Edinburgh New Philosophical Journal*, (April 1829), pp. 152–5.

(Translation of Geoffroy's March 23, 1829 paper to the *Académie des Sciences*, first appeared in *The Globe*.)

These three sources from the *Edinburgh New Philosophical Journal* have interested scholars of nineteenth-century evolutionism, as all three articles have been read as endorsing aspects of Lamarck's transformist theory. In this context, the articles have been interpreted as writings of a group of 'Edinburgh Lamarckians' – physicians and naturalists based in Edinburgh who sympathized with broadly Lamarckian ideas during the 1820s and 30s.[1] More recently, Pietro Corsi has re-examined these three articles, challenging scholarly consensus as to their authorship and placing them in the context of wider European debates about geology, natural history, and the species question.[2]

The first of the three articles was published in 1826 and included remarks that closely echo passages from Lamarck's transformist text *Philosophie zoologique* (1809). Initially attributed to the Lamarckian Scottish naturalist Robert Grant,[3] then to the natural history professor Robert Jameson,[4] the 1826 text is actually a translation of an extract from Albrecht Rengger's (1764–1835) *Beyträge zur Geognosie* (Contributions to Geognosy). Although Rengger did indeed endorse broadly Lamarckian tenets, Corsi notes that this text excised the more overtly materialist aspects of Lamarck's theory, making transformism less radical and thus more respectable. Rengger was a moderate liberal Swiss politician.

The second text, published in 1827, was previously attributed to Ami Boué (1794–1881), a German-speaking naturalist educated in Edinburgh. However, Corsi's research shows that this article, too, was a translation; the original was an article in the *Bulletin des sciences naturelles et de géologie* written by the journal's then-owner, Baron André Étienne d'Audebert de Férussac (1786–1836). Férussac was politically conservative and generally anti-transformist and was a staunch supporter of critical elements of the arguments espoused by Georges Cuvier, Lamarck's colleague and rival. Férussac did believe that natural laws could, over time, change both the earth and its living organisms. However, these were common arguments at the time and did not inherently indicate a belief in Lamarckism or species transformism.

The last text, published in 1829, is a translation made by Alexandre Bertrand (1795–1831) of a report delivered by the zoologist Étienne Geoffroy St-Hilaire (1772–1844) to the Académie des Sciences. Geoffroy's report was a summary of a paper by the naturalist François Désiré Roulin (1796–1824) on the changes that domesticated European animals underwent when transported to South America; it was initially run in the *Globe*, an opposition newspaper critical of the conservative Restoration government. As this suggests, this third text hints more at the links between politics and transformist theory; the *Globe* was not allowed to take explicit stances on political issues and used philosophical natural history like Geoffroy's to indicate support for liberal politics both within and without the sciences.

Together, these three articles indicate two important points: First, although they were not authored by Edinburgh Lamarckians, they show that journals such as the *Edinburgh New Philosophical Journal* were important conduits through which Edinburgh naturalists were kept up to date with ongoing transformist and biological debate in Europe. Second, they show that the belief in change over time in both Earth's history and in living organisms was not an inherently radical or necessarily "Lamarckian" position, even though Lamarck's name and ideas were frequently mentioned and commented on throughout these debates.

Notes

1 See Bill Jenkins, *Evolution Before Darwin: Theories of the Transmutation of Species in Edinburgh, 1804–1834* (Edinburgh: Edinburgh University Press, 2021).

2 Pietro Corsi, "Edinburgh Lamarckians? The Authorship of Three Anonymous Papers (1826–1829)." *Journal of the History of Biology* 54 (2021): 345–374.

3 Adrian Desmond, "Robert E. Grant: The social predicament of a pre-Darwinian transmutationist," *Journal of the History of Biology*, 17:2 (1984), pp. 189–223. Adrian Desmond, *The Politics of Evolution: Morphology, Medicine, and Reform in Radical London* (Chicago: University of Chicago Press, 1989).

4 James A. Secord, "Edinburgh Lamarckians: Robert Jameson and Robert E. Grant," *Journal of the History of Biology*, 24:1 (1991), pp. 1–18; Bill Jenkins, *Evolution Before Darwin. Theories of the Transmutation of Species in Edinburgh, 1804–1834* (Edinburgh: University of Edinburgh Press, 2021).

12

"OBSERVATIONS ON THE NATURE AND IMPORTANCE OF GEOLOGY"

[Albrecht Rengger]

Source: [Albrecht Rengger], "Observations on the Nature and Importance of Geology", *Edinburgh New Philosophical Journal*, (Oct. 1826), pp. 293–302

OBSERVATIONS ON THE NATURE AND IMPORTANCE OF GEOLOGY.

A CELEBRATED school of philosophy among the ancients, maintained that there was only one virtue. With as much, nay even more, propriety, it might be maintained, that there is only one science, at least one physical science. The various departments of this science are so framed, as, in some measure, to accommodate the incommensurability of nature to our capacity; and by connecting things that are homogeneous, they enable us to take a survey of natural phenomena; but, while we are occupied with a single department, we become sensible of its dependence on others, and are frequently at a loss to assign to each its peculiar province.

Of all the departments of physical science, geology is the most intimately connected with other branches, and stands in need of their assistance, or assists them more frequently than any other. This mutual relation, which contributes, in no small degree, to bestow a peculiar charm on geology, has, at the same time, a tendency to render it a difficult study.

When speaking of Geology, it must be understood to comprehend Oryctognosy as its foundation; the latter gives us a knowledge of the characters, the former of their combination. Whoever is in danger of mistaking one character for another, will never learn to read accurately; and he who continually devotes his attention to nothing but the characters, may, indeed, owing to the difficulty of recognising them, be very profitably engaged, but he will be frustrated as to the ultimate and most essential object of their study.

The terrestrial globe, whose structure, so far as it is exposed to our view, is the proper object of geognostic investigation, is the extensive workshop wherein the powers of nature, with which natural philosophy and chemistry are engaged, have operated, and are still operating. It is not therefore matter of surprise, though these two sciences have both a kindred affinity for geology, to which the latter is indeed so closely related, that geology may be considered as practical chemistry. In addition to this, geology has, with other departments of science, many points of contact, from which it may be allowable to select a single example.

Geometry, guided by simple principles, formed regular bodies from limited plane surfaces, and determined their peculiar properties, without foreseeing that models of them would be found in nature herself; but since observation has brought us acquainted with the regular figures of mineral bodies, they exhibit in relation to this science one of the most important applications, as well as one of the most unerring standards, by which they are distinguished.

When the geometer, by his measurements, proves that the figure of our earth may, like that of other planets, be determined by its revolutions, and hence draws conclusions regarding its original state of fluidity, we find, that the phenomena of geology lead to the same result. When he weighs its mass in a balance, whose arm is the semidiameter of the sun's orbit, we are unable to confirm his statement by immediate observation; but we obtain, in this way, a basis on which we can, in some measure, rest our conclusions regarding the internal structure of the earth.

If we contemplate its surface, with all its inequalities, it is geology alone that can give us a distinct representation of them. All local descriptions, not springing from this source, either leave behind them indistinct and erroneous conceptions, or are entirely fanciful. This surface being the habitation of our species, its figure and its changes must, therefore, be closely connected with the history of the human race; and though the most important of those changes may be far anterior to their origin, and to the period of history, we may yet, in more than one geognostic fact, find suggestions and disclosures, which cannot be unacceptable to the historical investigator. These facts concur with historical testimony, in representing the elevated platforms of Asia as the cradle of the human race, and in explaining their diffusion from that centre; and the traditions of deluges, found among all the nations of antiquity, are corroborated by the still existing traces of those violent events.

The monuments concealed in the bosom of the earth, and extending to the whole organic creation, are still more instructive. Between the dead and the living there yawns a chasm, indeed, which we can never overleap; but if any thing can lift the veil that hangs over the origin and progress of the organic world, it must be those remains of it, for the knowledge of which we are indebted to geology. So far as we have examined the crust of the earth, we have discovered in its structure and materials no transition from simple to compound. The order of time has established no relation, according to which the strata of simple rocks of the earliest formations are the simplest, while the newer are more and more compound; on the contrary, the oldest appear to be the most compound. In complete opposition to this, the organic world, in each of its two principal divisions, exhibits a series of formations from simple to compound; the simplest being the oldest. Thus we observe animal life commencing in infusory animals, without any discernible organs. Simple digestive organs are first visible in the polypi; in the echinodermata the organ of respiration first appears; in insects a system of nerves and muscles; in crustaceous animals circulation; and in the last two, simple organs of sense make their appearance. At the same time, generation preserves the peculiar character of organic beings; and after having accomplished its purpose, by mere division and dissolution, the particular generative organs develop themselves in distinct sexes. With the avertebral animals are conjoined the series of the vertebral, in which every system appears more perfect, and more closely connected. New organs of sense are unfolded, and the brain becomes the centre of feeling, perception and life, till in man it attains the highest state of perfection, and endows him with consciousness and rationality. Long ago, celebrated naturalists, relying upon these observations, attempted, with more or less success, to arrange the species of animals, sometimes according to a scale of gradation, and sometimes according to a reticulated form, without giving any distinct account of the meaning of such an arrangement. Should it, like the piling up of a collection of books, merely serve for a more convenient survey of innumerable creatures, without any reference to their origin? Or, do they intend, by means of such an arrangement, to express the design that hovered in the mind of Omnipotence, before he called these creatures into being? Or, have they originated in the way in which they appear in the scale of gradation, as if the hand of the Creator, like that of a human artist, perhaps, must first be exercised on simple formations, before it was capable of producing such as were compound?

Upon these questions, whose answer might contain no less than a key to the profoundest secrets of nature, Mr Lamarck, one of the most sagacious naturalists of our day, has expressed himself in the most unambiguous manner. He admits, on the one hand, the existence of the simplest infusory animals; on the other, the existence of the simplest worms, by means of spontaneous generation, that is, by an aggregation process of animal elements; and maintains, that all other animals, by the operation of external circumstances, are evolved from these in a double series, and in a gradual manner. On that account, the scale of gradation, according to which he arranges the animal kingdom, is, at the same time, the history of their origin; and the discovery of this truly natural method, the most important problem of the natural philosopher. Although it should not be forgotten, that this meritorious philosopher, more in conformity with his own hypothesis than is permitted in the province of physical science, has resigned himself to the influence of imagination, and attempted explanations, which, from the present state of our knowledge, we are incapable of giving, we nevertheless feel ourselves drawn towards it, and these notions of the progressive formation of the organic world, must be found more worthy of its first Great Author than the limited conceptions that we commonly entertain.

Geology can alone inform us, how far this successive course of development may have been followed by nature. When all the races of animals, whose remains are contained in the crust of the earth, have been better ascertained than at present, and their situations better known, when we have discovered at what period of the earth's formation any species of animals makes its appearance for the first time, we shall then be able to draw conclusions, more or less accurate, concerning the order of succession. The doctrine of petrifactions, even in its present imperfect condition, furnishes us with accounts that seem in favour of Mr Lamarck's hypothesis. We, in fact, meet with the more perfect classes of animals, only in the more recent beds of rocks, and the most perfect, those closely allied to our own species, only in the most recent; beneath them occur granivorous, before carnivorous, animals; and human remains, are found only in alluvial soil, in calcareous tuff, and in limestone conglomerates.

Geology does not inform us merely of the origin of animal species, but also of their destruction. Out of the vast number of animal remains, but few belong to species now living, and these only, in the most recent rock-formations; by far the greater number of their primitive structures are lost, and the older the beds of rock in which they make their appearance, so much the more do they deviate in their formation from the species now in existence. May this destruction, as is commonly received, have been the result of violent accidents, and destructive revolutions of the earth; or does it not rather indicate a great law of nature, which cannot be discovered by reason of its remote antiquity? Within the narrow circle of vision in which the organic world manifests itself to our observation, we observe individuals only going to destruction, and in opposition to that, great preparations making for the preservation of the species. But if all living perish, may no point of duration have been fixed for the species; or do we not rather, in these signs of a former world, discover a proof, that, from a change in the media in which organic creatures lived, and from powerful causes operating upon them, their power of propagation may be weakened, and at length become perfectly extinct? Is the continual decrease, then, which we observe among some species, a consequence of the various modes of destruction they experience from the hand of man, or may it not rather be produced by natural circumstances, and be a sign of the approaching old age of the species?

The distinction of species is undoubtedly one of the foundations of natural history, and her character is the propagation of similar forms. But are these forms as immutable as some distinguished naturalists maintain; or do not our domestic animals and our cultivated or artificial plants prove the contrary? If these, by change of situation, of climate, of nourishment, and by every other circumstance that operates upon them, can change their relations, it is probable that many fossil species to which no originals can be found, may not be extinct, but have gradually passed into others. As there are periodical movements of the heavenly bodies, that is, movements that are visible only after hundreds of years, so these are undoubtedly periodical changes in the organic world. If these have required intervals of time that are antecedent to all historical traditions, and to the duration even of the human race, the monuments concealed in the bosom of the earth can alone reveal them. We indeed observe that the Ibis, which was worshipped in ancient Egypt, and preserved as a mummy, is still the same in modern Egypt; but what are the few thousand years to which the mummy refers, in comparison with the age of the world, as its history is related by geology.

Geology likewise supplies us with instructive disclosures regarding the distribution of organic beings. If we, in all the regions and climates of the world, meet with a striking uniformity in the structure of the earth, we also, on the contrary, observe plants and animals of a most varied character scattered over its surface. As there are among the dicotyledons, that is, among the more perfect plants, no species, which are at the same time indigenous to the hot climates of the old and new world, so both halves of the globe in the same zone possess mammiferous animals, birds, reptiles, and insects peculiar to each. Species common to both are found only among the inferior gradations of organization, and species of a higher order are found only in those high northern latitudes, where the continents were undoubtedly at one time conjoined. From the combined results of organic geography, and the doctrine of petrifactions, it will at once follow, whether the ancient population of the terrestrial globe was distributed according to the same laws as at present. Even now, many of the petrifactions of cold climates, whose species and families are produced only in hot countries, indicate a great change in the temperature of their former situations, and phenomena, like that of the rhinoceros found on the shore of the Wilhui, and of the mammoth at the mouth of the Lena, are likewise indications of sudden changes in those places. Along with the distribution of species, we also acquire a knowledge of the distribution of individuals, and of their modes of life, from their fossil remains, because these remains, like living creatures, appear to us sometimes single, and dispersed at other times in numerous bodies, and closely crowded together.

The doctrine of petrifactions contains also the history of the organic world, as natural history contains its description. Like the coins, inscriptions, and works of art, which make us acquainted with the varied destiny of our own species, these monuments have been buried in the earth, and, by that means, have been secured against destruction. The Siberian and Chinese popular traditions of the mammoth living in the interior of the earth, are at least figuratively correct; and, in conjunction with the remains of a former world, bear evidence of an earlier state of things. The remains of all plants and classes of animals, whose structure permitted it, have been preserved in great abundance; and, although the distinction of species not unfrequently confronts us with unsurmountable obstacles, a knowledge of them must lead to important results; at least, if we admit that the various forms have been evolved from a primitive model, and that the species have arisen from an original generic form. But to perform what may be expected from it, the doctrine of petrifactions should keep pace with the improvement of botanical and zoological methods, and renounce names and distinctions which have no longer any meaning.

Independent also of this connection between the inorganic and the organic world, between geology, botany, and zoology, it is surely no unprofitable occupation for a rational being, to inquire what this earth upon which we live consists of, how it is constructed, what changes it may have suffered, and what it may still be destined to undergo. Whoever is still unsatisfied, whoever estimates the value of science, not by intellectual desires but by practical advantage, ought to recollect that there are few of the arts of life to which geology is not more or less applicable. It is one of the foundations of agriculture, which cannot flourish without a knowledge of the soil: it instructs us in the course and operation of water, whether we wish to prevent it from doing injury, or to turn it to advantage; it enables us to search out materials for our habitations and furniture, and the art of working mines, with which geology originated, and which in return yields its most valuable productions. We hence conceive that the study of geology brings us in continual contact with the most exalted scenes of nature, with all that can captivate our imagination, and fill our souls with vast conceptions, and thus explains the interest that is daily more and more excited by it, and which warrants the most sanguine expectations of its future progress.

Geology has shared the fate of all experimental sciences. Its first steps, for the most part directed by necessity, consisted of loose and superficial observations on those phenomena more immediately presented to our attention. But, as it is a peculiar prerogative of our nature to entertain a desire of tracing back causes, and explaining operations, theories of the earth were early indulged in; and these, although often absurd, were not without their use. Afterwards it was considered presumptuous, from those fragments of the earth's crust which we had looked upon rather than examined, to draw conclusions as to the formation of the earth, and to relate its history, as if we had been coeval with the events; and that true geology must be a collection, arrangement, and comparison of facts, and its theories only general observations. This view being generally admitted, geology may be said to have passed from the condition of childhood, and assumed its station among the sciences.

These theories are essentially different from those of other branches of physical science. When the natural philosopher makes mention of two electric fluids, or of a luminous matter, he is perfectly well aware that the causes of electrical or luminous phenomena might be different from what he imagines; yet these modes of expression are most convenient for producing unity and connection among the facts that have come under his observation. Geological theories are, on the contrary, of a purely historical character. Whether granite be a production of fire or water, is a matter of indifference in the explanation of its origin, if we are incapable of producing it either in the one way or the other; but whoever tells us that the present crust of the earth was once in a state of fusion, and that, upon cooling, it became a solid mass, exhibits an event which, like the heroic exploit of a Curtius or a Clœlia, should be received only upon the most indisputable testimony. Geological theories are, therefore, more exclusive than physical; hence a reason why geologists have always been more at variance than natural philosophers.

It is therefore the duty of the geologist to proceed cautiously with his conclusions. In return for that he is sufficiently indemnified by the nature of his study, which bears in the most distinguished manner the peculiar character of all physical science.

Geology obtains its materials from mineralogical geography, whose general results it selects and combines, in the same manner as state policy does with the results of civil geography. The advancement of the one, therefore, depends on the progress of the other; and although it may be advantageous to science, from time to time, to exhibit a correct view of its progressive advancement, as it is profitable for the traveller to stop sometimes and take a retrospective view of the country he has passed, geology has nevertheless to expect improvement principally from a patient and laborious investigation of single districts. There are but few who, by a glance, can determine general relations and throw light upon science, as there are but few travellers who are qualified to give any instructive information concerning the social condition of a country: On the contrary, any one provided with the necessary knowledge, may, by an accurate and detailed examination of a district, contribute, if not general views, facts that serve as a foundation for the great geological edifice. And, any one who reflects how much time and perseverance are necessary for examining the geognostic character of even a limited district, especially if its interior is not laid open by mines and natural sections, will agree with us, that this investigation, like that of the character and customs of a people, must chiefly be the work of an inhabitant.

13

"OF THE CHANGES WHICH LIFE HAS EXPERIENCED ON THE GLOBE"

[André Étienne d'Audebert de Férussac]

Source: [André Étienne d'Audebert de Férussac], "Of the Changes which Life has Experienced on the Globe" *Edinburgh New Philosophical Journal*, (1827), pp. 298–301

OF THE CHANGES WHICH LIFE HAS EXPERIENCED ON THE GLOBE.

Fossil remains of the animals which preceded man upon the earth are every day discovered on both continents; and every day are the documents regarding the history and successive changes of the various races that existed before the present, increased by new facts. This is equally the case with the vegetation which embellished the earth at that remote period, and with which those primitive animals were necessarily in close connection. New animals and vegetables have assumed the place of those that have been destroyed, and whose ancient existence is only revealed to us by their fossil remains. Thus, in the course of the ages that preceded the appearance of man upon the earth, its surface has successively changed its aspect, its verdure and its inhabitants; the seas have nourished other beings, the air has been peopled with other birds.

The remains of these various successions of animals and vegetables attest that they were at first much more uniform. The vegetables of the coal formation, for example, scarcely present any difference, whatever may be the latitude, the longitude, or the elevation at which they are found. Europe, Asia, and the two Americas, alike produced elephants, rhinoceroses, mastodons, &c. The differences which vegetables and animals exhibit at the present day, according to the various climates or situations in which they occur, have been gradually established under the predominating influence of a small number of natural causes, and constitute at length the order of distribution which life now presents at the surface of the earth.

Originally life extended from one pole to the other, and animated the whole of this surface. The frozen regions of the North, and the snow-clad summits of the Alps, were covered with the same verdure; and the forms of the pristine animals and vegetables presented either extraordinary types of which we have now no example, or species which belonged to families and genera still existing, but in most cases

DOI: 10.4324/9781003490548-15

only between the Tropics. As we approach nearer to the present times, we find in all places remains more and more resembling those of the plants and animals which now live in the same country. At a later period, the original races of animals and vegetables were gradually expelled from the north toward the south, from the summits to the plains, in proportion as the uniform mean temperature of the earth's surface yielded to more powerful causes, which brought about the establishment of climates. These gradual variations in the temperature, the lowering of the general level of the seas, the equally successive and gradual diminution of the energy of volcanic phenomena arising from the original igneous state of the earth, as well as of the strength and power of atmospheric phenomena, and of the tides – such were the regular, general, and continued natural causes of the modifications which life has undergone, and of almost all the changes that have been produced upon the earth's surface. The results of these first causes, such as the establishment of local influences over the temperature of the same climate, the formation of a multitude of particular basins, some containing salt, others fresh water; the pouring out of these lakes into one another, and into the great basin of the sea; the partial debacles which thence resulted; the ravages of the sea on the low parts of the continents at first, and then the formation of vast lagoons in the same places; lastly, the establishment of the general system of draining and watering, or of the hydrographic reticulation which covers the globe – such were the irregular, and more or less violent and perturbing secondary causes of the partial vicissitudes experienced by animal and vegetable life. The beings, which were unable to resist the influence of these various causes were destroyed and disappeared from the earth, with the circumstances for which they were created; new species appeared with new conditions of existence. But, in examining the series of fossil remains that are found buried in the strata of the globe, there is nowhere perceived a distinct line of demarcation between the different terms of that series, so as to prove that life has been once or oftener totally renewed on the earth. On the contrary, we discover in it a proof of the successive and gradual change which we have pointed out. Certain primitive types have indeed completely disappeared, but they are found existing at various epochs, and their remains are blended with those of more modern types; along with new species of types still existing, we find some of anterior epochs; certain genera that yet obtain are common to all the terms of the series; and toward the end of the series, we find the remains of some of our present species along with ancient types and extinct species. In consequence of the establishment of climates, life has almost entirely abandoned the polar countries, and the glaciers have usurped, on the high summits, the place of the verdure of primeval times. Palms, date-trees, cocoas, dracænæ, pandani, arecæ, the great reed, and the arborescent ferns, have forsaken our climates, together with the elephants, tigers, panthers, hippopotami, the gigantic tapirs, the rhinoceroses, palæotheria, anaplothæria, mastodons, and other extinct animals, as well as those enormous reptiles whose forms were so extraordinary. Sole masters, in those times, of the countries now subjected to the dominion of man, these animals are either entirely destroyed, or now live only between the tropics.

Man appears to have arrived upon the earth only after its surface was adapted to receive him, after the establishment of climates, and when a happy equilibrium among the elements had determined the permanency of the present state of things, or at least had rendered its variations almost imperceptible.

Such is a brief view of the changes which life has experienced at the surface of the globe, and of the causes which have produced those changes. Our theory, which is founded on all the facts that have been established, cannot but prevail over the systems hitherto proposed, for it is in harmony with the natural laws of order and permanency which rule the universe, and is, moreover, supported by the most accredited physico-mathematical theories; whereas those systems, founded upon perturbations of cataclysms, which science, facts and human reason equally reject, only increase the number of those imaginary conceptions which have been successively published for several centuries.

The above will suffice to shew, that there is no subject which, in all points of view, is more worthy to excite the interest and meditations of philosophers, and the investigations of geologists and naturalists.

14

"FOR THE CONTINUITY OF THE ANIMAL KINGDOM BY MEANS OF GENERATION FROM THE FIRST AGES OF THE WORLD TO THE PRESENT TIMES"

[Étienne Geoffroy St-Hilaire (trans. Alexandre Bertand)]

Source: [Étienne Geoffroy St-Hilaire (trans. Alexandre Bertand], "For the Continuity of the Animal Kingdom by Means of Generation from the First Ages of the World to the Present Times", *Edinburgh New Philosophical Journal*, (April 1829), pp. 152–5. (Translation of Geoffroy's March 23 1829 paper to the *Academie des Sciences*, first appeared in *The Globe*.)

Of the Continuity of the Animal Kingdom by means of Generation, from the first Ages of the World to the present Times.

On the relations of organic structure and parentage that may exist between the animals of the historic ages and those at present living, and the antediluvian and extinct species.

Are the animals whose remains occur buried in the earth, and which almost all belong to species or genera which are not observed in the living state, to be considered as having been the ancestors of those which now people the earth, and as having been modified by the influence of time, and of the changes that have supervened in the state of the globe? Or is the contrary opinion to be adopted? Are we to believe, that, after the occurrence of great cataclysms, new beings were produced by a new exertion of creative power, – in short, to make use of M. Geoffroy St Hilaire's expression, that *the six days work was resumed.* The solution of this great question is to form the subject of a series of memoirs to be laid before the French Academy of Sciences, of which the first only has been read.

The author of this memoir, M. Geoffroy St Hilaire, commenced with remarking, that science is not yet possessed of all the documents that might appear necessary for treating the question in a satisfactory manner. He even felt that he was not altogether excusable in meddling with it, at a period when the discussion which it involves might be considered premature; but a particular circumstance induced

DOI: 10.4324/9781003490548-16

him to enter upon it. "I had been reading," said he, "some important observations communicated to the Academy by Dr Roulin[a]. My mind being pre-occupied with old ideas respecting the antediluvian animals, there escaped me, in drawing up my report, a reflection which, to be rightly apprehended, would have required a greater developement. This has been remarked, and I have been enjoined to do justice to the subject."

M. Geoffroy St Hilaire believes in an uninterrupted succession of the animal kingdom, effected by means of generation, from the earliest ages of the world up to the present day. The ancient animals, indeed, whose remains have been preserved in the fossil state, are all, or at least almost all, different from those which now exist at the surface of the globe. But this is not a reason for thinking that they could not have been the ancestors of these latter. In the *first* place, the extinct species are united with the living species by the closest analogy. All have without difficulty entered into the prescribed limits of our great classifications; all, as being formed of analogous organs, *seem* to be nothing but modifications of the same being, of what is now called the vertebrate animal.

Viewing the animal creation in its aggregate from the commencement up to the present epoch, the author even thinks he can recognise in it a progressive series like the following: – *Ichthyosaurus, Plesiosaurus, Pterodactylus, Mesosaurus, Teleosaurus, Megalonyx, Megatherium, Anoplotherium, Palæotherium*, &c.; all animals that have been transformed in such a manner, that none of the genera which they form subsists at the present day. Through the medium of the mastodons, the author connects with these more ancient inhabitants of the globe the animals that succeeded them, and which are composed of species of the same genera, some extinct and antediluvian, and others still living. These latter are those which have been able to accommodate themselves without transformation, or at least by only undergoing very slight transformations, to the circumstances of the present world. These animals, some of which occur in the fossil, and others in the living state, are the elephants, rhinoceroses, some didelphides, hyenas, bears, &c.

M. Geoffroy St Hilaire cited, as the performance of an author who outstripped the age in which he lived, the work in which M. de Lamarck treats *Of the influence of circumstances upon the actions and habitudes of living bodies*, and reciprocally *on the influence of the actions and habitudes of living bodies upon the modification of their parts*.

The particular facts on which M. de Lamarck rests his grand idea, are far from being perfectly correct. Perhaps there is not even one of them that is not blemished by some inaccuracy; and yet the conclusion which he draws from them is true – such is the power of genius in foreseeing the great truths of nature. It was thus that Buffon perceived, by an inspiration of his genius, that the animals of the equatorial regions inhabit one of the continents to the exclusion of the other, although none of the proofs adduced by that celebrated man in support of his opinion would perhaps be admitted at the present day; and yet this proposition has become a law which has received entire confirmation from time.

On the subject of M. de Lamarck's opinion, the author quoted a remarkable passage from Pascal: "Animated beings," says that author, "were in their commencement nothing but formless and ambiguous individuals, whose constitution was originally decided by the permanent circumstances in the midst of which they lived."

To establish M. Geoffroy's opinion in a solid manner, the important point is to demonstrate that the differences of atmospheric constitutions may have been sufficiently great and powerful to bring the different species and genera, from the types which they originally presented, to what we now see them to be. Now, of this the author thinks no doubt can be entertained. Let attention be paid to the modifications which the species may still undergo, in consequence of a mere transportation from one latitude to another, – modifications which have been determined by Dr Roulin with respect to the animals transported from Europe to America. Let the important facts, in particular, which the study of monstrosity presents, be attended to, and there will appear nothing surprising in the modifications produced in the animal species, by the succession of ages, any more than in modifications induced in the agents under the influence of which animals are developed.

To determine the power of external causes in modifying the developement of living beings, was the real object of the experiments made by the author in the establishment of Anteuil, where chickens are reared that have been hatched under the influence of artificial heat.

The philosophical object of these inquiries the author now avows for the first time; he was obliged to conceal them at a period when science was under persecution. The experiments here alluded to are conclusive. M. Geoffroy St Hilaire, by varying the phenomena of heat, dryness, and motion, not only produced monstrosities at pleasure, but even produced a given species of monstrosity by means of a particular precaution. And let it not be objected that the monstrous species thus produced in an artificial manner, were incapable of being reproduced and perpetuated. Nature, aided by time, which he had not at his disposal, acting by more numerous and gentler modifications, could have done what will always be impossible in the most judiciously conducted experiments.

M. Geoffroy also spoke of the long debated question of the pre-existence of germs, and opposed to that theory the whole of our knowledge respecting monstrosities, and in particular the experiments above alluded to, in which he made the organization deviate at pleasure, and in a determinate direction from its natural course.

Note

a Vide account of Dr Roulin's observations in vol. vi. pages 190–193 of this Journal.

Editorial Headnote

15. Robert Knox, "Inquiry into the Origin and Characteristic Difference of the Native Races inhabiting the Extra-Tropical part of Southern Africa," *Memoirs of the Wernerian Natural History Society*, v.5, (Edinburgh: 1824), pp. 206–19.

Robert Knox (1791–1862) was a prominent Edinburgh anatomist. For a time he was the most sought-after teacher of anatomy in the city, and he often enrolled hundreds of students in his private classes. Having studied extensively in Europe, including with both Georges Cuvier and Geoffroy Saint-Hilaire, he brought the insights of transcendental anatomy back into his Edinburgh classroom and his published research. He later spelt out the main thrust of his ideas on this topic in his "Contributions to the Philosophy of Zoology," which appeared in *The Lancet* in 1855.[1] Following Geoffroy, he saw the origins of new species to lie in embryological developments and the adaptation of organisms to a changed climate, alterations in diet, and crossing of varieties. Knox was at one time reckoned to be the most famous transmutationist in Britain, although he later became more widely known for his involvement in the infamous Burke and Hare murders. Like many anatomists, he needed cadavers; Burke and Hare were happy to take his money, and Knox clearly did not care to ask too many questions. Knox, like many French Enlightenment thinkers, was also interested in the question of the origin and the mutability of the human races. He had presented papers on this topic from the 1820s in which, echoing the ideas of the German naturalist, anatomist, physiologist, and anthropologist Johann Friedrich Blumenbach, he argued that all the human races were of a common origin, descended from the Caucasian.[2] In this paper, he states that "the origin of the race, by which I mean the mode of descent, and separation from one or other of the more extended varieties of the human race, is one of the most interesting inquiries which the natural history of man presents."[3] He later wrote *The Races of Men* (1850, second revised edition 1862), in which he wrote extensively on the history of racial development. However, where Blumenbach had been clear that human races were not to be thought of in hierarchical terms, in this latter work, Knox did so and described what he took to be a biological racial hierarchy in which white Europeans were far superior to the non-white races.[4]

Notes

1 Robert Knox, "Contributions to the Philosophy of Zoology, with special reference to the Natural History of Man," *The Lancet*, July 14th 1855, pp. 24–6. This essay is reprinted in full Caden C. Testa and Piers J. Hale, *Evolution Before Darwin* (London: Routledge, 2025).

2 Robert Knox, "Inquiry into the Origin and Characteristic Difference of the Native Races inhabiting the Extra-Tropical part of Southern Africa," *Memoirs of the Wernerian*

Natural History Society, v.5, (Edinburgh: 1824), pp. 206–219. See headnotes and texts by Blumenbach in Caden C. Testa and Piers J. Hale, *Evolution Before Darwin* (Routledge: London, 2025); and Piers J. Hale and Caden C. Testa, *Evolution and Race* (Routledge: London, 2025); also Robert J. Richards, "Kant and Blumenbach on the Bildungstrieb: A historical misunderstanding," *Studies in the History and Philosophy of the Biological and Biomedical Sciences,* (2000), Vol.31, No.1, 11–32, on p. 20.

3 Knox, "Inquiry into the Origin and Characteristic Difference of the Native Races," p. 216.

4 On Blumenbach's views see Thomas Junker, "Blumenbach's theory of human races and the natural unity of mankind," in Nicolaus Rupke and Gerhard Lauer, (eds.), *Johann Friedrich Blumenbach: Race and Natural History, 1750–1850*, (London: Routledge 2019), pp. 96–112, on p. 104. On Knox see Evelleen Richards, "The moral anatomy of Robert Knox: the interplay between biological and social thought in Victorian Scientific Naturalism," *Journal of the History of Biology* (1989), 22, pp. 373–436. See also extracted in this collection: Robert Knox, *The Races of Men. A Fragment*, (Philadelphia, Lea & Blanchard, 1850).

15

INQUIRY INTO THE ORIGIN AND CHARACTERISTIC DIFFERENCE OF THE NATIVE RACES INHABITING THE EXTRA-TROPICAL PART OF SOUTHERN AFRICA

Robert Knox

Source: Robert Knox, "Inquiry into the Origin and Characteristic Difference of the Native Races inhabiting the Extra-Tropical part of Southern Africa", *Memoirs of the Wernerian Natural History Society*, v.5, (Edinburgh: 1824), pp. 206–19

INQUIRY INTO THE ORIGIN AND CHARACTERISTIC DIFFERENCES OF THE NATIVE RACES INHABITING THE EXTRA-TROPICAL PART OF SOUTHERN AFRICA.

By Robert Knox, M. D.
Member of the Wernerian Natural History Society, and of the Medico-Chirurgical Society of Edinburgh.

(*Read* 19*th April* 1823.)

The best apology I can offer for attempting the Inquiry I have now the honour to lay before the Society, is, that no similar one, relative to the country I speak of, has ever been instituted. Many travellers have described, with a minuteness proportioned to their talent for observation, the Peninsula of Southern Africa, and have given to the world valuable works on its natural history, political relations, &c.; but, so far as I know, none has ever viewed the savage races inhabiting this Peninsula in an anatomical point of view, and hence have arisen ill-founded conjectures, and positive errors, too numerous to be criticised. I have endeavoured to correct those more immediately connected with our inquiry, but have carefully avoided general criticism, as leading me from the subject. The Society will readily imagine the difficulties I experienced in avoiding frequent conjecture, and the impossibility of doing so altogether; but recourse has been had to conjecture rarely, and, though

DOI: 10.4324/9781003490548-17

some of the conclusions may, from insufficient data, or other causes, prove erroneous, I shall still feel sufficiently recompensed for my labour, should the facts I have collected be found serviceable to those engaged in writing the Natural History of Man.

The part of Africa situated to the south of the Tropic, contains at least three distinct races of men. Those met with in proceeding from Cape Point northerly, are intruders; they constitute the Anglo-Dutch colony of the Cape, and are a mixture of almost all the modern nations of Europe, the Dutch preponderating. The colonists dwelling in the remoter districts attain a gigantic size, owing, no doubt, to their descent from a race naturally tall; and in perfecting which, as to stature, much has been effected by climate, food, and other localities. This race, at present, extends from Cape Point northward to the banks of the Gariep, or Orange River, and eastward as far as the Keiskamma River. They have pushed before them, and partly exterminated, the race of Hottentots, or Bosjeman (for I shall consider them as the same), who are now found only in small numbers, either as servants to the colonists, or still preserving a sort of savage independence in that vast tract of almost desert country, extending from the chain of mountains out of which the Gariep and Great Kei rivers rise, westward to the shores of the Southern Atlantic. On the verge of the Tropic, and close to the western shore, the Damaras are found (a Negro race, as I have been assured), who extend towards Benguela and Congo; so that the Bosjeman race, if they exist much to the north of the Gariep River, must occupy a central stripe of Africa, bounded, on one hand, by the Damara and Benguela countries, and, on the other, by the Kaffer nations.

These latter extend from the Keiskamma eastward along the coast towards Inhambane; but previous to reaching this point, they become inland, and possess the mountain-country, which, there is every reason to suppose, may extend from the sources of the Gariep and Great Kei rivers, to the Equator. It would appear from the Journal of VAN REENEN, that the Temboo is the last of the Kaffre tribes found on the coast of Natal, and that beyond them, in about Lat. 26° S., are found the Hamboonas, a race totally different from the Kaffres. They are thus described: "Ce peuple a la peau jaunatre avec de longs cheveux fort epais et frisés, qui sont relevés sur le sommet de la tête en forme de turban[a]." At the point of coast where the Hamboonas disappear, the Negro race commences, and, extending around Sofala, Inhambane, and Mozambique, furnish the Portuguese with the most ample means of carrying on a traffic the most revolting to the feelings of mankind. The mountain-country, inhabited by the Kaffre race, cannot be of any great breadth, limited as it is on the east by the Negro country of Mozambique, and to the west by the vast and unknown deserts supposed to be inhabited by the wandering Macasses. It is unfortunate for our present inquiry, that two geographical problems of much interest remain to be solved. The Kaffre races have been conjectured to be Arabian, and the Bosjeman tribes have been considered by some of Chinese origin, by others of Egyptian; they have even been compared to the Troglodytes or Pygmies of HERODOTUS, that dwelt in the deserts to the south of Barca and Syrene. We shall presently see that the Kaffres are not Bedouin Arabs, and, moreover, that they cannot be traced to any European or Caucasian stock; but the question of origin, as regarding the Bosjeman, is of far more difficult solution. It were highly desirable, as well for geographical science as for the natural history of Man, that the extent of country occupied by the Kaffre race, from the Keiskamma northwards and eastwards, – in other words, the boundaries of Kaffraria Proper, – were correctly ascertained. We know that the race extends on either side the mountain-country stretching from the sources of the Gariep towards the Equator, inhabiting the valleys and slopes of the mountains: their progress westward was probably arrested by the existence of great central deserts, and by their natural dislike to such countries; whilst eastward, that is, towards the Indian Ocean, numerous Negro tribes were in possession of the country. That they have long been an inland race, is evident from their total ignorance of the use of boats or canoes: moreover, the Portuguese, in their ardent search after gold, received some severe checks from the black races inhabiting the mountains to the westward of their settlements: now, it is probable that these were Kaffre tribes, for Negroes are naturally timid, and easily subdued.

It is much more difficult to guess with any shew of probability at the extent of the Bosjeman nations, partly from their total dissimilarity to all surrounding tribes, and partly from the extremely defective state of our knowledge relative to the geography of Central Africa. It is well known that they originally extended to Cape Point, and have been found to the northward as far as Europeans have yet penetrated. Beyond this all is profound obscurity: How, then, are the aborigines of Southern Africa to be traced to the primitive races of the ancient world? In the absence of all historical detail relative to the affiliation of these races, and until some future Park shall make us acquainted with Central Africa, as well to the south as to the north of the Line, both being equally unknown, we shall bring together the results presented by anatomical inquiry, which, being a method founded on fixed and general physical laws, will, if it attain not the truth, at least approximate to it.

We may view the human race as derived originally from one stock, to which the arbitrary name of Caucasian has been given. This species, influenced by climate and civilization, assumed, at a very early period, five distinct forms, which have also been arbitrarily designated by the names of Caucasian, Mongolic, Ethiopian, American, and Malay[b]. We cannot hesitate referring the Kaffre nations to the Ethiopian, not so much from their position, as from their close resemblance; whilst the Bosjeman may be arranged, for the present, with the Mongolic, until further investigation shall have shewn a more intimate relation with some unknown African tribes, or until the race shall have been fairly traced through Central Africa to the Valley of the Nile, and thence to Asia, where we suppose all nations to have originated.

The cranium of the Kaffre is inferior in most of its measurements to the European. The temporal bones are flat and compressed, and the squamous suture often straight, instead of being semicircular, as in the European[c]. They are nice in the choice of their food, and reject fish, fowls, and generally whatever by the Levitical law is deemed unclean; yet they eat the inner parts of animals, as the intestines, stomach, lungs, &c. raw, torn from the animal just dead.

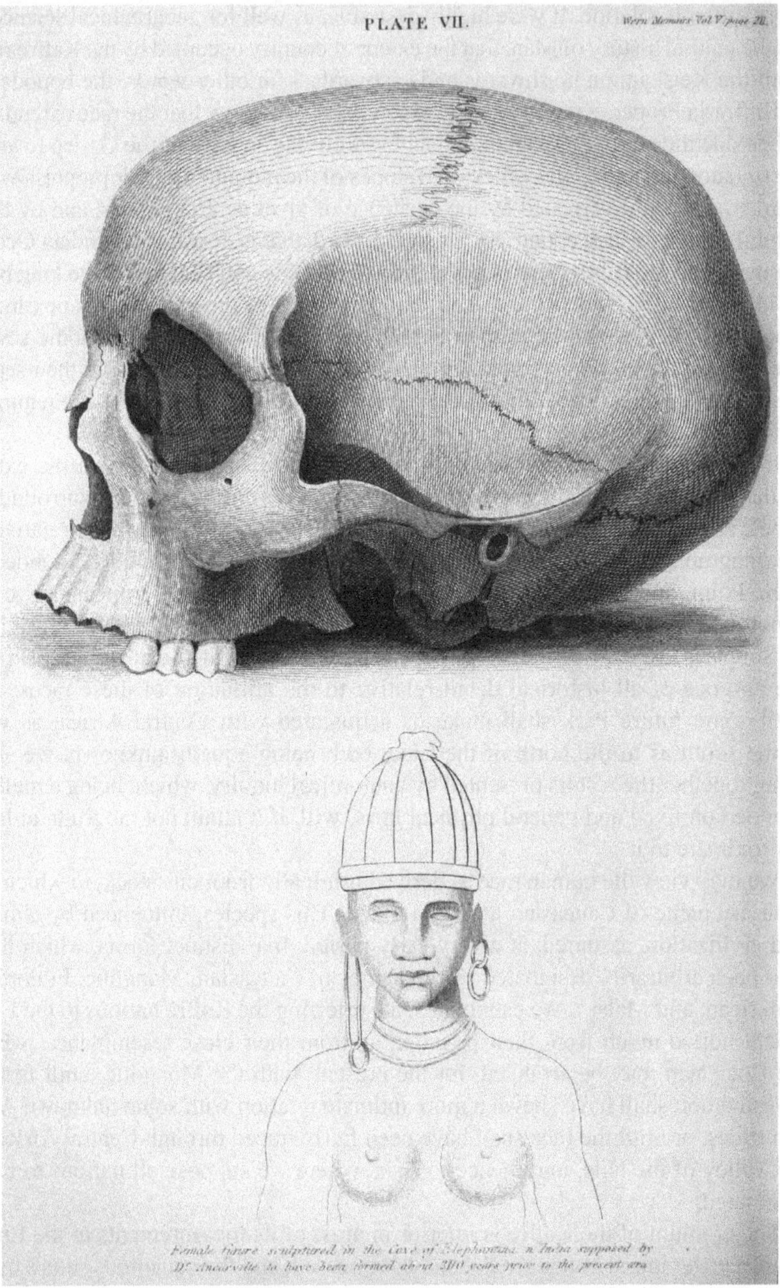

Female figure sculptured in the Cave of Elephantina in India supposed by Dr. Ancarville to have been formed about 2110 years prior to the present æra

	Var. Ethiop.
NEGRO.	KAFFRE, including Temboo, Briquas, Boshuanas, Red Kaffres, &c. &c.
Colour generally very dark.	Of a brown colour; some perfectly black.
Hair black, and, as it is called, woolly.	Hair uniformly black, crisp, and woolly, and scattered in small tufts over the scalp.
Head narrow, compressed on the sides.	Cranium narrow, elongated, resembling in outline the female European.
Forehead arched.	Forehead differing but little from the Negro, that is, rather retreating; very narrow, and not high.
Malar bones prominent; eyes full.	As in the Negro.
Nose thick, and little distinct from the lips.	In the greater number nearly as in the Negro; some have less of the Ethiopian physiognomy.
Upper jaws elongated forwards.	The osseous development of the upper jaw nearly as great as in the Negro.
The superior incisors obliquely projecting.	Hardly the case in the Kaffre.
Excessive fulness of the lips.	Nearly as in the Negro.
Retreating chin.	Not so much so as in the Negro.
Many bandy-legged. A certain difficulty in maintaining the upright position; the knees slightly bent, and the heels constantly tending to quit the ground. The gastrocnemii muscles placed too near the ham.	Never the case in the Kaffre. His lower extremities are finely proportioned, and frequently of Herculean strength; the upper extremities weak and disproportioned; whilst the lower limbs, pelvis, and loins, are superior even to the European.

From a careful consideration of the habits and configuration of this race, I believe them to be closely allied to the Negro, and that the observable differences are attributable to the different climates they inhabit. The Kaffres, in a word, are the Negroes of the mountains; they are Negroes changed by inhabiting an extra-tropical climate; like all mountaineers, they are hardy, bold, and fond of liberty. In intellect they are superior to the Negro, and I believe them capable of a very considerable degree of civilization. The remarkable disproportion observable between the upper and lower extremities of the Kaffre[d] may, perhaps, be owing to the very unequal degree of exertion to which these separate parts of the body are exposed. The Kaffre never labours, and hence his arms are weak; but he is strongly exercised in the chase and in predatory excursions, by which means his limbs become muscular, and occasionally altogether Herculean. The excess of nutrition gives rise, no doubt, to the enormously swelled legs to which many of them are liable, when, through indolence, or the infirmities of age, they are disinclined to continue the same life of activity. Circumcision and polygamy they practise with almost all African nations.

According to the relations of modern and ancient travellers, tribes similar to the Kaffre race are found scattered over Africa, apparently unconnected with each other. We are told of a nation of Negroes, called Nubee, inhabiting the country to the west of the Nile, near the confluence of the Abyssinian and true Nile: they are described as being mild in character, and of small features, though the nose be flat, and the hair woolly: they speak a soft sonorous language, differing in these points from their neighbours. Travellers describe the Ababdes, living to the east of the Nile, as being Blacks, with European features. But as few travellers have been anatomists, their accounts cannot altogether be depended on.

A very general belief prevails, that, by external means, and more particularly by pressure, the human cranium and form generally may be modified, and permanently altered, and that this may at last become hereditary. It is asserted, for example, that the more remarkable craniological differences amongst nations are occasioned by external pressure; that the flatness of the African nose arises from the same cause; that Negroes are bandy-legged, because they are carried during infancy on their nurses' back; and that the large feet of the Kaffre, and the small ones of the Bosjeman, are owing to an abundant supply of food to the former, and a deficient one to the latter[e]. All such assertions are constantly refuted by an appeal to facts. The feet and hands of the Kaffre and Bosjeman are strictly and

VAR. MONGOL.	
True MONGOL inhabiting the deserts of Central Asia.	AFRICAN BOSJEMAN, including the numerous Hottentot tribes, as Namaquas, &c. &c.
Colour yellow, or of an olive tinge.	Colour light yellow, or olive; it is difficult to describe, and varies in intensity amongst the Hottentots; but is pretty uniform amongst the true Bosjeman.
Hair black, coarse, straight, and scarce.	Hair black, usually short, though sometimes growing to a considerable length, and is scattered over the scalp in detached tufts, as In the Kaffre.
Head of a square form.	General outline of the head much resembles the Mongol, that is, broad and square.
Face broad, flat, depressed, the features being therefore, as it were, confluent.	Face having a considerable resemblance to the true Mongol; the lips are full.
Forehead smooth and flat.	As in the Mongol.
Nose small and flat.	
Aperture of the eye-lids narrow and linear.	As in the Mongol. The angle at the internal canthus of the eye is quite rounded away.
Chin somewhat projecting.	Chin very small and pointed, but not projecting.

beautifully proportional: though they be carried during infancy in the same manner as the Negro, they are never bandy-legged or deformed. The African nose is flat, independent of any mechanical skill exercised by the nurse; and human crania are fashioned all over the world by the hands of Nature, and not by man.

In addition to these characters of the races, I may remark, that in the Bosjeman there is the finest symmetry of the whole frame: their stature is remarkably diminutive; I should conjecture the average to be about 4 feet 6 inches in the male population. The females have remarkably prominent nates and elongated nymphæ; both these peculiarities are removed by a single cross with the Kaffre or European. The Bosjeman have uncommon powers of vision; this also is lost by intermarriage with another race. The cranium is finely formed, and thin; the nasal processes of the superior maxillary bones large and broad: these give to the Mongol and Bosjeman the appearance of breadth at the root of the nose. The foramina for the transmission of the hypoglossal nerves, are very large[f]. The cranium, viewed vertically, is nearly equal to the well-formed European head. The parietal bones bulge out very much, and the cranium is at this point broadest. It agrees with the Mongol in having the occipital foramen larger than in the other races. The floor of the orbit does not encroach so much on the cavity as in the Mongol; this affects the physiognomy to a considerable extent, in so far as regards the direction of the eyes.

The habits of the Bosjeman race have been described by most African travellers with sufficient accuracy, and I shall therefore limit my remarks to a few points of resemblance between the true Mongol and Bosjeman races. This consists, first, in the countries they inhabit, which, in either case, are vast elevated sandy deserts, nearly destitute of herbage and water: secondly, in the partiality both races have for horse-flesh as an article of diet, preferring it to every other sort of food; they strictly merit the appellation of *Hippophagi:* lastly, in the acuteness of their vision, which almost exceeds belief. I have found their sight to be equal to that of most Europeans when aided by excellent hand-telescopes of the best construction. The Bosjeman is ingenious, clever, and neat-handed; his powers of mimickry are great, his understanding good. He readily acquires languages, and his speed of foot is almost proverbial.

The origin of the race, by which I mean the mode of their descent, and separation from one or other of the more extended varieties of the human race, is one of the most interesting inquiries which the natural history of Man presents. To connect the Bosjeman with the Mongolin variety, we must step at once from the Peninsula of Southern Africa, to the great central deserts of Asia; the intermediate links are lost, – the intervening races unknown. History, though not altogether to be depended on in the consideration of events so remote, must not in the present instance be despised. Though surcharged and disfigured with fable, there is one fact to which such constant allusion is made, as almost to put it beyond a doubt, – I allude to the frequent descents of the northern Asiatic races on the southern states of Europe and Asia. The valuable monuments of antiquity, still preserved in the Cave of Elephantina, in Peninsular India, attest the predominating presence of the Mongol race, at a period removed from the birth of our Saviour by more than

2000 years[g], and that, at that time, the Mongol physiognomy bore the strongest resemblance to the present Chinese and Bosjeman races. The early introduction of the Mongol or Northern Asiatic races into Peninsular India is further attested by their influence on the modern Hindoo; for though the excellent BLUMENBACH assures us that the Hindoo cranium is quite equal in beauty and proportions to that of the Turk, and consequently refers the race to the Caucasian variety, yet, in the Hindoo heads I have examined, the development of the upper jaw has not been strictly Caucasian[h].

The vast antiquity of the Mongol hordes of Asia is further proved by the early establishment of the Chinese empire; and though I am fully persuaded of the still greater antiquity of those of Hindostan and of Egypt, yet many passages in HERODOTUS point out that the Mongolic tribes, with a rapidity even exceeding the Caucasian, rapidly assumed the form of great and warlike nations.

It would seem, then, from a hasty examination of historic records, of the remains of antiquity, and of the laws and religious ceremonies transmitted from generation to generation, that at a very early period the Mongol races penetrated into Europe[i] and southern Asia; and there is nothing improbable in the supposition, that they may have modified by their presence some of the races of central Africa.

Lest it might seem an omission that I have not described the Mongol race as having penetrated into the wilds of America, I shall here state, that the few Esquimaux crania examined by me, have proved to be strictly American, nor have I ever seen the slightest approach to the Mongol head in any native American race. I have only further to observe, that most of the opinions offered in this memoir are founded not on theory or conjecture, but on an anatomical examination of the crania of numerous races of men. For the opportunity of doing so, I am in a great measure indebted to Professor JAMESON, and to Doctors MONRO and BARCLAY, through whose politeness I was allowed to examine whatever specimens their valuable museums contained. I beg leave to offer my thanks to these gentlemen.

The annexed table gives a comparative view of the measurements of the head in several remarkable varieties of the human race.

EDINBURGH,
March 1. 1822.

	Hottentot.	*Cranium supposed to be from New Holland.*	*New Hollander.*	*Cranium supposed to be Malaic.*	*European Female.*	*Hottentot Venus, (from a cast).*	*Male Kaffre.*	*Male European.*	*Negress.*	*Turk.*	*Carib.*
Length of the face from the root of the nose to the edge of the alveolar process,	Inches. 2.3	Inches. 3	Inches. 2.5	Inches. 3	Inches. 2.8	Inches. 2.7	Inches. 2.8	Inches. 2.9	Inches. 2.6	Inches.	Inches.
From the anterior portion of the foramen occipitale to the incisive teeth of the upper jaw,	3.7	3.7	4	3.7	3.7		4.1	3.3	4	3.5	3.5
Breadth at the temples*,	3.6	3.8	3.8	4.1		3.5				3.5	3.4
Whole circumference,	19.7			20.1	19.7	19.8	21	20.7	19.6	19.6	18.3
Greatest breadth†,	5.2	5.4	5.7	5.8		5.3			5.1	5.1	5.4
Foramen occipitale‡,	1.3 by 1.5	1.5 by 1.1		1.3 by 1.5					1.3 by 1.1		

(*Continued*)

(Continued)

	Hottentot.	*Cranium supposed to be from New Holland.*	*New Hollander.*	*Cranium supposed to be Malaic.*	*European Female.*	*Hottentot Venus, (from a cast).*	*Male Kaffre.*	*Male European.*	*Negress.*	*Turk.*	*Carib.*
Length over the vertex, that is, from the ossa nasi to the posterior edge of the occipital foramen,	14.1	21.1	21	14.5	14.3	13.5	14.6	14.9	14.9		
Measurement from mastoid to mastoid over the vertex,	13	21	21	15.6	14	12.7	14.2	14.7	14.5		

* Not to be depended on.†

Varies very much; broadest, however, generally over or near the temporal bones; sometimes, as in most of the black races, in the middle of the parietal bones.

‡ Generally speaking, the foramen occipitale is largest in the Mongul, and smallest in the Negro. I have seen it very large in the Esquimaux, but am not sure that this is constant.

Notes

a I consider the Hamboonas (if they really exist) as descended from a race of shipwrecked Chinese or Malay mariners, modified by intermarriage with Negro or Kaffre tribes. It is, however, quite possible that they may be a race from Madagascar.

b I shall here take the liberty of remarking, that the Malay race seems to me an artificial variety, and has no existence in nature as a distinct race of men. I am inclined to refer the whole inhabitants of the innumerable islands scattered over the Great South Sea (including New Holland) to the American variety. In Southern India they have mixed with the Mongolic and Caucasian races, and form consequently a mixed or mulatto breed; but they may be traced tolerably pure as high as Sumatra, in which island several tribes are found much resembling the native Americans, and retaining all their customs, even to the pretended flattening of the head by artificial means. It is probable that in this race only are found the anthropophagous tribes.

c Some Kaffre skulls have been shewn me, in which the malar bones were remarkably large, and the face very broad; but this is not the usual Kaffre physiognomy. Such anomalies exist amongst all nations and races, but cannot be supposed sufficient to refute the general position, that each race has a form of cranium and face peculiar to itself.

d Of twenty-two ivory armlets, taken indiscriminately from the arms of Kaffres, four only were found which could admit the arm of a moderately muscular European.

e Blumenbach de Nat. Variet.

f These foramina are decidedly larger in the Black than in the White races, but more particularly in the Negro. They clearly indicate the passage of a proportionally large nerve.

g Compare the annexed drawing (Plate VII.) of the female figure sculptured in the Cave of Elephantina, and copied from the "Recherches sur l'Origine des Arts," with the beautiful and expressive portraits of the Bosjeman and Hottentot races by Mr Daniells in his "African Scenery."

h The inspection of a fine collection of skulls, collected on the Banks of the Ganges, and which, through the kindness of Professor Jameson, I was enabled to examine immediately on their arrival, has confirmed me in this opinion. In these skulls, which, with the exception of one Negro head, seem all to be of the Caucasian race, the cranium is quite equal to any European, and its longitudinal diameter shorter than in most: but there is a development and strength in the upper maxillary bone which, I should think, will not be found to exist in any Turkish, Syriac, or Jewish head.

i The peculiar Mongol face is very strongly marked in many families now inhabiting the Highlands of Scotland, and more particularly the Hebrides.

Editorial Headnote

16. Robert Knox, "Contributions to the Philosophy of Zoology with Special Reference to the Natural History of Man," *The Lancet*, July 14th 1855, pp. 24–6.

Robert Knox (1791–1862) was a prominent Edinburgh anatomist. For a time he was the most sought-after teacher of anatomy in the city, and he often enrolled hundreds of students in his private classes. Knox had studied extensively in Europe, including with both Georges Cuvier and Geoffroy Saint-Hilaire, and he brought the insights of transcendental anatomy back into his Edinburgh classroom and his published research. He later spelt out the main thrust of his ideas on this topic in his "Contributions to the Philosophy of Zoology," which appeared in *The Lancet* in 1855 and is reprinted in full here.[1] Following Geoffroy, he saw the origins of new species to lie in embryological developments and the adaptation of organisms to a changed climate, alterations in diet, and crossing of varieties. Knox, like many Enlightenment thinkers, was deeply interested in the question of the origin and mutability of the human races. He had presented papers on this topic from the 1820s in which, echoing the ideas of the German naturalist, anatomist, physiologist, and anthropologist Johann Friedrich Blumenbach, he argued that all the human races were of a common origin, descended from the Caucasian.[2] Later he wrote The *Races of Men* (1850, second revised edition 1862), in which he wrote extensively on the topic. However, where Blumenbach had been clear that human races were not to be thought of in hierarchical terms, in this later work Knox did so, and described what he took to be a biological racial hierarchy in which white Europeans were far superior to the non-white races.[3]

Knox was at one time reckoned to be the most famous transmutationist in Britain, although he later became more widely known for his involvement in the infamous Burke and Hare murders. Like many anatomists, he needed cadavers; Burke and Hare were happy to take his money, and Knox clearly did not care to ask too many questions.

Notes

1 Robert Knox, "Contributions to the Philosophy of Zoology, with special reference to the Natural History of Man," *The Lancet*, July 14th 1855, pp. 24–6. This essay is reprinted in full in this collection.

2 Robert Knox, "Inquiry into the Origin and Characteristic Difference of the Native Races inhabiting the Extra-Tropical part of Southern Africa," *Memoirs of the Wernerian Natural History Society,* v.5, (Edinburgh: 1824), pp. 206–219. See headnotes and texts by Blumenbach in Caden C. Testa and Piers J. Hale, *Evolution Before Darwin* (Routledge: London, 2024); and Piers J. Hale and Caden C. Testa, *Evolution and Race* (Routledge: London, 2024); also Robert J. Richards, "Kant and Blumenbach on the Bildungstrieb:

A historical misunderstanding," *Studies in the History and Philosophy of the Biological and Biomedical Sciences,* (2000), Vol. 31, No.1, 11–32, on p. 20.

3 On Blumenbach's views see Thomas Junker, "Blumenbach's theory of human races and the natural unity of mankind," in Nicolaus Rupke and Gerhard Lauer, (eds.), *Johann Friedrich Blumenbach: Race and Natural History, 1750–1850*, (London: Routledge 2019), pp. 96–112, on p. 104. On Knox see Evelleen Richards, "The moral anatomy of Robert Knox: the interplay between biological and social thought in Victorian Scientific Naturalism," *Journal of the History of Biology* (1989), 22, pp. 373–436. See also extracted in this collection: Robert Knox, *The Races of Men. A Fragment*, (Philadelphia, Lea & Blanchard, 1850).

16

"CONTRIBUTIONS TO THE PHILOSOPHY OF ZOOLOGY WITH SPECIAL REFERENCE TO THE NATURAL HISTORY OF MAN"

Robert Knox

Source: Robert Knox, "Contributions to the Philosophy of Zoology with Special Reference to the Natural History of Man", *The Lancet*, July 14th 1855, pp. 24–6

CONTRIBUTIONS TO THE PHILOSOPHY OF ZOOLOGY,

WITH SPECIAL REFERENCE TO

THE NATURAL HISTORY OF MAN.

BY ROBERT KNOX, M.D., F.R.S. EDIN.,

LECTURER ON ANATOMY, AND CORRESPONDING MEMBER OF THE ACADEMY OF MEDICINE OF FRANCE.

(*Continued from vol. i. p.* 627.)

> "A magnificent temple is a laudable monument of national taste and religion; and the enthusiast who entered the dome of St. Sophia might be tempted to suppose that it was the residence, or even the workmanship, of the Deity. Yet how dull is the artifice, how insignificant is the labour, if it be compared with the formation of the vilest insect that crawls upon the surface of the temple!"
>
> – GIBBON. *Decline and Fall.*

PART I.

Introduction. – The philosophy of zoology can be based only on anatomy. Men, it is true, of vast genius had attempted, and with some success, to formule the laws regulating or presiding over the formation of animal bodies – the laws, in fact, regulating animal forms – by a bold synthesis, independent of all analysis. Such were the attempts of Aristotle, and of Newton. Buffon, availing himself, no doubt,

DOI: 10.4324/9781003490548-18

to a certain extent of the labours of Daubenton and others, but more especially influenced by the comprehensive views which a deep and constant meditation of Nature taught him to adopt, made an effort in in the same direction, sufficiently powerful to attract the attention of the Sorbonne and the Vatican. But so long as the philosophy these great men taught or proposed was without a basis in anatomical research, so long was it entitled only to the name of "hypothesis," unsupported by proof, by an appeal to intuitive evidence, or by any analysis of the material composing the essence of animal forms. Their views, then, were, by all inquirers into truth, regarded – and this, in the nature of things, could not well be otherwise – merely in the light of ingenious hypotheses, grouping together phenomena of which the greater part were avowedly mysterious and inexplicable.

So long as the laws of zoology were confined to, and embraced within, the doctrine of final causes, all philosophic minds were of course sensible that the philosophy of natural history, including, of course, zoology, could have no existence; for these doctrines placed it of necessity within that ultimate appeal, the First Cause, where all science ceases. The principle, or rather the phenomena, of life, so mysterious in their nature, so difficult to investigate, led no doubt to this view – to this ultimate appeal. It was in vain that gifted men, as Fontana and others, mathematicians and geometricians, pointed out that final causes were not causes, but effects; the doctrine maintained its ground, and does so still. The successive philosophies of the day, as they were called – the theological. namely, and the metaphysical – gave the doctrine sufficient support to overrule the natural philosopher, and to remove the phenomena of life, at least, from his field of inquiry. The reason was this: the field of inquiry included MAN, whose obvious connexion with the zoological world could not be mistaken, could not be denied. But man's history had been made the subject of elaborate works by historical and theologico-historical writers; the animal and vegetable worlds had been described as subservient to him; whatever trenched on histories which, taken literally, are clearly and simply fabulous, gave alarm to powerful classes of men, deeply and intensely interested in obstructing all inquiry into truth, and Goëthe, Oken, and Geoffroy St. Hilaire would have shared the fate of Galileo, but for the accidentally altered circumstances of the European world. The French revolution had occurred – that mightiest of all human events. The artificial bonds of society were loosened. Men for the first time ventured to think. The mass of mankind rose incredibly in intellectual calibre for a short space; it attempted to soar with science into the loftier, purer, and calmer atmosphere of the thinking world; its natural grossness brought it rapidly to the earth, where it now lies, a prey to every imposture, the struggle serving only to rivet each chain more securely than before. In the struggle, science escaped, not unobserved by the natural enemies of truth, repudiating alike the deceiver and the deceived. But great as was the genius of Goëthe, Oken, and Geoffroy, it would have proved unavailing to stem, even for an instant, the vast turbid current of folly and fable which passed even to later times for a history of this globe and its living inhabitants, but for the accidental appearance of one man, who unconsciously submitted to thinking men, not a hypothesis, not a theory, but a demonstration, irrefutable as those of Euclid – a demonstration of the fact, that all previous histories of the globe and its living inhabitants were simply fables. That man was George Cuvier, an anatomist. Palissy, it is true, had attempted this demonstration long before, but Palissy was a potter, not an anatomist. Buffon threw out a bold conjecture, and was instantly silenced by the Church of Rome. Cuvier's demonstration could not be so disposed of; it admitted of no refutation. The most subtle Jesuit, ever ready with a refutation even of Newton, quailed before the organic remains, attesting the existence of a former world, of a vast, if not of an infinite, antiquity.

It is worthy of remark, that Goëthe, Oken, Geoffroy, and Cuvier, however much they might differ on some points, agreed in this: that the basis of Philosophic Zoology was to be sought for in Anatomy. Goëthe saw this at once, even when a mere student; so did Oken, Cuvier, and Geoffroy. Comparative anatomy was the starting-point of all, but from this their course was different, divergent; Goëthe, by the transcendant force of his genius, was straightforward, cautious, and philosophical; Oken, theoretical, transcendental, and mythical; Geoffroy halted between the systems; Cuvier refusing the aid of embryology, preferred a perpetual recurrence to the miraculous interposition of supreme power – a frank confession that with him science and philosophy no longer guided his pen.

The discovery of the signification of the fossil remains (*ossemens fossiles*) led Cuvier to the conclusion, startling to the thinking mind, the philosophic, – namely, that there had been three or four successive creations of the animal and vegetable worlds. As he grew older, and his intellect stronger, he himself felt the antagonism existing in such a view to sound philosophy, and he committed to paper a direct contradiction to those who had ascribed such an opinion to him. But still later he returned somewhat to his original views – the views of his early years. Goëthe, Oken, Geoffroy, never changed; as at first, so at last: one creation was their motto; unity of organization; unity of type; unity of plan, was their theory. Not so expressed it is true, and variously modified by each, but still based on one great idea. Unity of plan, unity of design, observable, traceable, demonstrable, in all forms which live, which have lived, or which may hereafter come into being.

I commenced these inquiries in 1810, and since that moment have never lost sight of them. My first endeavour was to reduce the structure of the fore-foot of the horse to the corresponding parts in man; to compare and to discover the *corresponding* structures. I felt sure that they must have been formed on one plan; it was a deep conviction, arrived at by no reasoning – instinctive. As might be supposed, my success at that time was not great. The *rudimentary* fingers in the horse surprised me. Why rudimentary fingers? I said to myself. I fancied that all Nature's works were perfect. The theory of arrested development, at which Meckel and the German school caught so strongly, did not satisfy me, and never did, even when forced to teach it for want of a higher generalization. It is a doctrine I am now prepared to refute. The laws of deformation are as regular as the laws of formation. My next effort (1812) was to compare the organs of sight in man with the same in all other animals. Other structures followed, in all the natural families from man downwards. Satisfied, by a constant appeal to structure, that one great plan regulated all – that there could not be two or more plans – two or more creations – I had not yet seen any of the founders of the philosophy of zoology. This happened in 1821. Acquainted with what I had done for comparative anatomy, no introduction was required to those whom I most desired to see – Cuvier, Geoffroy, and De Blainville: Goëthe and Oken were in Germany, and thus I never met the real founders of all philosophical anatomy. But my own dissections furnished me with data sufficient to prevent me coinciding with Geoffroy in his views of unity of the organization; whilst deep instinct, which never errs, told me that of the four creations maintained by Cuvier, three at least must be superfluous; to the geometrician I leave the refutation of the one creation, intending not to meddle with that which seems beyond the reach of human thought. My object in this free inquiry is merely to ascertain, if possible, if there really be a philosophy in zoology – if its basis can be shown to rest on science – if zoological knowledge forms a part of science, or merely an amusing mode of contrasting the human with the non-human. As I view it, a history of zoology requires for its basis an inquiry into –

1. The laws of species and of natural families. Do species and natural families exist, or are these distinctions merely human contrivances resulting from imperfect observation? Is mankind composed of one great natural family, or of several? and if of one, how many species, as permanent varieties are called, enter into its composition? Are species and natural families permanent and eternal, or is it merely a question of time? What part do embryonic forms play in Nature's grand scheme? Do they merely shadow forth a unity of plan, or do they, rather, pourtray the endless varieties of forms which animal matter may in time and space assume? In this view, the embryo is the most perfect of forms, embracing within it *the possible* of the past, the present, and the future. Individual adult forms simply show the development of that which can exist in unison with the existing order of things. The modern rhinoceros and elephant, tapir and sloth, are simply the developed forms of a natural family which once existed under other forms. They are not the mere hereditary descendants of the fossil world, as Geoffroy thought, but forms new in space and in time, but pre-existing in every embryo of their natural families. The embryo, then, is that which Nature perfects, that to which she looks for the continuation of *varied life* as it exists and has existed on the globe. The future is wrapt up in the same category. The embryo of any species of any natural family contains within it during its phases of development, all the forms or species which that natural family can assume or has assumed in past time. In the embryo and the young individual of any species of the natural family of the Salmonidæ, for example, you will find the characteristics of the adult of all the species. The same, I believe, holds in man; so that, were all the existing species of any family to be accidentally destroyed, saving one, in the embryos and young of that one will be found the elements of all the species ready to reappear to repeople the waters and the earth, the forms they are to assume being dependent on, therefore determined by, the existing order of things. With another order will arise a new series of species, also foreseen and provided for in the existing world. There can be no such

thing as arrested development, nor a gradual development of all forms towards perfection. Nature's works are perfect, first and last; and the hypothesis which supports the idea of a development tending to perfection is simply another expression of the doctrine of a final cause. The so-called "arrest of development" in the individual is a doctrine admitting of the easiest refutation. Development can only mean development, and nothing else. This may be either individually and specifically progressive, or zoologically retrogressive – i. e., back towards other forms; but always development, notwithstanding – a retrogressive development. The laws of deformation are as constant and regular as the laws of formation, which we call perfect, as belonging to the existing world; this is all. But to enter on this inquiry, a preliminary investigation is necessary. Zoological forms either obey general laws or they do not; they have been formed on one great plan, or on several plans. If the latter be maintained, it ought to be shown in what the plans differ; if the former, the scheme or plan ought to be explained and its existence demonstrated, if possible. If no such plan exist, there exist no laws, which, indeed, in that case, are not required, the scientific basis on which zoology rests being removed, it would cease to be a science. Whether life was coeval with the globe or not, is a question not essential to the present inquiry. At its appearance it of necessity obeyed those laws which philosophy is called on to investigate. In their essence these laws can never alter: this is what science teaches. No fossil remains have ever been discovered contradictory of the theory that one great scheme or plan has at all times existed. In a strictly philosophical sense, there could be only one creation; but the real difficulty is to determine what were the zoological forms of that creation. Were they *species* – that is, *races*, or did the individuals represent natural families embracing many species? However this may be, one thing is certain: many races or species have ceased to be, whilst others, which were not then, now live. These are new species merely as regards man, for, in point of fact, distinct species exist not in Nature. They are not included in the great scheme which fills up all voids, all differences, all distinctions. The gap – absence of the link connecting one species, as we call them, with another – man assumes as a positive fact in Nature – an intentional part of her plan; but it is not so, as science will soon demonstrate. Some persons have troubled themselves with a refutation of the hypothesis of the convertibility of one species into another. It is wholly a labour of supererogation, for no strictly scientific man ever entertained such an hypothesis in the sense assumed by them.

My first step, then, is to establish the doctrine of one great plan or type for all animals; a unity of plan – that is, a type – a type "which exists everywhere, but is nowhere to be found." A material archetype is the invention of a mind sunk in materiality and in error. No such can exist, since a part can never represent a whole.

A type being proved to exist, there arises next the question of varieties in zoological forms, embraced within the type. The application of these two inquiries includes all natural history as a science; if applicable to one they are applicable to all forms, man included. The law called "the arrest of development hypothesis" explains nothing, and involves increasing contradictions in terms. For many years

I have preferred using the terms "persistence of embryonic forms," although this phrase also is open to manifest objections. The persistence of the webbing of the human fingers to the adult state has been described as "an arrest of development." I should have thought it rather "a retrogressive development" towards another form. Is the presence of a third and a fourth head of the biceps muscle in man "an arrest of development"? Is the presence of a supra-condyloid process in the humerus and the passage of the humeral artery and median nerve through the passage thus formed an arrested development? Is it even a persistence of embryonic forms? These doctrines are nearly the reverse of those I hold. In the embryo I see a perfect, not an imperfect being; its developments, transmutations, metamorphoses, follow certain laws; they equally tend to perfection of the individual species as a species, of the individual as an individual. But their developments must not be called imperfect, inferior, or arrested developments, merely because they happen to tend towards, and to produce, inferior forms of organization, as we esteem them. The brain of the human embryo, like all its other structures, no matter to what species it belongs, embraces every possible form which man assumes, or has assumed, on the earth. The white races are not the more fully developed, and the negro the more imperfectly developed, species of one common natural family. The development of each is perfect in its way – equally so. The negro brain is not an imperfectly developed white brain, but a specifically formed brain, developed according to its own specific laws. When in the white races we find developments resembling the negro, and *vice versâ*, these are not arrested developments, but retrogressive, if you will, yet perfect in their kind. To instance a class of animals, supposed, erroneously, however, to be less complex in their organization than man, lower in the scale, less perfect – the class Fishes; and selecting amongst these, as, perhaps, most familiar to my readers, the natural family Salmonidæ, – if in this natural family, divisible into three sub-families, I find the young of any of the species of any of the sub-families strictly to resemble each other, and yet the grown *individual* of any of the species materially to differ from any other, it by no means follows that the one is a species less developed than another. In the embryo – in the young – all are alike; each offers in itself the specific characters found in all. The embryo salmo, for example, is of no species; it is of all. Take from it one set of characters, it becomes the salmo; deduct another, you have the forelle; subtract a third, you have the trout. In every embryo, of every species, we have *the possible* of all forms appertaining to that natural family at least. The destruction, then, by geological or other phenomena, of all the species, saving one, of any natural family, would not necessarily extinguish that family, since, in the embryo of the remaining species, there exist the elements for the reappearance, perhaps under modified forms, of all the extinct species. As of fishes, so of man: one natural family – one embryonic form, equal to the production of all species in accordance with the essential conditions of existence in time and space.

Editorial Headnote

17. [John Gordon], "The Doctrines of Gall and Spurzheim", *Edinburgh Review* 25, (June 1815), pp. 227–68.

This essay by John Gordon (1786–1818), an extra-mural lecturer on anatomy and physiology in Edinburgh, was published anonymously in the *Edinburgh Review*. Gordon's intent was to undermine the scientific standing of both Franz Joseph Gall (1758–1828) and Johann Gaspar Spurzheim (1776–1832) in a review of two of their recently published works. What Gall called 'cerebral physiology,' which Spurzheim later popularized as 'phrenology,' was controversial because it claimed to offer a material science of mind and morals. It was thus inevitably seen as challenging to religious orthodoxy, and therefore, Gordon was eager to disassociate such ideas from the practice of science. Gordon aimed to leave his readers in no doubt on this point, concluding his essay with the following words:

> The writings of DR GALL and SPURZHEIM, have not added one fact to the stock of our knowledge, respecting either the structure or the functions of man; but consist of such a mixture of gross errors, extravagant absurdities, downright misstatements, and unmeaning quotations from Scripture, as can leave no doubt, we apprehend, in the minds of honest and intelligent men, as to the real ignorance, the real hypocrisy, and the real empiricism of the authors.

A year after the publication of this review, as the historian John van Wyhe noted, "In 1816 Spurzheim travelled to Edinburgh where, before a packed lecture theatre, he delivered his well-known refutation of Gordon with the *Edinburgh Review* in his hand."[1]

Despite Gordon's efforts to distance the works of Gall and Spurzheim from respectable science, more than one historian has argued that it was in their work that we find the origins of scientific naturalism.[2]

Notes

1 John van Wyhe, "History of Phrenology on the Web," www.historyofphrenology.org.uk/texts/gordon.htm. Accessed 26th November 2023.

2 John van Wyhe, *Phrenology and the Origins of Scientific Naturalism* (London: Routledge, 2004); Roger Cooter, *The Cultural Meaning of Popular Science. Phrenology and the Organization of Consent in Nineteenth-Century Britain* (Cambridge: Cambridge University Press, 1985). For an extract from and commentary upon Combe and *Constitution of Man* see Caden C. Testa and Piers J. Hale, *Evolution Before Darwin* (London: Routledge, 2025); see also their essay in that volume, Caden C. Testa and Piers J. Hale, "Mapping evolutionary ideas in Britain in the first half of the nineteenth century."

17

"THE DOCTRINES OF GALL AND SPURZHEIM"

[John Gordon]

Source: [John Gordon], "The Doctrines of Gall and Spurzheim", *Edinburgh Review* 25 (June 1815), pp. 227–9, 247–54, 268

ART. X. *Anatomie et Physiologie du Système Nerveux en général, et du Cerveau en particulier; avec des Observations sur la Possibilité de reconnoître plusieurs Dispositions intellectuelles et morales de l'Homme et des Animaux par la Configuration de leurs Têtes*. Par F. J. GALL & G. SPURZHEIM. Premier vol. 4°. pp. 352. avec dix-sept planches fol. Paris. 1810. Deux vol. prem. part. pp. 212. avec quinze planches fol. Paris, 1812.

The Physiognomical System of Drs Gall *and* SPURZHEIM, *&c.* Illustrated with nineteen copper-plates. 2d edit. 8vo. pp. 581. London, 1815.

OUR readers will here recognize, without any difficulty, the same man of skulls, whom we had occasion to take notice of, more than twelve years ago. Long before this time, we should have looked for his craniological death. But he seems somewhat cleverer than most of his tribe; and has discovered the inestimable secret, that a man's reputation, as well as his health, may often be prolonged, by a little well-timed locomotion. So far, therefore, from suffering the demise we had confidently anticipated, he has gone on, it would appear, exercising his calling with indefatigable industry, and extending his fame. He has uttered his prelections, we are credibly informed, from one end of the Continent to the other; – imprinted his doctrines in two quarto volumes – defended them in a third, and illustrated the whole by a folio of engravings; – challenged all the anatomists of the Imperial Institute, conjunctly and severally, to a trial of skill in cerebral dissection; – gauged all the prominent *craniums* in Germany; – and ascertained the solid contents of every celebrated head in France.

VILLERS's *letter to* CUVIER[a] had, we confess, left very little doubt in our minds as to the real merits and the real views of this formidable personage; and the present publications have not only confirmed our original judgment with respect to him, but led us to extend the same opinion, without the slightest modification, to the partner he has since assumed, DR J. G. SPURZHEIM. We look upon the whole

DOI: 10.4324/9781003490548-19

doctrines taught by these two modern peripateties, anatomical, physiological, and physiognomical, as a piece of *thorough quackery* from beginning to end; and we are persuaded, that every intelligent person who takes the trouble to read a single chapter of the volumes before us, will view them precisely in the same light.

There are a certain number of individuals, however, in every community, who are destined to be the dupes of empirics; so it would be rather matter of SURPRISE, if these itinerant philosophers did not make some proselytes wherever they come. How many disciples Dr Spurzheim may have already collected from this class in England, by his English book, or his Lectures in the metropolis, we do not feel very anxious to inquire; because we are quite certain we should find, that they are more than ten times the number he has seduced, from the same proportion of persons, in any other nation in Europe. Great Britain is a field for quacks to fatten in; they flock to it from all quarters of the world; and England is the sweetest corner of the pasture. Well has the learned and most witty historian of Mrs John Bull's indisposition remarked, 'There is nothing so impossible in nature, but mountebanks will undertake; nothing, so incredible, but they will affirm.' As truly might he have added, that there is nothing mountebanks can undertake, which John Bull will not think possible; nothing they can affirm, which he will not believe. It may seem a little inconsistent, that the 'most thinking' people in the world should thus be the most credulous; but still it is comfortable to think, that theirs is not the credulity of ignorance, but the credulity of an honest and unsuspicious nature.

That DRS GALL and SPURZHEIM, however, should have brought over any of the better informed in the island, particularly from among those with whom anatomy and physiology are either favourite or professional pursuits, into a belief of any of the amazing absurdities they are bold enough to teach, is, we should hope, a thing really impossible. There is nothing, indeed, in the shape of reasoning, calculated to mislead, in their whole writings; not one clever sophistry to captivate, nor even an occasional successful induction to redeem; – nothing but a perpetual substitution of assertion for demonstration, and conjecture for fact. Were they even to succeed in shaking off the suspicion of *mala fides*, which we apprehend is inseparably attached to their character, we should not hesitate to say, that we do not know any writers, who, with a conceit so truly ludicrous, and so impudent a contempt for the opinions and labours of others, are so utterly destitute of every qualification necessary for the conduct of a philosophical investigation.

We have two objects in view in a formal *exposé* and exposure of the contents of the volumes before us. The first is, to contradict directly various *statements*, in point of fact, made by Drs GALL and SPURZHEIM with unparalleled boldness and effrontery, which persons perfectly satisfied of the general absurdity of their *opinions* may not have the same opportunities of refuting as ourselves: The second, and by far the most important, to save the purses of our readers if possible, before it be too late, by satisfying that curiosity which might otherwise lead them to purchase the books themselves, or attend the lectures of these cunning craniologers.

[...]

By a proper gauging of the head without, they can tell to a trifle how much it contains within. They have always found, that the larger the cranial part of the head measured upon the outside, skin and all, the greater the quantity of brain lodged in the cranial cavity. So it is the simplest process in the world: Shave a man's head, and you have the 'measure of his mind' in a moment: Multiply the length by the breadth, and the product by the thickness; and his philosophy and feeling will come out to the fraction of an inch. The remark is as old as it is said to be true, that no real hero is a hero to his *valet de chambre*. Let all those whom it may concern, now remember, that no man can be a pretended philosopher to his barber.

Were we disposed to be captious, we should perhaps object to this application of the Solid Measure; because it hath appeared to us, that human heads are rather like unto hazel nuts, whereof many that be equally large to look upon, do yet possess, some a thicker, some a thinner shell; some a smaller, some a larger kernel. But we will acquiesce implicitly for the present in the proposition, (familiar to physiologists long before the age of GALL or SPURZHEIM), that there is in most instances a general correspondence between the size of the cranium and the quantity of cerebrum; that large heads usually contain large brains, and small heads small brains. Nevertheless, we again deny, that there is any constant correspondence, or any connexion whatever, between the dimensions of a man's head and his intellect and inclinations, either in kind or degree. We do not even subscribe to the ancient proverb, '*Large head, little wit;*' although this obviously goes to establish, by vulgar observation, a species of connexion, directly the reverse of that maintained by our authors; and although the maxim seems to derive no small credit from the fact, (for such we affirm it to be), that idiots in general have *uncommonly* large heads. We are aware, that it is sometimes necessary to unsay 'old sayings;' and this we believe to be the case in the present instance. Much less are we disposed to allow any weight in this discussion to the familiar banter of *thick skull;* for it admits of no application to the inquiry, unless GALL should choose to interpret it as meaning, that when the sides of the cranium are thick, the cavity of the cranium is small, and the brain small in proportion, and consequently the understanding poor. But we demur at this explanation; and humbly submit, that the jeer in question is founded entirely on a false hypothesis; the vulgar very naturally, but very erroneously conceiving, that when there is more than usual difficulty in forcing sense into a man, there must be a greater than usual thickness of wall to keep it out. GALL and SPURZHEIM, in fact, in affirming that the vigour of intellect is always proportional to the size of the head, seem to have been desirous of trying how far their effrontery might be carried. If they succeed in convincing a single individual of common parts and observation, that this assertion is a truth, they will find little difficulty, we apprehend, in persuading mankind in general, that they hear by their eyes and see by their ears.

But we will even indulge these gentlemen so far as to admit, that the relation they assert is true. Will it, therefore, follow, that the greater size of the brain is the *cause* of the peculiarities in the intellect or inclinations of the individual?

Certainly not. Can it be possible that the great DRS GALL and SPURZHEIM have not observed, in the course of their multifarious inquiries into nature, that phenomena may coincide, without being related to each other as cause and effect? Were it established, that all great mathematicians had black eyes, and all poets blue ones, would any sensible man, from this alone, think of ascribing the mathematical talent in the one case, or the poetical genius in the other, to the colour of the iris. In the course of our own experience, we have observed, that persons who have a lurking affection for port wine, have uniformly a certain redness of nose; and yet we are far from conceiving ourselves warranted to infer from this, that the nasal hue is the *cause* of the vinous partiality. Some, on the contrary, are disposed to maintain, that it is rather the *effect;* but this we hold to be quite wicked and calumnious. Again, it is a remark which we have never found to fail, that all great lawyers have long and very mobile fingers, '*digiti prehensiles*,' as LINNÆUS would have called them, with a remarkably smooth cuticle or epidermis on the palms of their hands. Shall we therefore conclude, that this length and flexibility of finger, and this exceeding smoothness of palm, are the cause of eminence in the law? No; this may be a case of mere coincidence; nay, the professional eminence may, indirectly, be the *cause* of some of these phenomena: – But this is dangerous ground.

[...]

In the two first plates, then, of *Spurzheim's* work,[b] there will be found different views of the surface of the brain, when deprived of its membranes; one from above, another from below, and two lateral views. On different points of the surfaces thus represented, will be seen engraved, the various numbers, from one to thirty-three; and on turning to the table of explanation, the reader will discover, that each of these numbers refers to the *organ* of the faculty corresponding to it, in the enumeration of the special faculties formerly given; – number V, for example, marking the *organ* of *combativeness;* and number X, of *self-esteem;* and so on. Number I. makes its appearance on every surface of the laminated part of the cerebellum; and all the other numbers are scattered over the convolutions of the brain proper; – none, however, appearing on the lower surface of the middle or posterior lobes; and none on that surface of the hemispheres, which looks towards the falx of the dura mater. It deserves, also, to be remarked, that not a single number is to be seen on the medulla oblongata, or the annular protuberance, or the peduncles either of the cerebellum or brain proper, or the mamillary eminences, or the infundibulum; or, in short, any of the parts which lie between the middle lobes.

These numbers or figures, as they are said to refer to the different organs, are of course situated on particular parts of these organs; but they cannot of themselves, it is obvious, mark the whole dimensions of an organ; and no attempt has been made to do this in the engravings, by the usual contrivance of dotted lines, such as are to be seen in all maps and diagrams. The plates, therefore, leave the reader in utter ignorance as to the superficial extent, the internal direction, and the total bulk, of any one organ. Whether the *organ* of *Hope* goes upwards or downwards, backwards or forwards; whether the *organ* of *Order* stands quite clear of that of *Tune;* whether the *organ* of *Combativeness* does not intertwine with the *organ* of *Destructiveness;* whether the *organ* of *Wit* does not run the *organ* of *Imitation* through the body; whether one might not scoop out the *organ* of *Covetiveness* from end to end, as a cheesemonger with his wimble does a bit of Stilton, and yet not interfere in the least with the organ either of *Benevolence* or of *Veneration;* – these, and many other questions of equal importance, would be in vain determined, by an inspection of the engravings alone. The reader, therefore, will naturally turn to the text, for more minute and precise information on these points; and there we have the pleasure of telling him, he will find the following very satisfactory *detail.*

'It is indeed true, that the organs are not confined to the surface of the brain; they extend from the surface to the great swelling of the occipital hole (medulla oblongata), and probably include even the commisures; for the whole mass of the brain constitutes the organs.' – *Spurzheim*, p. 239.

Such is the trash, the despicable trumpery, which two men, calling themselves scientific inquirers, have the impudence gravely to present to the physiologists of the nineteenth century, as specimens of reasoning and induction.

But let us now attend a little to the ingenious methods, by which these gentlemen have discovered the precise situation of each of these *organs;* and to the external signs by which they assure us that it may be known during life, whether any particar *organ* be unusually large or small. This leads to the most important part of their doctrines perhaps; the practical part, as it may be called; the *physiognomy, craniology* or *cranioscopy;* the part which teaches us how to find out, by the shape of the head, whether a man loves his children or kills them, whether he steals or is very benevolent; whether he has much hope or little; whether he believes in a God, or is a freethinker; in short, what he is remarkable for, in knowledge, talent, or dispositions; or whether he be remarkable for any thing at all.

In the *first* place, then, our readers must know, that, according to DRS GALL and SPURZHEIM, when any portion of the surface of the brain happens to be more prominent than usual, there is always a corresponding eminence on that part of the osseous shell which covers it; and not only is this prominence obvious, whatever structure the shell may chance to have at this part, but it is invariably of sufficient size to be perceptible by sight or touch on the living head, with whatever soft textures the part may happen to be covered; whether with tendinous substance, cellular substance, adipose substance, muscular fibres, or common integuments with the hair; or two or three of these, or all of them together. Now, as it has been already shown, that the brain is the general *organ* of all the faculties, that each faculty has a portion of the cerebrum appropriated to itself, and that the perfection or vigour of each faculty is proportionate to the size of its *organ;* it obviously follows, that the aforesaid bumps on the head are infallible signs of peculiar energy in some of the special faculties. When, therefore, any uncommon bump of this sort presents itself on the head of an individual, it is only necessary to ascertain what faculty that person is remarkable for; and thus the particular part of the brain, which constitutes the *organ* of that faculty, is at once demonstrated. By a steady application of this method, to the skulls or busts of the celebrated dead, and to the heads of the living, unshaven as well as shaven, in all classes of society, and in different nations; in establishments for education, in hospitals for idiots and madmen, in houses of correction, and in prisons; DRS GALL and SPURZHEIM have had the merit of ascertaining the exact site of the thirty-three organs. Nay, of such amazingly precise application does this method seem to be susceptible, that although our authors found it impossible to define the superficial boundaries of the different organs on the brain itself, they have marked out the limits of each organ with the utmost distinctness, on the outside of the whole head. Accordingly, in the frontispiece to SPURZHEIM's book, a most extraordinary and engaging spectacle presents itself. The human head there appears in different aspects, cleanly shaven, with the whole of its upper surface divided into regions, like the maps of revolutionized France. The regions are of different sizes, from a quarter of an inch, perhaps, to two inches; and of very various shapes, some being rhomboidal, others elliptical, others almost circular, some like a half-moon, some like the point of a scymitar, others distinctly heart-shaped, but by far the greater number, as it appears to our eye, like the scales of a salmon magnified. The facilities which these diagrams must afford, in the application of the intellectual topography to practice, must be very great. When we discover any unusual projection or depression, any remarkable specimen of hill and dale, on the head of a stranger, we have only to consult a copy of the frontispiece quietly; and, without a moment's delay, we shall be enabled to decide, whether we ought to approach him as an honest man, or shun him as a knave.

If we must reply gravely to all this, which it requires some effort to do, we have briefly to observe, that not one of the assertions are true, and that not one step of the reasoning is correct.

It is not true, that there are ever such eminences on the surface of the brain, accompanied with projections of the cranium, as GALL and SPURZHEIM have affirmed. It is true, no doubt, and we presume it was known to anatomists before our authors were born, that the different regions of the brain, differ in their relative proportions in almost every individual. But the difference, whether it be confined to one dimension, or extend to all, is at the utmost very inconsiderable; seldom we believe amounting to half an inch, and never, we are confident, exceeding an inch, over an extent of six inches; and very often it is so small, as just to be perceptible, and no more. In this last case, as might be expected, it is difficult to discover any corresponding difference in the proportional capacity of the cerebral cavity of the skull; but in all varieties which exceed the ratio of a quarter of an inch in five or six inches, this cavity is in general obviously larger in the same proportion. Now, our anatomical readers are not to be informed, surely, that the two surfaces of the bones which form the cerebral cavity of the cranium are not everywhere parallel to each other. They know very well, that there are often considerable depressions within, where the corresponding surface without, does not exhibit the slightest appearance of projection, but is quite flat, or even hollow; and that there are often large prominences without, where there are no corresponding concavities within; so that when the outer surface of the walls of the cerebral cavity are compared with a mould in Paris plaster or in wax, of the cavity itself, the greatest difference is perceptible between them. All anatomists know, too, that this difference is not the same in degree, in any two skulls; the distance of the two surfaces of the bones from each other is so various. When, therefore, any particular region of the brain is considerably larger than usual, compared with other parts, although the corresponding region of the cerebral cavity be always proportionally larger also, this increased capacity within, is far from being uniformly accompanied with any greater prominence of the osseous shell without. Such correspondence does often occur, no doubt, particularly in those cases where the greater proportional size of the particular regions of the brain approaches to the maximum we have stated; but even in these instances it is not constant. Moreover, we must observe, that, admitting it were uniform and invariable, still as the prominences and depressions on the outer surface often vary, without any corresponding variety within, the examination of the outer surface alone never can lead to any certain conclusions as to the proportional dimensions of the brain.

The whole of what we have now stated, the ingenuous Drs Gall and Spurzheim must not only have been aware of from the beginning, but must have felt to be strictly true. We looked, therefore, with perfect confidence for a denial of its truth, somewhere in the volumes before us; for there is nothing sufficiently certain in nature, which these gentlemen will not call in question, if it be hostile to their views. Were the fact in any degree unfavourable to their system, that heavy bodies fall to the ground, we should, without doubt, have these fearless craniologers demonstrating, by 'incontestable facts,' by 'reasoning thereupon,' and by the 'refutation of all objections,' that such a phenomenon never can happen, and, in point of fact, never has happened since the world began. But we were not a little surprized to find, that not only are the facts, in the present instance, not contradicted, but that no other attempt has been made to evade them than is to be found in the following short and unintelligible sentence; –

'It is not necessary to appreciate any minute differences of size, 'in order to determine the development of the organs.' *Spurzheim*, p. 234.

We venture to affirm, therefore, that such prominences on the head, as Gall and Spurzheim have described, indicating certain eminences of the brain within, and uniformly accompanying some peculiarity of intellect or inclinations in the individual, never have been observed; and that all they have been so good as to write on this subject, is a mere fiction. Were it worth our while, we could even undertake to show, without much difficulty, that this piece of invention is inconsistent with itself, in various circumstances, and that it presumes a degree of blindness and ignorance in those to whom it is addressed, which it was really very cruel in Drs Gall and Spurzheim to suppose. How could these gentlemen, for example, think so poorly of the eyesight of their readers, as to imagine, that by the aid of their beautiful engravings they could fail to discover, that some of the prominences in the skull which they describe, are said to be caused by elevations on portions of the brain, which are not even in contact with the skull at these parts? Then, in point of extravagance, we do think, that since the integuments of every sort, covering the skull, seem to present so little impediment to the exercise of their acute vision and their erudite touch, in the discovery of the bumps, it would have made very little difference to them, and been vastly more convenient for their *customers*, if they had affirmed, that they could discover a man's character through his nightcap, or his hat, or a wig of *four stories*, or even through both hat and wig, at the distance of twenty miles, provided they had a good telescope, and the weather were clear.

IV. We are so heartily tired of the mass of nonsense we have been obliged to wade through, that we could now most willingly have done. But the *Anatomical Discoveries* of DRS GALL and SPURZHEIM yet remain to be considered; and these are on no account to be passed over in silence. It appears to us, that in this department they have displayed more quackery than in any other; and their bad faith is here the more unpardonable, that it was so much more likely to escape detection.

[...]

We must needs indulge ourselves with a summary paragraph too. The writings of DRS GALL and SPURZHEIM, have not added one fact to the stock of our knowledge, respecting either the structure or the functions of man; but consist of such a mixture of gross errors, extravagant absurdities, downright mistatements, and unmeaning quotations from Scripture, as can leave no doubt, we apprehend, in the minds of honest and intelligent men, as to the real ignorance, the real hypocrisy, and the real empiricism of the authors.

Notes

a See our Number for April 1803.

b We refer to this work alone, because it is the latest; and because, in the large French work, with folio engravings, the conjoint production of GALL and SPURZHEIM, neither the number nor the place of the *organs* accords exactly with the representation now given of them by SPURZHEIM himself. Let GALL and SPURZHEIM settle this point between them.

Editorial Headnote

18. George Combe, *The Constitution of Man Considered in Relation to External Objects* (Edinburgh: John Anderson, & London: Longman & Co., 1828), pp. 143, 145–56, 160, 164–6.

While not a book that spoke explicitly to the question of zoological speciation, George Combe's *Constitution of Man* was important for a number of other related reasons. Originally published in 1828, *The Constitution of Man* eventually became one of the most popular scientific works of the nineteenth century, even though there were many who denied that the ideas it advanced were truly scientific. Combe's work popularized the work of Franz Joseph Gall (1758–1828), who had developed a system of mental analysis based on the idea that the shape of the brain moulded to the shape of the skull, and thus that it was possible to read a person's character from the shape of their head. Gall called his analysis "Die Schädellehre." Gall's system was popularized initially by Gall's one-time assistant Johann Gasper Spurzheim (1776–1832) as "phrenology." Combe first encountered phrenological ideas as a result of reading an account of a lecture that Spurzheim had given in Edinburgh and quickly became enamoured by the topic. He was subsequently instrumental in the founding of the Edinburgh Phrenological Society and the publication of the Society's journal, *The Phrenological Journal and Miscellany*, (later renamed *Phrenological Journal and Magazine of Moral Science*). The Society quickly drew a large number of adherents, who met regularly to discuss the methods, insights, and broader social implications that might be drawn from this new science. Beyond this, though, Combe's *The Constitution of Man* brought phrenology to a much broader audience.

As the historian of science and medicine Roger Cooter has pointed out in his own compelling study of the history of phrenology, Combe's book sold vastly more copies than either Darwin's *Origin of Species* or the anonymously published popular evolutionary text *The Vestiges of the Natural History of Creation.*[1] Phrenology thus quickly became a national phenomenon, especially after the publication of a number of subsidized cheap editions from 1835. While Combe's book was not explicitly an evolutionary text, Combe did rely on what he called the "law of hereditary descent," by which he described how attributes of parents might variously be passed on to their offspring. Combe further suggested not only that human morality and character were material phenomena that reflected the physiology of the brain, but that it was possible to mould and develop one's own character (and thus one's physiology) through one's actions. The historian of science, John van Wyhe, has made the case that phrenology was instrumental in ushering in the development of scientific naturalism.[2] Phrenology was also foundational to later developments in the science of the mind.[3]

Although originally published in 1828, the book was republished in revised and expanded editions across the rest of the century. In 1832, two cheap editions were produced as a result of being subsidized by William Ramsay Henderson, who donated five thousand pounds for the "advancement of Phrenology."[4]

Notes

1 Roger Cooter, *The Cultural Meaning of Popular Science. Phrenology and the organization of Consent in Nineteenth-Century Britain* (Cambridge: Cambridge University Press, 1985). For more on the publication of the *Vestiges of the Natural History of Creation* and an extract from it, see Caden C. Testa and Piers J. Hale, *Evolution Before Darwin* (London: Routledge, 2025), including the contextual essay by the editors, "Mapping evolutionary ideas in Britain in the first half of the nineteenth century." The exhaustive book-length study of *Vestiges* and its publication is James Secord, *Victorian Sensation. The Extraordinary Publication, Reception, and Secret Authorship of The Vestiges of the Natural History of Creation* (Chicago: University of Chicago Press, 2000).

2 John van Wyhe, *Phrenology and the Origins of Victorian Scientific Naturalism* (London: Ashgate, 2004). For the full text of *Constitution of Man* and major reviews of it, see John van Wyhe (ed.), *Combe's Constitution of Man and Nineteenth Century Responses*. 3 vols. (London: Thoemmes Press, 2004).

3 It is worth noting that Robert Chambers referred to Gall's work in giving an account of mind in his popular 1844 evolutionary work *The Vestiges of the Natural History of Creation* (London: George Routledge & Sons, 1887), p. 251.

4 James Secord, *Visions of Science. Books and Readers at the Dawn of the Victorian Age* (Chicago: University of Chicago Press, 2015). pp. xiii.

18

THE CONSTITUTION OF MAN CONSIDERED IN RELATION TO EXTERNAL OBJECTS

George Combe

Source: George Combe, *The Constitution of Man Considered in Relation to External Objects* (Edinburgh: John Anderson, & London: Longman & Co., 1828), pp. 143, 145–56, 160, 164–6

018 "George Combe"
Image in the public domain.

Physiologists, in general, are agreed, that a vigorous and healthy constitution of body in the parents, communicates existence, in the most perfect state, to the offspring[a], and many observers of mankind, as well as medical authors, have remarked, also, the transmission, by hereditary descent, of mental talents and dispositions.

[...]

Phrenology reveals the principle on which these phenomena take place. Mental talents and dispositions are determined by the size and constitution of the brain. The brain is a portion of our organised system, and, as such, is subject to the organic laws, by one of which its qualities are transmitted by hereditary descent. This law, however faint or obscure it may appear in individual cases, becomes absolutely undeniable in nations. When we place the collection of Hindoo, Charib, Negro, New Holland, North American, and European skulls, possessed by the Phrenological Society, in juxtaposition, we perceive a national form and combination of organs in each actually obtruding itself upon our notice, and corresponding with the mental characters of the respective tribes; the cerebral development of one tribe is seen to differ as widely from that of another, as the European mind does from that of the New Hollander. Here, then, each Hindoo, Chinese, New Hollander, Negro, and Charib, obviously inherits from his parents a certain general type of head; and so does each European. If, then, the general forms and proportions are thus so palpably transmitted, can we doubt that the individual varieties follow the same rule, modified slightly by causes peculiar to the parents of the individual? The differences of national *character* are equally conspicuous as those of national *brains*, and it is surprising how permanently both endure. It is observed by an author in the *Edinburgh Review*, that "the Vicentine district is, as every one knows, and has been for ages, an integral part of the Venetian dominions, professing the same religion, and governed by the same laws, as the other continental provinces of Venice; yet the English character is not more different from the French, than that of the Vicentine from the Paduan; while the contrast between the Vicentine and his other neighbour, the Veronese, is hardly less remarkable." – No. lxxxiv. p. 459.

If, then, form, size, and constitution of brain, are transmitted from parents to children, if these determine natural mental talents and dispositions, which in their turn exercise the greatest influence over the happiness of individuals through the whole of life, it becomes extremely important to discover according to what laws this transmission takes place. Three principles present themselves to our consideration, at the first aspect of the question. Either, in the *first* place, the constitution and qualities of brain, which the parents themselves inherited at birth, are transmitted absolutely, so that the children, sex following sex, are exact copies, without variation or modification, of the one parent or the other; or, *secondly*, the natural and inherent qualities of the father and mother combine, and are transmitted in a modified form to the offspring; or, *thirdly*, the qualities of the children are determined jointly by the constitution of the stock, and by the faculties which predominate in power and activity in the parents, at the particular time when the organic existence of each child commences.

Experience shews that the *first* cannot be the law; for, as often mentioned, a real law of nature admits of no exceptions, and it is well established, that the minds of children are *not exact* copies, without variation or modification, of those of the parents, sex following sex. Neither can the second be the law, because it is equally certain that the minds of children, although *sometimes, are not always*, in talents and disposition, perfect modifications of those of the father and mother. If this law prevailed, no child would be a copy of the father, none a copy of the mother, nor of any collateral relation, but each would be invariably a compound of the two parents, and all the children would be exactly alike, sex alone excepted. Experience shews, that this cannot be the law. What, then, does experience say to the *third* idea, that the mental character of each child is determined by the particular qualities of the stock, combined with those which predominated in the parents when its existence commenced.

I have already adverted to the influence of the stock, and shall now illustrate that of the condition of the parents, when existence is communicated.

A strong illustration, in the case of the lower animals, appeared in the Edinburgh Review, No. lxxxiv. p. 457.

"Every one conversant with beasts," says the reviewer, "knows, that not only their natural, but that many of their acquired qualities, are transmitted by the parents to their offspring. Perhaps the most curious example of the latter fact may be found in the pointer.

"This animal is endowed with the natural instinct of winding game, and stealing upon his prey, which he surprises, having first made a short pause, in order to launch himself upon it with more security of success. This sort of *semicolon* in his proceedings, man converts into a *full stop*, and teaches him to be as much pleased at seeing the bird or beast drop by the shooter's gun, as at taking it himself. The staunchest dog of this kind, and of the original pointer, is of Spanish origin, and our own, is derived from this race, crossed with that of the foxhound, or other breed of dog, for the sake of improving his speed. This mixed and factitious race, of course, naturally partakes less of the true pointer character; that is to say, is less disposed to stop, or at least he makes a shorter stop at game. The *factitious pointer is, however, disciplined, in this country, into staunchness; and, what is most singular*, THIS QUALITY IS, IN A GREAT DEGREE, INHERITED BY HIS PUPPY, who may be seen earnestly standing at swallows or pigeons in a farm-yard. For intuition, though it leads the offspring to exercise his parent's faculties, does not instruct him how to direct them. The preference of his master afterwards guides him in his selection, and teaches him what game is better worth pursuit. On the other hand, the pointer of pure Spanish race, unless he happen to be well broke himself, which in the south of Europe seldom happens, produces a race which are all but unteachable, according to our notions of a pointer's business. They will make a stop at their game, as natural instinct prompts them, but seem incapable of being drilled into the habits of the animal, which education has formed in this country, and has rendered, as I have said, in some degree, capable of transmitting his acquirements to his descendants."

"Acquired habits are hereditary in other animals besides dogs. English sheep, probably from the greater richness of our pastures, feed very much together; while Scotch sheep are obliged to extend and scatter themselves over their hills, for the better discovery of food. Yet the English sheep, on being transferred to Scotland, *keep their old habit of feeding in a mass*, though so little adapted to their new country; so do their descendants; and the English sheep is not thoroughly naturalized into the necessities of his place *till the third generation.* The same thing may be observed as to the nature of his food, that is observed in his mode of seeking it. When turnips were introduced from England into Scotland, *it was only the third generation* which heartily adopted this diet, the first having been starved into an acquiescence in it."

In these instances, long continued impressions on the parents appear to have at last effected change of disposition in the offspring.

"We have seen," says an author whom I have already quoted, "how wonderfully the *bee* works – according to rules discovered by man thousands of years after the insect had followed them with perfect accuracy. The same little animal seems to be acquainted with principles of which we are still ignorant. We can, by crossing, vary the forms of cattle with astonishing nicety; but we have no means of altering the nature of an animal, once born, by means of treatment and feeding. This power, however, is undeniably possessed by the bees. When the queen-bee is lost, by death or otherwise, they choose a grub from among those who are born for workers; they make three cells into one, and, placing the grub there, they build a tube round it; they afterwards build another cell, of a pyramidal form, into which the grub grows: they feed it with peculiar food, and tend it with extreme care. It becomes, when transformed from the worm to the fly, not a worker, but a queen-bee." – *Objects, Advantages, and Pleasures of Science*, p. 33. It is difficult to conceive that man will ever possess such a power as this last.

Man, however, as an organized being, is subject to laws similar to those which govern the organization of the lower animals. Dr PRITCHARD, in his Researches into the Physical History of Mankind, has brought forward a variety of interesting facts and opinions on this subject of transmission of hereditary qualities in the human race. He says, "Children resemble, in feature and constitution, both parents, but, I think, more generally the father. In the breeding of horses and oxen, great importance is attached, by experienced propagators, to the male. In sheep, it is commonly observed that black rams beget black lambs. In the human species, also, the complexion chiefly follows that of the father; and I believe it to be a general fact, that the offspring of a *black father* and white mother is *much darker* than the progeny of a *white father* and a black mother." – Vol. ii. p. 551. These facts appear to me to be referable to both causes. The stock must have had some influence, but the mother, in all these cases, is not impressed by her own colour, because she does not look on herself; while the *father's* complexion must strikingly attract her attention, and may, in this way, give the darker tinge to the offspring[b].

Dr PRITCHARD states the result of his investigations to be, *First*, That the organization of the offspring is always modelled according to the type of the *original structure* of the parent; and, *secondly*, "That changes, produced by external causes in the appearance or constitution of the individual are temporary; and, in general, acquired characters are transient; they terminate with the individual, and have no influence on the progeny." – Vol. ii. p. 536. He supports the first of these propositions by a variety of facts occurring "in the porcupine family," "in the hereditary nature of complexion," and, "in the growth of supernumerary fingers or toes, and corresponding deficiencies." "MAUPERTUIS has mentioned this phenomenon; he assures us, that there were two families in Germany, who have been distinguished for several generations by six fingers on each hand, and the same number of toes on each foot," &c. He admits, at the same time, that the *second* proposition is of more difficult proof, and that an opinion contrary to it "has been maintained by some writers, and a variety of singular facts have been related in support of it." But many of these relations, as he justly observes, are obviously fables.

In regard to the foregoing propositions, I would observe, that a manifest distinction exists between transmission of monstrosities, or mutilations, which constitute additions to, or abstractions from, the natural lineaments of the body, and transmission of a mere tendency in particular organs to a greater or less development of their natural functions. This last appears to me to be influenced by the state of the parents, at the time when existence is communicated to the offspring. On this point Dr PRITCHARD says, "The opinion which formerly prevailed, and which has been entertained by some modern writers, among whom is Dr DARWIN, that at the period when organisation commences in the ovum, that is, at or soon after the time of conception, the structure of the fœtus is capable of undergoing modification from impressions on the mind or senses of the parent, does not appear altogether so improbable. It is contradicted, at least, by no fact in physiology. It is an opinion of very ancient prevalence, and may be traced to so remote a period, that its rise cannot be attributed to the speculations of philosophers, and it is difficult to account for the origin of such a persuasion, unless we ascribe it to facts which happened to be observed." – P. 556.

A striking and undeniable proof of the effect on the character and dispositions of children, produced by the form of brain transmitted to them by hereditary descent, is to be found in the progeny of marriages between Europeans, whose brains possess a favourable development of the moral and intellectual organs, and Hindoos, and native Americans, whose brains are inferior. All authors agree, and report the circumstance as singularly striking, that the children of such unions are decidedly superior in mental qualities to the native, while they are still inferior to the European parent. Captain FRANKLIN says, that the half-breed American Indians "are upon the whole a good looking people; and, where the experiments have been made, have shewn much expertness in learning, and willingness to be taught; they have, however, been sadly neglected." – P. 86. He adds, "It has been remarked, I do not know with what truth, that half breeds shew more personal courage than the pure breeds." Captain BASIL HALL, and other writers on South

America, mention, that the offspring of native American and Spanish parents, constitute the most active, vigorous, and powerful portion of the inhabitants of these countries; and many of them rose to high commands during the revolutionary war. So much is this the case in Hindostan, that several writers have already pointed to the mixed race there, as obviously destined to become the future sovereigns of India. These individuals inherit from the native parent a certain adaptation to the climate, and from the European parent a higher development of brain, the two combined constituting their superiority.

Another example of the same law occurs in Persia. In that country, it is said that the custom has existed for ages among the nobles, of purchasing beautiful female Circassian captives, and forming alliances with them as wives. It is ascertained that the Circassian form of brain stands comparatively high in the development of the moral and intellectual organs[c]. And it is mentioned by some travellers, that the race of nobles in Persia is the most gifted in natural qualities, bodily and mental, of any class of that people; a fact diametrically opposite to that which takes place in Spain, and other European countries, where the nobles intermarry constantly with each other, and set the organic laws altogether at defiance.

The degeneracy and even idiocy of some of the noble and royal families of Spain and Portugal, from marrying nieces, and other near relations, is well known; and defective brains, in all these cases, are observed.

[...]

I do not present these views as ascertained phrenological science, but as inferences strongly supported by facts, and consistent with known phenomena. If we suppose them to be true, they will greatly strengthen the motives for preserving the *habitual* supremacy of the moral sentiments and intellect, when, by doing so, improved moral and intellectual capacities may be conferred on offspring.

[...]

Every phrenologist knows, that the New Hollanders, Charibs, and other savage tribes, are distinguished by great deficiencies in the moral and intellectual organs[d]. If, however, it be true, that considerable development of intellectual organs is indispensable to the comprehension of science, and the practice of virtue, it would, on the present supposition, be impossible to raise the New Hollanders, as a people, one step higher in capacity for intelligence and virtue than they now are. We might cultivate each generation up to the limit of its powers, but there the improvement, and a low one it would be, would stop; for the next generation, being produced with brains equally deficient in the moral and intellectual regions, no principle of increasing amelioration would exist. The same remarks are applicable to every tribe of mankind. If we assume modern Europeans as the standard, then, if the law of hereditary descent were abrogated, every deficiency that at this moment is attributable to imperfect or disproportionate development of brain, would be irremediable, and continue as long as the race existed. Each generation might be cultivated till the summit level of its capacities was attained, but there each succeeding generation would remain. When we contrast with this prospect the very opposite effects flowing from the law of hereditary transmission of qualities in an increasing ratio, the whole advantages are at once perceived to be on the side of the latter constitution. According to this rule, the children of the individuals who have obeyed the organic, the moral, and the intellectual laws, would start from the highest level of their parents, not only in acquired knowledge, but in consequence of that very obedience, they would inherit an enlarged development of the moral and intellectual organs, and thereby enjoy an increasing capability of discovering and obeying the Creator's institutions. This improvement will, no doubt, have its limits; but it may probably extend to that point at which man will be capable of placing himself in harmony with the natural laws. The effort necessary to *maintain* himself there, will still provide for the activity of his faculties.

Notes

* Parents frequently live again in their offspring. It is quite certain that children resemble their parents, not only in countenance and the form of their body, but also in their mental dispositions, in their virtues and vices, &c.[2]

a Very young hens lay small eggs, but a breeder of fowl will never set these to be hatched because the animals produced would be feeble and imperfectly developed. They select the largest and freshest eggs and endeavour to rear the healthiest stock possible.

b Black hens lay dark-coloured eggs.

c In Mr W. ALLAN's picture of the Circassian Captives, the form of the head is said to be a copy from nature, taken by that artist, when he visited the country. It is engraved by Mr JAMES STEWART with great beauty and fidelity, and may be consulted as an example of the superiority of Circassian development of the brain.

d This fact is demonstrated by specimens in most Phrenological Collections.

Part 3

THE DEVELOPMENTAL HYPOTHESIS IN MID-VICTORIAN BRITAIN

Editorial Headnote

19. Charles Lyell, *Principles of Geology*, Vol. 2. (London, John Murray, 1832), pp. 1–19, 20–1.

Charles Lyell's *Principles of Geology* is an important work in this collection for a number of reasons beyond its signal contribution to the field of geology. Firstly, it was in volume two of this three-volume work that Lyell included a lengthy and detailed rejection of the transmutationist argument that the French naturalist Jean-Baptiste Lamarck had made in his 1809 publication *Zoological Philosophy*. Lyell was motivated to comment on the transmutationist ideas of Lamarck in particular (although he also mentioned Lamarck's colleague and fellow transmutationist at the French National Museum in Paris, Étienne Geoffroy Saint-Hilaire) because he feared that many people might be led to read his own gradualist and historical account of earth history as implying a gradual, historical, and progressive natural history. In order to undermine this kind of conclusion, Lyell insisted that earth history was cyclical and not progressive, but still felt it necessary to give a blow-by-blow attack on Lamarck's transmutationist understanding both of the earth and of the species that lived upon it.

In the process of seeking to deny Lamarck's conclusions, Lyell felt it necessary to give a full and accurate account of his opponent's views and thereby provided the fullest and most explicit account of Lamarck's views published in England at that time. Indeed, Darwin and Herbert Spencer were only two of Lyell's readers who found that the Frenchman's argument was more convincing than Lyell's rebuttal. As the historian Anne Dewitt has recently noted, the atheist and free-thinking press made ready use of science to advance their cause, and one such effort utilized Lyell's account of Lamarck in their own advocacy of transmutation as grounds for unbelief.[1]

Note

1 Anne Dewitt, "Evolution and Political Revolution in Blackwood's Periodical Poetry," *Victorian Periodicals Review*, Vol. 56, no.2 (Summer 2023), pp. 158–181, on p. 164. See [William Chilton], "Theory of Regular Gradation XLV," *Oracle of Reason, September 30, 1843 p. 332–34*. As James Secord notes, Lyell's geology had been appropriated to undermine biblical literalism by other atheist and freethinking writers, such as in Henry Hetherington's *The Freethinker's Information for the People*, 1, no.1–2 (1842): pp. 1–16. The series "The Theory of Regular Gradation" also later appropriated the *Vestiges of the Natural History of Creation* to the atheist cause. See James Secord, *Victorian Sensation. The Extraordinary Publication, Reception, and Secret Authorship of The Vestiges of the Natural History of Creation* (Chicago: University of Chicago Press, 2000), p. 313, n.33.

19

PRINCIPLES OF GEOLOGY

Charles Lyell

Source: Charles Lyell, *Principles of Geology*, Vol. 2. (London, John Murray, 1832), pp. 1–19, 20–1

PRINCIPLES OF GEOLOGY.

CHAPTER I.

In our first volume we treated of the changes which have taken place in the inorganic world within the historical era, and we must next turn our attention to those now in progress in the animate creation. In examining this class of phenomena, we shall treat first of the vicissitudes to which *species* are subject, and afterwards consider the influence of the powers of vitality in modifying the surface of the earth and the material constituents of its crust.

The first of these divisions will lead us, among other topics, to inquire, first, whether species have a real and permanent existence in nature; or whether they are capable, as some naturalists pretend, of being indefinitely modified in the course of a long series of generations? Secondly, whether, if species have a real existence, the individuals composing them have been derived originally from many similar stocks, or each from one only, the descendants of which have spread themselves gradually from a particular point over the habitable lands and waters? Thirdly, how far the duration of each species of animal and plant is limited by its dependance on certain fluctuating and temporary conditions in the state of the animate and inanimate world? Fourthly, whether there be proofs of the successive extermination of species in the ordinary course of nature, and whether there be any reason for conjecturing that new animals and plants are created from time to time, to supply their place?

Before we can advance a step in our proposed inquiry, we must be able to define precisely the meaning which we attach to the term species. This is even more necessary in geology than in the ordinary studies of the naturalist; for they who deny that such a thing as a species exists, concede nevertheless that a botanist or zoologist may reason as if the specific character were constant, because they confine their observations to a brief period of time. Just as the geographer, in constructing his maps from century to century, may proceed as if the apparent places of the fixed stars remained absolutely the same, and as if no alteration was brought about by the precession of the equinoxes, so it is said in the organic world, the stability of a species may be taken as absolute, if we do not extend our views beyond the narrow period of human history; but let a sufficient number of centuries elapse, to

DOI: 10.4324/9781003490548-22

allow of important revolutions in climate, physical geography, and other circumstances, and the characters, say they, of the descendants of common parents may deviate indefinitely from their original type.

Now, if these doctrines be tenable, we are at once presented with a principle of incessant change in the organic world, and no degree of dissimilarity in the plants and animals which may formerly have existed, and are found fossil, would entitle us to conclude that they may not have been the prototypes and progenitors of the species now living. Accordingly, M. Geoffroy St. Hilaire has declared his opinion, that there has been an uninterrupted succession in the animal kingdom effected by means of generation, from the earliest ages of the world up to the present day; and that the ancient animals whose remains have been preserved in the strata, however different, may nevertheless have been the ancestors of those now in being. Although this notion is not generally received, we feel that we are not warranted in assuming the contrary, without fully explaining the data and reasoning by which we conceive it may be refuted.

We shall begin by stating as concisely as possible all the facts and ingenious arguments by which the theory has been supported, and for this purpose we cannot do better than offer the reader a rapid sketch of Lamarck's statement of the proofs which he regards as confirmatory of the doctrine, and which he has derived partly from the works of his predecessors, and in part from original investigations.

We shall consider his proofs and inferences in the order in which they appear to have influenced his mind, and point out some of the results to which he was led while boldly following out his principles to their legitimate consequences.

The name of species, observes Lamarck, has been usually applied to 'every collection of similar individuals, produced by other individuals like themselves[a].' This definition, he admits, is correct, because every living individual bears a very close resemblance to those from which it springs. But this is not all which is usually implied by the term species, for the majority of naturalists agree with Linnæus in supposing that all the individuals propagated from one stock have certain distinguishing characters in common which will never vary, and which have remained the same since the creation of each species.

In order to shake this opinion, Lamarck enters upon the following line of argument. The more we advance in the knowledge of the different organized bodies which cover the surface of the globe, the more our embarrassment increases, to determine what ought to be regarded as a species, and still more how to limit and distinguish genera. In proportion as our collections are enriched, we see almost every void filled up, and all our lines of separation effaced; we are reduced to arbitrary determinations, and are sometimes fain to seize upon the slight differences of mere varieties, in order to form characters for what we choose to call a species, and sometimes we are induced to pronounce individuals but slightly differing, and which others regard as true species, to be varieties.

The greater the abundance of natural objects assembled together, the more do we discover proofs that everything passes by insensible shades into something else; that even the more remarkable differences are evanescent, and that nature has, for the most part, left us nothing at our disposal for establishing distinctions, save trifling and, in some respects, puerile particularities.

We find that many genera amongst animals and plants are of such an extent, in consequence of the number of species referred to them, that the study and determination of these last has become almost impracticable. When the species are arranged in a series, and placed near to each other, with due regard to their natural affinities, they each differ in so minute a degree from those next adjoining, that they almost melt into each other, and are in a manner confounded together. If we see isolated species, we may presume the absence of some more closely connected, and which have not yet been discovered. Already are there genera, and even entire orders, – nay, whole classes, which present an approximation to the state of things here indicated.

If, when species have been thus placed in a regular series, we select one, and then, making a leap over several intermediate ones, we take a second, at some distance from the first, these two will, on comparison, be seen to be very dissimilar; and it is in this manner that every naturalist begins to study the objects which are at his own door. He then finds it an easy task to establish generic and specific distinctions; and it is only when his experience is enlarged, and when he has made himself master of the intermediate links, that his difficulties and ambiguities begin. But while we are thus compelled to resort to trifling and minute characters in our attempt to separate species, we find a striking disparity between individuals which we know to have descended from a common stock, and these newly-acquired peculiarities are regularly transmitted from one generation to another, constituting what are called *races*.

From a great number of facts, continues the author, we learn that, in proportion as the individuals of one of our species change their situation, climate, and manner of living, they change also, by little and little, the consistence and proportions of their parts, their form, their faculties, and even their organization, in such a manner, that everything in them comes at last to participate in the mutations to which they have been exposed. Even in the same climate a great difference of situation and exposure causes individuals to vary; but if these individuals continue to live and to be reproduced under the same difference of circumstances, distinctions are brought about in them which become in some degree essential to their existence. In a word, at the end of many successive generations, these individuals, which originally belonged to another species, are transformed into a new and distinct species[b].

Thus, for example, if the seeds of a grass, or any other plant which grows naturally in a moist meadow, be accidentally transported, first to the slope of some neighbouring hill, where the soil, although at a greater elevation, is damp enough to allow the plant to live; and if, after having lived there, and having been several times regenerated, it reaches by degrees the drier and almost arid soil of a mountain declivity, it will then, if it succeeds in growing and perpetuates itself for a series of generations, be so changed that botanists who meet with it will regard it as a particular species[c]. The unfavourable climate in this case, deficiency of nourishment, exposure to the winds, and other causes, give rise to a stunted and dwarfish race, with some organs more developed than others, and having proportions often quite peculiar.

What nature brings about in a great lapse of time we occasion suddenly by changing the circumstances in which a species has been accustomed to live. All are aware that vegetables taken from their birth-place and cultivated in gardens, undergo changes which render them no longer recognizable as the same plants. Many which were naturally hairy become smooth or nearly so; a great number of such as were creepers and trailed along the ground, rear their stalks and grow erect. Others lose their thorns or asperities; others again, from the ligneous state which their stem possessed in hot climates, where they were indigenous, pass to the herbaceous, and, among them, some which were perennials become mere annuals. So well do botanists know the effects of such changes of circumstances, that they are averse to describe species from garden specimens, unless they are sure that they have been cultivated for a very short period.

'Is not the cultivated wheat,' (*Triticum sativum*) asks Lamarck, 'a vegetable brought by man into the state in which we now see it? Let any one tell me in what country a similar plant grows wild, unless where it has escaped from cultivated fields? Where do we find in nature our cabbages, lettuces, and other culinary vegetables, in the state in which they appear in our gardens? Is it not the same in regard to a great quantity of animals which domesticity has changed or considerably modified[d]?' Our domestic fowls and pigeons are unlike any wild birds. Our domestic ducks and geese have lost the faculty of raising themselves into the higher regions of the air, and crossing extensive countries in their flight, like the wild ducks and wild geese from which they were originally derived. A bird which we breed in a cage cannot, when restored to liberty, fly like others of the same species which have been always free. This small alteration of circumstances, however, has only diminished the power of flight, without modifying the form of any part of the wings. But when individuals of the same race are retained in captivity during a considerable length of time, the form even of their parts is gradually made to differ, especially if climate, nourishment, and other circumstances, be also altered.

The numerous races of dogs which we have produced by domesticity are nowhere to be found in a wild state. In nature we should seek in vain for mastiffs, harriers, spaniels, greyhounds, and other races, between which the differences are sometimes so great, that they would be readily admitted as specific between wild animals; 'yet all these have sprung originally from a single race, at first approaching very near to a wolf, if, indeed, the wolf be not the true type which at some period or other was domesticated by man.'

Although important changes in the nature of the places which they inhabit modify the organization of animals as well as vegetables, yet the former, says Lamarck, require more time to complete a considerable degree of transmutation, and, consequently, we are less sensible of such occurrences. Next to a diversity of the medium in which animals or plants may live, the circumstances which have most influence in modifying their organs are differences in exposure, climate, the nature of the soil, and other local particulars. These *circumstances* are as varied as are the characters of species, and, like them, pass by insensible shades into each other, there being every intermediate gradation between the opposite extremes. But each locality remains for a very long time the same, and is altered so slowly that we can only become conscious of the reality of the change, by consulting geological monuments, by which we learn that the order of things which now reigns in each place has not always prevailed, and by inference anticipate that it will not always continue the same[e].

Every considerable alteration in the local circumstances in which each race of animals exists, causes a change in their wants, and these new wants excite them to new actions and habits. These actions require the more frequent employment of some parts before but slightly exercised, and then greater development follows as a consequence of their more frequent use. Other organs no longer in use are impoverished and diminished in size, nay, are sometimes entirely annihilated, while in their place new parts are insensibly produced for the discharge of new functions[f].

We must here interrupt the author's argument, by observing that no positive fact is cited to exemplify the substitution of some *entirely new* sense, faculty, or organ, in the room of some other suppressed as useless. All the instances adduced go only to prove that the dimensions and strength of members and the perfection of certain attributes may, in a long succession of generations, be lessened and enfeebled by disuse; or, on the contrary, be matured and augmented by active exertion, just as we know that the power of scent is feeble in the greyhound, while its swiftness of pace and its acuteness of sight are remarkable – that the harrier and stag-hound, on the contrary, are comparatively slow in their movements, but excel in the sense of smelling.

We point out to the reader this important chasm in the chain of the evidence, because he might otherwise imagine that we had merely omitted the illustrations for the sake of brevity, but the plain truth is, that there were no examples to be found; and when Lamarck talks 'of the efforts of internal sentiment,' 'the influence of subtle fluids,' and the 'acts of organization,' as causes whereby animals and plants may acquire *new organs*, he gives us names for things, and with a disregard to the strict rules of induction, resorts to fictions, as ideal as the 'plastic virtue,' and other phantoms of the middle ages.

It is evident, that if some well authenticated facts could have been adduced to establish one complete step in the process of transformation, such as the appearance, in individuals descending from a common stock, of a sense or organ entirely new, and a complete disappearance of some other enjoyed by their progenitors, that time alone might then be supposed sufficient to bring about any amount of metamorphosis. The gratuitous assumption, therefore, of a point so vital to the theory of transmutation, was unpardonable on the part of its advocate.

But to proceed with the system; it being assumed as an undoubted fact, that a change of external circumstances may cause one organ to become entirely obsolete, and a new one to be developed such as never before belonged to the species, the following proposition is announced, which, however staggering and absurd it may seem, is logically deduced from the assumed premises. "It is not the organs, or, in other words, the nature and form of the parts of the body of an animal which have given rise to its habits, and its particular faculties, but on the contrary, its habits, its manner of living, and those of its progenitors have in the course of time determined the form of its body, the number and condition of its organs, in short, the faculties which it enjoys. Thus otters, beavers, water-fowl, turtles, and frogs, were not made web-footed in order that they might swim; but their wants having attracted them to the water in search of prey, they stretched out the toes of their feet to strike the water and move rapidly along its surface. By the repeated stretching of their toes, the skin which united them at the base acquired a habit of extension, until in the course of time the broad membranes which now connect their extremities were formed.

In like manner the antelope and the gazelle were not endowed with light agile forms, in order that they might escape by flight from carnivorous animals; but having been exposed to the danger of being devoured by lions, tigers, and other beasts of prey, they were compelled to exert themselves in running with great celerity, a habit which, in the course of many generations, gave rise to the peculiar slenderness of their legs, and the agility and elegance of their forms.

The cameleopard was not gifted with a long flexible neck because it was destined to live in the interior of Africa, where the soil was arid and devoid of herbage, but being reduced by the nature of that country to support itself on the foliage of lofty trees, it contracted a habit of stretching itself up to reach the high boughs, until its fore-legs became longer than the hinder, and its neck so elongated, that it could raise its head to the height of twenty feet above the ground."

Another line of argument is then entered upon, in farther corroboration of the instability of species. In order it is said that individuals should perpetuate themselves unaltered by generation, those belonging to one species ought never to ally themselves to those of another: but such sexual unions do take place, both among plants and animals; and although the offspring of such irregular connexions are usually sterile, yet such is not always the case. Hybrids have sometimes proved prolific where the disparity between the species was not too great; and by this means alone, says Lamarck, varieties may gradually be created by near alliances, which would become races, and in the course of time would constitute what we term species[g].

But if the soundness of all these arguments and inferences be admitted, we are next to inquire, what were the original types of form, organization, and instinct, from which the diversities of character, as now exhibited by animals and plants, have been derived? We know that individuals which are mere varieties of the same species, would, if their pedigree could be traced back far enough, terminate in a single stock; so according to the train of reasoning before described, the species of a genus, and even the genera of a great family, must have had a common point of departure. What then was the single stem from which so many varieties of form have ramified? Were there many of these, or are we to refer the origin of the whole animate creation, as the Egyptian priests did that of the universe, to a single egg?

In the absence of any positive data for framing a theory on so obscure a subject, the following considerations were deemed of importance to guide conjecture.

In the first place, if we examine the whole series of known animals, from one extremity to the other, when they are arranged in the order of their natural relations, we find that we may pass progressively, or at least with very few interruptions, from beings of more simple to those of a more compound structure; and in proportion as the complexity of their organization increases, the number and dignity of their faculties increase also. Among plants a similar approximation to a graduated scale of being is apparent. Secondly, it appears from geological observations, that plants and animals of more simple organization existed on the globe before the appearance of those of more compound structure, and the latter were successively formed at later periods: each new race being more fully developed than the most perfect of the preceding era.

Of the truth of the last-mentioned geological theory, Lamarck seems to have been fully persuaded; and he also shews that he was deeply impressed with a belief prevalent amongst the older naturalists, that the primeval ocean invested the whole planet long after it became the habitation of living beings, and thus he was inclined to assert the priority of the types of marine animals to those of the terrestrial, and to fancy, for example, that the testacea of the ocean existed first, until some of them, by gradual evolution, were *improved* into those inhabiting the land.

These speculative views had already been, in a great degree, anticipated by Delamétherie in his Teliamed, and by several modern writers, so that the tables were completely turned on the philosophers of antiquity, with whom it was a received maxim, that created things were always most perfect when they came first from the hands of their Maker, and that there was a tendency to progressive deterioration in sublunary things when left to themselves –

––– omnia fatis
In pejus ruere, ac retrò sublapsa referri.

So deeply was the faith of the ancient schools of philosophy imbued with this doctrine, that to check this universal proneness to degeneracy, nothing less than the re-intervention of the Deity was thought adequate; and it was held, that thereby the order, excellence, and pristine energy of the moral and physical world had been repeatedly restored.

But when the possibility of the indefinite modification of individuals descending from common parents was once assumed, as also the geological generalization respecting the progressive development of organic life, it was natural that the ancient dogma should be rejected, or rather reversed; and that the most simple and imperfect forms and faculties should be conceived to have been the originals whence all others were developed. Accordingly, in conformity to these views, inert matter was supposed to have been first endowed with life; until in the course of ages, sensation was superadded to mere vitality: sight, hearing, and the other senses, were afterwards acquired; and then instinct and the mental faculties; until, finally, by virtue of the tendency of things to *progressive improvement*, the irrational was developed into the rational.

The reader, however, will immediately perceive, that if all the higher orders of plants and animals were thus supposed to be comparatively modern, and to have been derived in a long series of generations from those of more simple conformation, some further hypothesis became indispensable, in order to explain why, after an indefinite lapse of ages, there were still so many beings of the simplest structure. Why have the majority of existing creatures remained stationary throughout this long succession of epochs, while others have made such prodigious advances? Why are there still such multitudes of infusoria and polypes, or of confervæ and other cryptogamic plants? Why, moreover, has the process of development acted with such unequal and irregular force on those classes of beings which have been greatly perfected, so that there are wide chasms in the series; gaps so enormous, that Lamarck fairly admits we can never expect to fill them up by future discoveries?

The following hypothesis was provided to meet these objections. Nature, we are told, is not an intelligence, nor the Deity, but a delegated power – a mere instrument – a piece of mechanism acting by necessity – an order of things constituted by the Supreme Being, and subject to laws which are the expressions of his will. This nature is *obliged* to proceed gradually in all her operations; she cannot produce animals and plants of all classes at once, but must always begin by the formation of the most simple kinds; and out of them elaborate the more compound, adding to them successively, different systems of organs, and multiplying more and more their number and energy.

This Nature is daily engaged in the formation of the elementary rudiments of animal and vegetable existence, which correspond to what the ancients termed *spontaneous generations*. She is always beginning anew, day by day, the work of creation, by forming monads, or 'rough draughts' (ébauches), which are the only living things she ever gives birth to *directly*.

There are distinct primary rudiments of plants and animals, and *probably* of each of the great divisions of the animal and vegetable kingdoms[h]. These are gradually developed into the higher and more perfect classes by the slow, but unceasing agency of two influential principles: first, *the tendency to progressive advancement* in organization, accompanied by greater dignity in instinct, intelligence, &c.; secondly, *the force of external circumstances*, or of variations in the physical condition of the earth, or the mutual relations of plants and animals. For as species spread themselves gradually over the globe, they are exposed from time to time to variations in climate, and to changes in the quantity and quality of their food; they meet with new plants and animals which assist or retard their development, by supplying them with nutriment, or destroying their foes. The nature also of each locality is in itself fluctuating, so that even if the relation of other animals and plants were invariable, the habits and organization of species would be modified by the influence of local revolutions.

Now, if the first of these principles, *the tendency to progressive development*, were left to exert itself with perfect freedom, it would give rise, says Lamarck, in the course of ages, to a graduated scale of being, where the most insensible transition might be traced from the simplest to the most compound structure, from the humblest to the most exalted degree of intelligence. But in consequence of the perpetual interference of the *external causes* before mentioned, this regular order is greatly interfered with, and an approximation only to such a state of things is exhibited by the animate creation, the progress of some races being retarded by unfavourable, and that of others accelerated by favourable, combinations of circumstances. Hence, all kinds of anomalies interrupt the continuity of the plan, and chasms, into which whole genera or families might be inserted, are seen to separate the nearest existing portions of the series.

Such is the machinery of the Lamarckian system; but our readers will hardly, perhaps, be able to form a perfect conception of so complicated a piece of mechanism, unless we exhibit it in motion, and shew in what manner it can work out, under the author's guidance, all the extraordinary effects which we behold in the present state of the animate creation. We have only space for exhibiting a small part of the entire process by which a complete metamorphosis is achieved, and shall, therefore, omit the mode whereby, after a countless succession of generations, a small gelatinous body is transformed into an oak or an ape. We pass on at once to the last grand step in the progressive scheme, whereby the orang-outang, having been already evolved out of a monad, is made slowly to attain the attributes and dignity of man.

One of the races of quadrumanous animals which had reached the highest state of perfection, lost, by constraint of circumstances, (concerning the exact nature of which tradition is unfortunately silent,) the habit of climbing trees, and of hanging on by grasping the boughs with their feet as with hands. The individuals of this race being obliged for a long series of generations to use their feet exclusively for walking, and ceasing to employ their hands as feet, were transformed into bimanous animals, and what before were thumbs became mere toes, no separation being required when their feet were used solely for walking. Having acquired a habit of holding themselves upright, their legs and feet assumed insensibly a conformation fitted to support them in an erect attitude, till at last these animals could no longer go on all fours without much inconvenience.

The Angola orang, *Simia troglodytes*, Linn., is the most perfect of animals, much more so than the Indian orang, *Simia Satyrus*, which has been called the orang-outang, although *both* are *very inferior* to man in corporeal powers and intelligence. These animals frequently hold themselves upright, but their organization has *not yet* been sufficiently modified to sustain them habitually in this attitude, so that the standing posture is very uneasy to them. When the Indian orang is compelled to take flight from pressing danger, he immediately falls down upon all fours, shewing clearly that this was the original position of the animal.

Even in man, whose organization, in the course of a long series of generations, has advanced so much farther, the upright posture is fatiguing and can only be supported for a limited time, and by aid of the contraction of many muscles. If the vertebral column formed the axis of the human body, and supported the head and all the other parts in equilibrium, then might the upright position be a state of repose; but as the human head does not articulate in the centre of gravity; as the chest, belly, and other parts, press almost entirely forward with their whole weight, and as the vertebral column reposes upon an oblique base, a watchful activity is required to prevent the body from falling. Children which have large heads and prominent bellies can hardly walk at the end even of two years, and their frequent tumbles indicate the natural tendency in man to resume the quadrupedal state.

Now, when so much progress had been made by the quadrumanous animals before mentioned, that they could hold themselves habitually in an erect attitude, and were accustomed to a wide range of vision, and ceased to use their jaws for fighting, and tearing, or for clipping herbs for food, their snout became gradually shorter, their incisor teeth became vertical, and the facial angle grew more open.

Among other ideas which the natural *tendency to perfection* engendered, the desire of ruling suggested itself, and this race succeeded at length in getting the better of the other animals, and made themselves masters of all those spots on the surface of the globe which best suited them. They drove out the animals which approached nearest to them in organization and intelligence, and which were in a condition to dispute with them the good things of this world, forcing them to take refuge in deserts, woods and wildernesses, where their multiplication was checked, and the progressive development of their faculties retarded, while in the mean time the dominant race spread itself in every direction, and lived in large companies where new wants were successively created, exciting them to industry, and gradually perfecting their means and faculties.

In the supremacy and increased intelligence acquired by the ruling race, we see an illustration of the natural tendency of the organic world to grow more perfect, and in their influence in repressing the advance of others, an example of one of those disturbing causes before enumerated, that *force of external circumstances*, which causes such wide chasms in the regular series of animated beings.

When the individuals of the dominant race became very numerous, their ideas greatly increased in number, and they felt the necessity of communicating them to each other, and of augmenting and varying the signs proper for the communication of ideas. Meanwhile the inferior quadrumanous animals, although most of them were gregarious, acquired no new ideas, being persecuted and restless in the deserts, and obliged to fly and conceal themselves, so that they conceived no new wants. Such ideas as they already had remained unaltered, and they could dispense with the communication of the greater part of these. To make themselves, therefore, understood by their fellows, required merely a few movements of the body or limbs – whistling, and the uttering of certain cries varied by the inflexions of the voice.

On the contrary, the individuals of the ascendant race, animated with a desire of interchanging their ideas, which became more and more numerous, were prompted to multiply the means of communication, and were no longer satisfied with mere pantomimic signs, nor even with all the possible inflexions of the voice, but made continual efforts to acquire the power of uttering articulate sounds, employing a few at first, but afterwards varying and perfecting them according to the increase of their wants. The habitual exercise of their throat, tongue and lips, insensibly modified the conformation of these organs, until they became fitted for the faculty of speech[i].

In effecting this mighty change, "the exigencies of the individuals were the sole agents, they gave rise to efforts, and the organs proper for articulating sounds were developed by their habitual employment." Hence, in this peculiar race, the origin of the admirable faculty of speech; hence also the diversity of languages, since the distance of places where the individuals composing the race established themselves, soon favoured the corruption of conventional signs[j].

[...]

CHAPTER II.

[...]

THE theory of the transmutation of species, considered in the last chapter, has met with some degree of favour from many naturalists, from their desire to dispense, as far as possible, with the repeated intervention of a First Cause, as often as geological monuments attest the successive appearance of new races of animals and plants, and the extinction of those pre-existing. But, independently of a predisposition to account, if possible, for a series of changes in the organic world, by the regular action of secondary causes, we have seen that many perplexing difficulties present themselves to one who attempts to establish the nature and the reality of the specific character. And if once there appears ground of reasonable doubt, in regard to the constancy of species, the amount of transformation which they are capable of undergoing, may seem to resolve itself into a mere question of the quantity of time assigned to the past duration of animate existence.

Before we enter upon our reasons for rejecting Lamarck's hypothesis, we shall recapitulate, in a few words, the phenomena, and the whole train of thought, by which we conceive it to have been suggested, and which have gained for this and analogous theories, both in ancient and modern times, a considerable number of votaries.

In the first place, the various groups into which plants and animals may be thrown, seem almost invariably, to a beginner, to be so natural, that he is usually convinced at first, as was Linnæus to the last, "that genera are as much founded in nature as the species which compose them[k]." When, by examining the numerous intermediate gradations, the student finds all lines of demarcation to be in most instances obliterated, even where they at first appeared most distinct, he grows more and more sceptical as to the real existence of genera, and finally regards them as mere arbitrary and artificial signs, invented like those which serve to distinguish the heavenly constellations for the convenience of classification, and having as little pretensions to reality.

Doubts are then engendered in his mind as to whether species may not also be equally unreal. The student is probably first struck with the phenomenon, that some individuals are made to deviate widely from the ordinary type by the force of peculiar circumstances, and with the still more extraordinary fact, that the newly-acquired peculiarities are faithfully transmitted to the offspring. How far, he asks, may such variations extend in the course of indefinite periods of time, and during great vicissitudes in the physical condition of the globe?

[...]

His first opinions are now fairly unsettled, and every stay at which he has caught has given way one after another; he is in danger of falling into any new and visionary doctrine which may be presented to him; for he now regards every part of the animate creation as void of stability, and in a state of continual flux. In this mood he encounters the Geologist, who relates to him how there have been endless vicissitudes in the shape and structure of organic beings in former ages – how the approach to the present system of things has been gradual – that there has been a progressive development of organization subservient to the purposes of life, from the most simple to the most complex state – that the appearance of man is the last phenomenon in a long succession of events – and, finally, that a series of physical revolutions can be traced in the inorganic world, coeval and coextensive with those of organic nature.

These views seem immediately to confirm all his preconceived doubts as to the stability of the specific character, and he thinks he can discern an inseparable connexion between a series of changes in the inanimate world, and the capability of species to be indefinitely modified by the influence of external circumstances. Henceforth his speculations know no definite bounds; he gives the rein to conjecture, and fancies that the outward form, internal structure, instinctive faculties, nay, that reason itself, may have been gradually developed from some of the simplest states of existence, – that all animals, that man himself, and the irrational beings, may have had one common origin; that all may be parts of one continuous and progressive scheme of development from the most imperfect to the more complex; in fine, he renounces his belief in the high genealogy of his species, and looks forward, as if in compensation, to the future perfectibility of man in his physical, intellectual, and moral attributes.

Notes

a Phil. Zool. tom. i. p. 54.
b Phil. Zool. tom. i. p. 62.
c Ibid.
d Phil. Zool. tom. i. p. 227.
e Phil. Zool. tom. i. p. 232.
f Phil. Zool. tom. i. p. 234.
g Phil. Zool p. 64.
h Animaux sans Vert., tom. i. p. 56, Introduction.
i Lamarck's Phil. Zool., tom. i. p. 356.
j Ibid. p. 357.
k Sir J. Smith's Introduction to Botany.

Editorial Headnote

20. Carpenter, W. B., *Principles of General and Comparative Physiology, Intended as an Introduction to the Study of Human Physiology and as a Study Guide to the Pursuit of Natural History.* London: John Churchill, 1839, pp. x, 129–39, 414–24, 463–4.

In 1872, the Unitarian physiologist William Benjamin Carpenter (1813–1885) delivered the Presidential Address to the assembled British Association for the Advancement of Science.[1] Carpenter was thus, at this time, arguably the foremost representative of the British scientific establishment. However, as the historian of science Alison Winter has shown, his rise in science was anything but inevitable and depended as much on social, political, and religious concerns as on the quality of his scientific investigations.[2] Carpenter had studied in Bristol, then at University College London, before graduating with his MD degree from Edinburgh in 1839. His *Principles of General and Comparative Physiology* followed upon his award-winning graduation thesis, but was reviewed as being a controversial and heterodox work. Winter contends that, but for his careful management of the situation, in which he successfully appealed to established men of science for testimonials, his career might have taken a very different trajectory. Carpenter's belief that the universe was created by a beneficent Creator, who had created through the operation of the physical secondary causes of natural law, fit well with contemporary developments in science, but what was controversial was that he extended this line of thinking to the science of mind. Thus, in this work, he made the claim that physiology was subject to the same overarching laws as the physical science, and that this was as true for the mind as for the body.

The suggestion of a physiological study of the mind was read as leading towards materialist and, therefore, also to atheist ideas. This reading of Carpenter's work was facilitated by the prevalence of phrenological ideas that could also be read in this way, as well as the fact that physiology was still under the shadow of the early nineteenth-century debate between the physiologist William Lawrence and the surgeon John Abernathy over the relationship between mind and body. Lawrence had argued that life was an inherent quality of the organic world and that this was as true for mental phenomena as for other aspects of physiology. In contrast, Abernathy had argued that life – and mind – were superadded to the material world.[3] Lawrence's position was thought by many critics to be monistic, reductive, materialist, and even atheistic, and Carpenter's work was initially interpreted as making a similar case.

However, Carpenter was ultimately successful in defending himself and his science from such disreputable associations. In later decades, his work in physiology became a model for practitioners, and – indicative of how much society had changed (and how successful he had been in establishing himself as a devout and orthodox scholar) – by the 1870s, he was able to publish extensively on mental physiology without negative comment.[4]

Notes

1 William Benjamin Carpenter, "Inaugural Address of Dr. William Carpenter, F.R.S., President," *Nature*, 15th August 1872, pp. 206–12.

2 Alison Winter, "The Construction of Orthodoxies and Heterodoxies in the Early Victorian Life Sciences," in Bernard Lightman (ed.), *Victorian Science in Context* (Chicago: University of Chicago Press, 1997), pp. 24–50.

3 L.S. Jacyna, "Immanence or Transcendence: Theories of Life and Organization in Britain 1790–1835," *ISIS*, (1983), 74: 311–29.

4 William Benjamin Carpenter, *Principles of Mental Physiology, with their applications to the training and discipline of the mind and the study of its morbid conditions* (New York: D. Appleton & Co, 1874).

20

PRINCIPLES OF GENERAL AND COMPARATIVE PHYSIOLOGY, INTENDED AS AN INTRODUCTION TO THE STUDY OF HUMAN PHYSIOLOGY AND AS A STUDY GUIDE TO THE PURSUIT OF NATURAL HISTORY

W. B. Carpenter

Source: W. B. Carpenter, *Principles of General and Comparative Physiology, Intended as an Introduction to the Study of Human Physiology and as a Study Guide to the Pursuit of Natural History* (London: John Churchill, 1839), pp. x, 129–39, 414–24, 463–4

020 "William Benjamin Carpenter"

Image in the public domain.

PRINCIPLES

OF

GENERAL AND COMPARATIVE PHYSIOLOGY,

INTENDED AS AN

INTRODUCTION TO THE STUDY OF HUMAN PHYSIOLOGY,

AND AS A

GUIDE TO THE PHILOSOPHICAL PURSUIT OF NATURAL HISTORY.

BY
WILLIAM B. CARPENTER,
MEMBER OF THE ROYAL COLLEGE OF SURGEONS, LONDON;
LATE PRESIDENT OF THE ROYAL MEDICAL AND ROYAL PHYSICAL SOCIETIES, AND PELLOW OF THE ROYAL BOTANICAL SOCIETY, EDINBURGH; AND LECTURER ON FORENSIC MEDICINE IN THE BRISTOL MEDICAL SCHOOL.

LONDON:
JOHN CHURCHILL, PRINCES STREET, SOHO;

MACLACHLAN AND STEWART, AND CARFRAE AND SON, EDINBURGH; FANNIN AND CO., DUBLIN; W. STRONG, AND PHILP AND EVANS, BRISTOL.

MDCCCXXXIX.

ANALYSIS OF THE CONTENTS.

N.B. The Numbers refer to the Paragraphs.

INTRODUCTION. – ON ORGANISED STRUCTURES.

I. – *Preliminary Remarks*, 1–7.

On the Nature and Objects of the Science of Physiology. – Difficulties in the way of its advancement. – True mode of pursuing it.

II. – *Of Organised Structures in General*, 8–20.

The object of this Section, taken in connection with Chapter I. (Book I), is to show the relations between Life, or Vital Action, and Organised Structure. The argument may be thus briefly stated. – The *properties* of any aggregation of Matter depend upon the mode in which its ultimate molecules are combined and arranged, and have not an existence separate from the matter itself (150). The simplicity of our notion of the properties of inorganic matter depends upon the facility with which we may become acquainted with them, through the command which we possess over the agencies by whose operation they are manifested; and their *evident uniformity* of action enables us at once to refer them to definite and comprehensive laws (4, 141).

These *Laws* simply express the *conditions of action* of the material bodies which they concern (147, 148). All Physical Phenomena result from the excitement of the *physical properties* of matter by external agents; and, when these are not in operation, no change takes place (1, 9, 10, 144). In like manner, Vital Actions result from the excitement of the *vital properties* of certain forms of matter by external agents or *stimuli;* and are not manifested, or called into play, without the influence of these (10, 144, 155). These Vital properties exist only in *organised* tissues, and stand in the same relation to them as do the Physical properties to matter in general (140). They may be regarded as essentially dependent on the peculiar state in which the component particles exist. This state can only be induced by an action of *Organisation* effected upon inorganic matter by a pre-existing structure; and the change thus operated developes properties that previously lay dormant, the material particles not being, until then, in the condition required to exhibit them (10, 151, 152). In proportion as the properties thus called into exercise differ from those common to matter in general, does the organised tissue which possesses them, exhibit peculiarities of structure and composition unlike those presented by other forms of matter (11, 19, 159). These peculiarities are such that spontaneous decomposition of the elements has a constant tendency to take place, especially in the most highly organised structures; but this is kept in check, in the living body, by the continual renovation which is characteristic of Vital Action (18, 214). So long as this takes place, the vital properties are retained; but if any considerable alteration of structure or composition occur, they are perverted or altogether destroyed (152–4). All Vital Action, therefore, is dependent on two conditions; – an organised structure possessed of properties not manifested by inorganic matter; and a stimulus by which these properties are excited to action. But many of the changes concerned in the maintenance of Life are of a strictly physical character; and it is by these that the connection is effected between the Organism and the external world (159–165).

BOOK I.

GENERAL PHYSIOLOGY.

CHAPTER I.

ON THE NATURE AND CAUSES OF VITAL ACTIONS.

140. THERE is no department of philosophy around which so much unnecessary mystery has been cast, as the investigation of the character and laws of the Phenomena of Life. And this veil of mystery will long continue to baffle the curiosity of those who have not received, either directly or indirectly, the mental training which the pursuit of Physical Science affords. To designate any of the actions of a living body, – whether its reception of food from without, its conversion of that food into the materials of its own structure, the movements of its parts upon one another or of the whole fabric, its production of other beings like itself, – as *vital*,

or as effected by the *vital principle*, has long been regarded as placing a sufficient check upon further enquiry. The history of Physical science shows, however, that it was once labouring under the same restraint, and that until the true objects of investigation were understood, scarcely any advance was made. Thus, in past ages all the phenomena of the movements of the heavenly bodies were attributed to the operation of some vague "principle of motion," the laws of which it was considered impracticable to attain. In like manner, the simple optical fact – that when the sun's light passes through a hole, the bright image, if formed at a considerable distance from it, is always round, instead of imitating the figure of the aperture, – was attributed by Aristotle to the "circular nature" of the sun's light; whilst the simple consideration, that the rays of light travel in straight lines, would, if properly applied, have explained this phenomenon, not only as regards the sun, but in the case of any other round luminous body placed at a sufficient distance. To refer all the operations of Life which cannot be explained by physical laws, to a "Vital Principle," is in reality to proceed just as unphilosophically as the ancients did in the cases just quoted: since a strict examination into their character will show that, although not *identical with* physical phenomena, they are *analogous to* them; being the results of the operation of the vital properties which are peculiar to organised matter, just as the physical changes of inorganic matter result from the operation of its mere physical properties. The characters and laws of these Vital properties are as open to investigation as those which give rise to the phenomena of Gravity, Electricity, or Magnetism; although more difficult of attainment, owing to the intricacy of the combinations in which the results are presented to our observation (§ 4).

141. Few terms have been employed in a greater variety of significations, or more frequently without any definite meaning at all, than the word *Life*. The older philosophers regarded it as a distinct entity or substance residing in certain forms of matter, and the cause both of their organisation, and of the actions exhibited by them.[a] We have seen that the tendency to rest satisfied with vague hypotheses of this kind, operated in the retardation of Physical Science; and had it not been for the comparative prominence and simplicity of its phenomena, it is not improbable that even at the present day we should hear employed in *it* such terms as the "Vital Principle," "Organic Agent," or any other equally unphilosophical refuge of those physiologists who neglect the substance to grasp at the shadow. To the term *principle* no very definite meaning can be attached. It has been remarked that "this word, characteristic of a less advanced state of science, has been generally employed (as the final letters of the alphabet are used by algebraists) to denote an unknown element, which, when thus expressed, is more conveniently analysed." Thus, it has been customary to speak of the principle of gravity, the principle of electricity, or of the principle of magnetism, as the unknown causes of certain phenomena that are as yet imperfectly comprehended. In so far, however, as the laws of these phenomena are understood, they terminate in referring all the results to simple properties of matter, from which they may be deduced by demonstrative reasoning, just as geometrical theorems from the postulates on which they are founded. Thus, the law of gravitation, – which is only an expression of the property inherent in masses of matter, of attracting others, and of being attracted

by them, – joined with those of motion, which are equally simple, explains all the movements of the heavenly bodies, in whose phenomena we may witness their operation uncontrolled by any other agencies, and of which the astronomer is thus enabled to predict the perturbations as well as the regular motions. And it is probable that all the phenomena of electricity and magnetism will, ere long, be generalised to the same extent, – that is to say, will be reduced to an expression of equal simplicity; and that a still higher generalisation will then include all those now alluded to. It has been in the Science of Life that the term *principle* has been most used, and most abused. It must be admitted that the conditions of vital phenomena are not yet determined with sufficient precision, to enable us to refer all observed facts, through the medium of general laws, to simple vital properties; and there might be no peculiar objection to the use of the term "Vital principle," as a convenient expression for the sum of the unknown powers which are developed by the action of these properties. But care must be taken not to rest satisfied in its use.

142. The terms *Vital Principle*, and *Life*, are commonly employed almost synonymously, to imply the controlling agent by which the phenomena of living beings are directed, if not immediately produced. Thus, it is frequently said that the action of Physical and Chemical laws is modified or entirely checked by the living principle. Now if we come to analyse this expression, we shall find it to mean one of two things; – either that the living principle is a distinct intelligent agent, capable of harmonising all the actions of Physics and Chemistry, and of rendering them subservient to its government, employing them in fact as subordinate agents to execute its mandates; – or that the actions in question result from the mixed operation of those properties which organised structures possess in common with inorganic matter, and of those which are peculiar to the former. In the first sense it means *every thing*, for to the Vital principle all meaner agents must then acknowledge their subordination; in the latter case, it means *nothing* more than would be better expressed in other language, free from cavil or misapprehension.

143. The doctrine of a Vital Principle is not only quite unnecessary to explain facts, but is totally unsupported by the analogies of Nature, and by what we know of the Divine Government in general. No reflecting mind has any doubt that this earth and its inhabitants form a system, of which every part is perfectly adapted to the rest, (so that we might almost call it an *organised* one, if the idea of a particular structure were not involved in the term,) and of which all the actions and changes, however in appearance contrary, have one common tendency, the ultimate happiness of the creatures of Infinite Benevolence. It cannot be regarded as an improbability that the other spheres and systems, – whose countless multitudes, revealed by the aid of science, impress our minds with the nearest conception of infinity of which our finite comprehension is capable, – are peopled with beings, if not similar in structure with ourselves, at least equally worthy of the Creator's care. In the government of our own planet, itself but a point in the vast universe, we are able to recognise, to a small extent, the laws by which its physical changes are guided; and we discern faint glimmerings of those by which the

moral condition of sentient beings is controlled. So far as we can understand the mutual adaptation of these laws, we everywhere see them working to the same end; and we entertain the highest anticipations of that beauty and harmony which will be revealed to us, when our imperfect glimmerings of knowledge shall be extended and corrected by the light of Eternal Truth. Should we not consider it degrading to the dignity of Infinite Wisdom to suppose that at the creation of each world, He had found it necessary to delegate to a subordinate the control over its working, – instead of at once impressing upon its elements those simple properties, from whose mutual actions, foreseen and provided for in the laws according to which they operate, all the varieties of change which it was His intention to produce, should necessarily result?

144. The harmony of means and ends is shown in the structure of the universe, not less than in the adaptation of the parts of a single organised being to one another. And if the actions of the former can be reduced to simple and general laws, which are but expressions of the Divine Will, there is nothing absurd or unphilosophical, or derogatory to the dignity of living beings, in the belief that those of the latter may be ultimately placed on the same footing. For if we come to enquire into the function of any single organ, or, in other words, into the nature of the changes produced by it, we find that it may be referred to the property of the structure, manifested or called into action by a *stimulus* of some kind, to which it is expressly fitted to respond. This is evidently the case even in the inorganic world. The process of evaporation, for example, will not take place when fluid is exposed to an atmosphere already saturated with moisture; since one of the conditions of the action is a dry air capable of dissolving watery vapour, or, still better, a vacuum into which it may freely rise. In the same manner, the electrical properties of matter are not manifested by one mass alone; to exhibit electric or magnetic attractions and repulsions, two substances are necessary. In machines constructed to take advantage of the physical properties of matter, and to bring them into useful operation, a *stimulus* to their action is required, in some means which shall develope these *properties*, and thus create *powers*. Thus, the power of gravitation is called into exercise, when the clock-weight is wound up; that which results from elasticity, when the main-spring of a watch is coiled within the barrel; that of the expansibility of gases generated by combustion, when a cannon-ball is propelled by the ignition of gunpowder. In the Steam-engine we have a still closer parallel with the mechanism of organised structures, since this apparatus consists of a *number of parts*, having functions which are totally distinct in themselves, and yet all tending to a common purpose. In the construction of the steam-engine, (at least in the usual forms of it), advantage is taken of *two* of the properties of water, and means are provided to bring these into operation *against* one another, yet still with a common end. Thus, heat is applied to the boiler to generate steam, the expansion of which is employed as one motive force; whilst cold is applied to the condenser, to produce a vacuum by the condensation of that steam, against which vacuum the expansive force may act with greater advantage than against the elastic air. Heat and cold, then, properly applied, may be regarded as the two *stimuli* which are essential to excite the machine to action; and without

these, however perfect might be its structure, however beautiful and harmonious the adaptation of its parts, it would for ever remain dormant.

145. It must not be imagined for a moment, however, that any intention is here implied of identifying the structure and actions of an organised being, with those of a steam-engine or any other such piece of human mechanism. The latter is framed to take advantage of the properties with which the Creator has endowed all forms of matter, inanimate as well as animate. But the actions of a living being incontestably show that beside these properties, there are others which are exclusively confined to organised tissues; and if we compare these with the general structure of the fabric, we shall find that its mechanism is adapted to bring them into the most advantageous relation with one another. Thus, the circulating system is a piece of apparatus which acts chiefly upon physical principles, the fluid impelled by the heart moving through its ramifying canals, just as water ejected from a forcing pump might traverse elastic tubes of similar construction; and its object is to bring the alimentary materials which have been absorbed, into contact with the tissues whose nutrition they are to supply. The powers which move the blood may altogether result from vital operations; yet the motion itself is strictly conformable to physical laws. Different Schools of Physiologists have erred in opposite extremes, with regard to the agency of these laws, or, in other words, of the physical properties of matter, in the production of vital phenomena; some attributing all the actions of living beings to the immediate operation of the vital properties of their structure; others maintaining that they are of a purely physical nature. The truth appears, with regard to this, in common with so many disputed questions, to lie in the mean between the opposing extremes; and it will be the object of much of the present work to show where the boundary line may be most naturally drawn.

146. If the application of the term *Life* to some imaginary agent which is the immediate cause of vital phenomena, be found useless or injurious, it may reasonably be enquired what is to be understood by it. If we regard as a *living* being, an organised structure which we observe growing and moving and resisting decay, it is evidently no improper use of the term to designate by it the *sum of all the actions* performed by such a being, from its first production to its final dissolution. Observation of these actions leads us to arrange them, as has been already stated (§ 6), into certain groups termed Functions; and analysis of the functional changes exhibited by living beings, terminates in referring them all to certain properties possessed by their component structures; which properties stand in the same relation to organised tissues, as do those of gravitation, electricity, &c., to matter in general. Their existence must, for the present at least, be regarded as ultimate facts in physiology. They are called into action by *stimuli* of various kinds, adapted to excite each of them to its own peculiar operations. Now, although the adaptation of the various functions to one another, and the manifest tendency of all the vital processes to a common end, would appear, at first sight, to favour the idea of a presiding power by which the whole is regulated, a little consideration will show that these *really* imply no more than an original adaptation of the structure and properties of the organs by which they are performed. Every tissue has its own laws

of development, (some of which have been glanced at in the Introduction, III – V); but all these laws are subservient to one general principle, – that every organised structure is produced by a previously-existing organism, no living being ever taking its origin from spontaneous combinations of inorganic matter (§ 516). Our enquiry leads us back, therefore, to the first creation of each species; and here we may again revert to the character of Physical laws, as illustrating the more obscure nature of those of Vitality.

147. The term *Law* of Nature, as already employed, expresses the *conditions of action* of the properties of matter. The Divine Creator of the universe "has, by creating his materials, endued with certain fixed qualities and powers, impressed them in their origin with the *spirit* not the *letter* of his law, and made all their subsequent combinations and relations inevitable consequences of this first impression."[b] In other words, the unchangeableness of His nature is manifested by his continued action in the material creation, according to the same plan by which He at first adjusted the relations of its parts. Our belief in the uniformity of Nature, which leads us to seek for a common cause when a number of similar phenomena are presented to our observation, is based, not only upon experience, but upon the conviction which every believer in the existence of the Deity feels of His immutability. If it were otherwise, we should be led by *analogy* only to infer the existence of law and order where none is evident; but the mind which is once satisfied of the existence of a Creator, possesses a moral *certainty* that to him must belong a Consummate Wisdom, which shall contrive the attainment of every end by the best adapted means, – an Omnipotence, which shall have all these means at full command, – and an Omniscience, which shall foresee in every action not only its immediate but its remotest consequences. To imagine, therefore, that the plan of the Universe, once established with a definite end, could require alteration during the continuance of its existence, is at once to deny the perfection of the Divine attributes; whilst, on the other hand, to suppose, as some have done, that the properties first impressed upon matter would *of themselves* continue its actions, is to deny all that revelation teaches us regarding our continued dependence on the Creator. Let it be borne in mind, then, that when a *law* of Physics or of Vitality is mentioned, nothing more is really implied than a simple expression of the *mode* in which the Creator is *constantly* operating on inorganic matter, or on organised structures.

[...]

CHAPTER XIV.

SUBORDINATE LAWS REGULATING THE EXERCISE OF THE REPRODUCTIVE FUNCTION. – DISTINCTION OF SPECIES. – PROPAGATION OF SPONTANEOUS OR ACQUIRED PECULIARITIES.

542. When we contemplate the immense number of diversified forms which the study of the organised creation brings under our notice, and witness these forms

perpetuated, as it would seem, by the process of reproduction, so as to constitute distinct races, the question naturally arises whether all these had a different origin; or whether the characters of any of them have been so modified in the course of time, as to lead to the belief in the diversity of origin of those which were at first really identical. When it can be *shown* that two races have had a separate origin, they are regarded as of *different species;* and, in the absence of proof, this is *inferred* when we see some peculiarity of organisation, characteristic of each, so constantly transmitted from parent to offspring, that the one cannot be supposed to have lost, or the other to have acquired it, through any known operation of physical causes. It cannot be regarded as an unimportant question to the *naturalist* to ascertain what these constant distinctions are; whilst it is an investigation of high interest in a *physiological* point of view, to trace the modifying influence of external circumstances upon the structure and functions of living beings, and to enquire how far the result of such influences may be transmitted hereditarily, so that the difference produced by them may be perpetuated. Where races which have originally sprung from a common stock present marked differences, they are spoken of as *varieties;* and the variety may be *transient*, from its peculiarity manifesting a tendency to disappear, – or *permanent*, where it continues to be transmitted without change. The uncertainty of the limits of species is daily becoming more and more evident; and every naturalist is aware that a very large number of races are usually considered as having a distinct origin, when they are nothing more than permanent varieties of a common stock. Whilst the exertions of the enterprising discoverer are adding to our already enormous list of species, from the unexplored resources of foreign lands, the skill of the horticulturist and of the breeder is exerted to produce new varieties of species already in our catalogues: and it has unfortunately too often happened, that a new specific name has been invented for the latter as well as for the former; and that a mere hybrid or transient variety has thus taken the rank of a species, to the confusion of all true principles of arrangement. The philosophic naturalist, on the other hand, aims to reduce the number of species by investigating the degree of variation which each is liable to undergo, the forms it assumes at different periods of its existence, the permanent characters by which it may be distinguished during its whole life, the habits which are natural to it, the degree in which these may be changed by the influence of circumstances; – and, in fine, he endeavours to become acquainted with the *whole* natural history of a reputed species, before separating it from another to which it may be closely allied.

543. Many examples may be given of the success with which this mode of investigation is now being prosecuted. The belief which is gaining ground that many diversified forms of the simpler Cryptogamia may arise from similar germs developed under different circumstances, has already been noticed (§ 65–8). The same may, perhaps, be surmised without improbability of the Infusorial Animalcules; and with respect to these, patient observation has already done much in reducing the number of species amongst the forms previously known (whilst the improved powers of the microscope have revealed many new ones), by showing

that the same individual may present very diversified appearances at different times, owing to the variable distention of its digestive cavities, and the changes which it undergoes in the process of *fissiparous* reproduction (§ 528). Among the higher plants, the experiments of Mr. Herbert on the primrose, cowslip, oxslip, and polyanthus (which he proves to be all varieties of one species), are sufficient evidence of the important results which would probably accrue from a similar investigation in other quarters. In Zoology, again, the very interesting paper of Mr. Gray[c] may be referred to, as proving the great influence of external circumstances in modifying the form of shells; it is there shown, among other instances, that what have been regarded as *six* distinct species of *Murex* (§ 100) are in reality but different states of *one;* and Mr. Stutchbury has been equally successful in reducing the number of species of Patella, Cypræa, and Oliva, by attending to the changes of form which each individual undergoes in the progress of its development. Many instances might be related in proof of the uncertainty of reputed specific distinctions among higher classes. Insects have been seen presenting the characters of different species on the two sides of the body; and it is now certain that an erroneous multiplication of species among Birds, especially in the migrating tribes, has been occasioned by their change of plumage at different seasons. And finally, to return to the Vegetable kingdom, the uncertainty of all principles of arrangement founded upon arbitrary characters has been demonstrated by the fact recently published,[d] that the flowers and pseudo-bulbs of three distinct *genera* of Orchideous plants have been produced by the same individual.[e]

544. It has been formerly stated (§ 72) that the Naturalist endeavours to simplify the acquirement and pursuit of his science, by the adoption of easily-recognised external characters as the basis of his classification; but these can only be safely employed, when indicative of peculiarities in internal structure which are found to be little subject to variation, and which are not liable to be affected by the influence of physical causes. The colour of flowers, for example, is liable to so much alteration from the influence of soil and climate, that it is seldom regarded as of itself any test of the unity or diversity of species: in moths and butterflies, on the other hand, the uniform appearance of particular spots on the wings is held sufficient to constitute a specific character, because it is never known to vary; and it would probably be found associated, if the examination were pushed far enough, with some unequivocal differences in the configuration of internal organs. Amidst all these difficulties attending the discrimination of species from structural characters alone, it is not unreasonable to enquire if there are any other means of effecting the object with greater certainty. This subject has been fully considered by Dr. Prichard in his elaborate work on the Physical History of Man; all that can be here considered are the laws regulating the intermixture of species, and the propagation of hereditary or acquired peculiarities.

545. The conclusions which have now been attained on the first of these points, and which (if stated in a sufficiently general form) are equally applicable to both the Animal and Vegetable kingdoms, may be regarded as one of the most valuable tests which naturalists possess. In plants, the stigma of the flower of one species

may be fertilised with the pollen of an allied species; and, from the seeds produced, may be raised plants of an intermediate character. But these *hybrid* plants will not long continue the race; for, although they may ripen their seed for one or two generations, they will not continue to reproduce themselves beyond the third or fourth. But, if the intervention of one of the parent species be used, its stigma being fertilised with the pollen of the hybrid, or *vice versâ*, a mixed race may be kept up for some time longer; but it will then have a manifest tendency to return to the form of the parent whose intervention has been employed. Where, on the other hand, the parents were themselves only varieties, the hybrid is only another variety, and its powers of reproduction are rather increased than diminished; so that it may continue to propagate its own race, or may be used for the production of other varieties, almost *ad infinitum*. In this way many beautiful new varieties of garden flowers have been obtained, especially among such species as have a natural tendency to change their aspect.[f] Amongst animals, the limits of hybridity are more narrow, since the hybrid is totally unable to continue its race with one of its own kind; and although it may be fertile with one of its parent species, the progeny will of course be nearer in character to the pure blood, and the race will ultimately merge into it.[g] In animals, as among plants, the mixed offsprings originating from different races within the limits of the same species, generally exceed in vigour, and in the tendency to multiply, the parent races from which they are produced, so as often to gain ground upon the older varieties, and gradually to supersede them. Thus, the mixture of the European races with the Hindoo and South American has produced tribes of such superior characters of body, and of such rapid tendency to multiplication, that there is reason to believe that they will ultimately become the dominant powers in the community.[h] The general principle, then, is that beings of distinct species, or descendants from stocks originally different, cannot produce a mixed race which shall possess the capability of continuing itself; whilst the union of varieties has a tendency to produce a race superior in energy and fertility to its parents.

546. In examining into the characters of the different species of Plants and Animals with which different regions on the earth's surface are peopled, the naturalist soon becomes aware that there are many kinds which are restricted to particular localities, whilst others are diffused extensively or even universally over the globe; – that there are some spots (especially insular ones), of which the aboriginal inhabitants are almost entirely different from those elsewhere found; – and yet that amongst these there will always be found species holding the same rank with regard to the remainder, and thus representing each other in different countries. Thus, the species of plants and animals originally inhabiting the eastern and western hemispheres were probably almost entirely different, until the agency of man changed their geographical distribution; and almost the same may be said of the species north and south of the Equator. On the other hand *man*, and his constant attendants the dog and the fly, exist in every quarter of the globe. Again, we find in New Holland no quadrupeds which do not belong to the order *Marsupialia* or *Monotremata* (§ 75), with the exception of a dog which is believed to have been

introduced by man, and to have run wild; and none of these species are found elsewhere. The greater part of the plants also belong to new genera; and those included in the genera already known constitute distinct species, – with scarcely any exception but among the Cryptogamia, the distribution of which seems more extended than that of flowering plants. The Flora of insular situations, if at a great distance from land, contains very few species which occur elsewhere. Thus, among the flowering plants of St. Helena, which is so far removed even from the western shores of Africa, there have been found, out of 61 native species, only *two* or *three* which exist in any other part of the globe. From these and many similar facts it appears fair to conclude, that every species of plant and animal had originally a distinct locality, from which it has been dispersed, according to the capabilities possessed by its structure of adapting itself to changes in its external conditions, its own locomotive powers, and the degree in which it is subject to external agencies. "What is a rare plant," says Decandolle, "but one which is so organised that it can only live in a particular locality, and which perishes in all others; such a plant is incapable of assuming different forms. What, on the other hand, is a common plant? It is one robust enough to exist in very different localities, and under very different circumstances, and which will therefore put on many different forms."[i] Plants, then, are liable to run into varieties in proportion as they are more robust, more common, or more cultivated; and some native species are, from this cause, domesticated with greater difficulty than many exotics. Precisely the same may be said of animals; those which have the power of adaptation to differences of temperature, food, &c. are most universally diffused; while those which can only exist within narrower limits of variation are restricted to the neighbourhood of their original locality.

547. It becomes a most interesting question, then, to determine what are the changes which may be produced by the influence of external circumstances, and how far these are hereditarily transmissible. On this subject, a few facts may be stated which will give an insight into the nature of the enquiry; but it is one which deserves more attention than it has yet received, since it is not only essential to the correctness of all Natural-history classifications, but is connected with some of the highest questions in Physiological science. One of the most obvious distinctions, where it is well marked, is that of size; and yet a little examination will show that it is one most open to fallacy. Thus, a plant only a few inches high in a poor dry soil, may become much larger in a damp rich one; and this is a very common effect of cultivation. On the other hand, by *starvation* naturally or artificially induced, plants may be dwarfed, or reduced in stature: thus, the *Dahlia* has been diminished from six feet to two; the Spruce Fir, from a lofty timber tree to a pigmy bush; and many of the trees of plains become more and more dwarf as they ascend mountains, till at length they exist as mere underwood. That a similar influence will be productive (within narrower limits, however,) of corresponding effects in the animal kingdom, no one can be ignorant; and a very curious illustration is given by Mr. Gray of the effect of external conditions upon the size of Mollusca, in the fact that there is so much difference of size between individuals of *Bulimus*

rosaceus on the coast and on the mountains of Chili, that the latter have been described as distinct species. He also mentions that the *Littorina petræa* found on the sea side of Plymouth Breakwater acquires, from its superior exposure to light and heat, and probably also from the greater supply of nutriment which it obtains, twice the size which is common to individuals living on the north side within the harbour.[j] It is interesting to remark that these great variations occur in animals which, from their fixed condition, and the preponderance of their nutritive system, have most alliance with the vegetable kingdom; and it seems probable that a diminution of the vital stimuli, which in them only reduces the growth, would be fatal to other tribes whose animal powers are more active, and which have therefore greater means of suiting their external conditions to their bodily constitution.

548. Other modifications in the form and relative size of individual parts are very common in Vegetables, where the tissues are so simple, and the different organs so much alike in elementary constitution. Thus, cultivation often converts a single flower into a double one, by the metamorphosis of its stamens into petals, or by the development of a row of petals previously abortive, or by the change of the small tubular florets of a composite flower (like those composing the *disk* or *eye* of the Dahlia) into flat expanded florets which constitute the ray. Cultivation has a similar effect in obliterating the spines, prickles, and thorns, from the surface of many plants; a change which was fancifully, but not improperly, termed by Linnæus "the taming of wild fruits." The instances of such alterations effected by external agency in the vegetable kingdom, are almost innumerable; but it is very difficult to say how far the varieties thus created may become permanent by their hereditary transmission. The usual principle is, that propagation by seeds will only reproduce the *species*, the *race* not being continued with any certainty. In most plants which have been much altered by cultivation – such as the Apple, the Cabbage, or the Dahlia – the seeds, if dropped on a poor soil, will produce plants which approximate to the original type of the species; whilst from the seeds of the *Cerealia* (corn-grains), which are believed to have been originally grasses of some very different aspect, no other forms are ever produced which might assist in the solution of the curious problem of their origin. It is not improbable that, as among animals, varieties which arise from some peculiarity in the constitution of the being itself, are more liable to be reproduced in the offspring, than those which are simply the result of external agencies. It is evident, at least, that here also the capability of undergoing such modifications is that which renders the species most truly valuable to man.

549. Amongst animals, the various breeds of domestic cattle, of the horse, dog, &c. afford abundant evidence of the modifying influence of external conditions; since there is no doubt that they have originated from single stocks, and that their peculiarities have been engrafted, as it were, upon their specific characters. Between the Shetland pony and the Arabian racer, for example, or between the Newfoundland dog and the Italian greyhound, there would seem much greater difference than between the Lion and Tiger (the sculls of which are so much alike that even Cuvier was not always able to distinguish them), or between various other species of the Feline tribe, which, from the incapability of domestication, have not been exposed

to such influences. That these domesticated races, however different their external characters, have a common origin, is proved by the fact that, whenever they return to a state of nature, – as is the case with the dogs introduced by the Spaniards into Cuba, and the horses and wild cattle which now overspread the plains of South America, – the differences of breed disappear, and a common form is possessed by all the individuals. It is not a little curious, too, that instincts which must have remained dormant for many generations during the domesticated condition of the race, should re-appear when this change takes place in its habits; thus, among the wild horses of South America there is the same tendency to associate in herds under the protection of a leader, as among those of Asia whose ancestors have never been reduced to subjection. "It seems reasonable to conclude," as Mr. Lyell has justly remarked, "that the power bestowed on the horse, the dog, the ox, the sheep, the cat, and many species of domestic fowls, of supporting almost every climate, was given expressly to enable them to follow man throughout all parts of the globe, in order that he may obtain their services and they our protection." "Unless some animals had manifested in a wild state an aptitude to second the efforts of man, their domestication would never have been attempted. If they had all resembled the wolf, the fox, and the hyæna, the patience of the experimentalist would have been exhausted by innumerable failures before he at last succeeded in obtaining some imperfect results; so, if the first advantages derived from the cultivation of plants, had been elicited by as tedious and costly a process as that by which we now make some slight additional improvement in certain races, we should have remained to this day in ignorance of the greater number of their useful qualities."

550. How all the varieties of breeds have been produced which are now so striking, is a question much more easily asked than replied to satisfactorily. That peculiarities of structure sometimes arise independently of external agencies can scarcely be doubted; thus, it is by no means uncommon to find individuals of the human species with six fingers and six toes; and such peculiarities are more likely to be continued hereditarily than those which have been acquired. Sometimes advantage has been taken by man of accidental varieties of this kind, for some purpose useful to him, and he has exerted his skill to perpetuate them. The following example is of comparatively recent occurrence. In the year 1791 one of the ewes on the farm of Seth Wright in the state of Massachusets, produced a male lamb, which, from the singular length of its body and the shortness of its legs, received the name of the *otter* breed. This physical conformation, incapacitating the animal from leaping fences, appeared to the farmers around so desirable that they wished it continued. Wright determined on breeding from this ram, and the first year obtained only two with the same peculiarities. The following years he obtained greater numbers; and, when they became capable of breeding with one another, a new and strongly-marked variety, before unknown to the world, was established.[k] This shows the influence which the circumstance of a scanty population may have formerly had in the production of varieties, both in the human and other species. At the present time, any peculiarity which may occasionally arise speedily merges by intermixture, and returns to the common standard; but it may be imagined that, in the older ages of the world,

some race in which a peculiarity existed, may have been so far separated from the rest as to necessitate frequent union among its members, so that the character would be rendered still more marked instead of disappearing; and, being propagated for a few generations, would be rendered permanent. Acquired peculiarities, on the other hand, are seldom reproduced in the offspring, unless they have a relation with the natural habits and physical wants of the species; but, when this relation exists, they may be transmitted as regularly as the specific character. Thus, in dogs, the relative perfection of the organs of sight and smell, perhaps also of hearing, varies much in different breeds, and their mode of hunting their prey undergoes a corresponding change; but in these cases no new instinct is developed, the difference merely consisting in the relative proportion of those already existing; and the new peculiarities have an intimate relation to the habits of the animal in a wild state.[1] It is impossible not to recognise in many acquired habits, however, something more than a relation to the instincts necessary for the preservation of the species; they evidently arise, in part at least, from the connection of the race with man. This is more particularly exemplified in the instance of the breed of shepherds' dogs, which often display an extraordinary hereditary sagacity respecting their peculiar vocation; as well as in cases which have been frequently mentioned, where the descendants of dogs to which peculiar tricks have been taught, have displayed an unusual aptitude for learning the same. It may then be considered that the capability of undergoing such modifications, is a part of the psychical as well as structural character of the dog, even in a wild state; and that his relation to man may have as important an influence on *his* hereditary propensities, as the supply of their physical wants has on animals of other species. The same may, perhaps, be said of the horse, in the races of which we find peculiar habits transmitted from parent to offspring, which are the pure results of human instruction. It is from the want of this relation towards either the natural habits of the species or their subserviency to man, that habits acquired by other animals do not become hereditary. Thus, pigs have been taught to hunt and point game with great activity and steadiness, and other learned individuals of the same species have been taught to spell; but these acquirements have in no instance been transmitted to the offspring, not being the result of the development or modification of any instinctive propensity naturally existing. In like manner, however artificially the forms of domesticated animals may have been altered in all the individuals of successive generations, the usual character of the species and variety is maintained in each one of the offspring; unless, as sometimes happens, this alteration happens to coincide with natural varieties of the species. Thus, instances are on record in which dogs, that have been deprived of their tails by accident or design, have produced puppies with a similar deficiency; but as breeds of tail-less dogs have spontaneously arisen, there would be a stronger tendency to the perpetuation of the acquired peculiarity, than when no such peculiarity naturally occurred. It has also been asserted, however, that cats deprived of their tails will often produce one or two tail-less kittens at each birth; and that a cat which had its tail distorted by accident, has been known to transmit the deformity to some of its offspring. These are certainly exceptions to the general rule, but must not be left out of view; there can be no doubt that much has yet to be learned of the influence of the state of the parent upon the development of the offspring; and that, though credulity and the love of the marvellous have been the occasion of many strange fictions being transmitted to us, we are by no means justified in rejecting the doctrine without further enquiry.[m]

551. Any one who takes an extensive survey of the psychical as well as corporeal peculiarities of the human race must discover that both are susceptible of a higher degree of education than those of any other tribe of animals; and it is in consequence of this, that man has surmounted the obstacles interposed by his naked and defenceless condition, and found the means of existence in every part of the globe. And in general it may be observed, that the more difficulties are presented by circumstances to the supply of his instinctive wants, the more are his intellects called into exercise for their gratification. Thus, the conditions of civilised life are more calculated to excite the dormant energies of his mind, than the pastoral habits of the Nomade tribes, scarcely now advanced beyond patriarchal simplicity, or the easily satisfied wants of the Indian hunter or the Polynesian fisherman. If, again, this power of self adaptation had been confined to the *mind* of man, whilst his body continued unable to resist changes in its external conditions, or to perform those actions which his new circumstances might require, his race must as necessarily have ceased long ago to exist, except in spots peculiarly favoured by Nature, as if, with his present organisation, he had been made dependent upon those mere instincts, which are just capable of maintaining his life when supplied by the ministration of others. The educability of man's bodily frame is in fact scarcely less remarkable than that of his psychical powers. Although each of his organs of sensation is naturally inferior in acuteness to the corresponding organ of some other animal, it may be rendered by constant practice so far superior to the usual standard as to convey a degree of information greater than that which brutes can attain. Thus, the experienced seaman announces with confidence the proximity of land, or the aspect and direction of a vessel, which the ordinary voyager cannot discern; and the watchful ear of the North American Indian distinguishes the tread of friends or foes when his civilised companion is unconscious of their neighbourhood. That these acquired powers are sometimes propagated as hereditary instincts, seems probable when we remember that, among some savage nations of North America and New Holland, precisely the same notion of *direction* is manifested, as is evinced, in a degree scarcely more remarkable, by the lower animals; individuals frequently traversing pathless forests for the first time without swerving in the least from the direct line towards the point at which they are aiming. No one, who has sufficient opportunity of observation, can doubt that the intellectual faculties which have been developed by cultivation, are generally transmitted to the offspring in an improved state; so that the descendant of a line of educated ancestors will probably have a much higher capacity for instruction than the child that springs from an illiterate race.

[...]

600. It has been one object of the foregoing pages to show that *vital* properties are as essentially connected with certain forms of matter, as are those usually denominatd *physical* with matter under its more common aspects. One more question yet remains. It is possible that the physical and vital properties of matter, which are at present our ultimate facts or axioms, may be included within a more general

expression common to both? On this subject we can only speculate; but the probability appears decidedly in the affirmative. It has already been remarked that the rapid progress of generalisation in the physical sciences renders it probable that, ere long, a single formula shall comprehend all the phenomena of the inorganic world (§ 141); and it is not, perhaps, too much to hope for a corresponding simplification in the laws of the organised creation, although its progress is necessarily retarded by the many obstacles which the nature of the subject presents to the philosophic enquirer. Every step which we take in the progress of generalisation, increases our admiration of the beauty of the adaptation, and the harmony of the action, of the laws we discover; and it is in this beauty and harmony that the contemplative mind delights to recognise the wisdom and beneficence of the Divine Author of the Universe. This, in fact, is one of the highest results to which the exercise of our intellectual faculties should lead; and we cannot but believe that the Creator, in endowing us with these faculties, intended that they should conduct us nearer to the conception of his Infinite mind. But, at the same time, the vastness of the prospect thus disclosed can scarcely fail to impress us with the most humbling consciousness of our own insignificance.

601. If, then, we can conceive that the same Almighty *fiat* which created matter out of nothing, impressed upon it one simple law which should regulate the association of its masses into systems of almost illimitable extent, controlling their movements, fixing the times of the commencement and cessation of each world, and balancing against each other the perturbing influences to which its own actions give rise, – should be the cause, not only of the general uniformity, but of the particular variety of their conditions, governing the changes in the form and structure of each individual globe protracted through an existence of countless centuries, and adjusting the alternation of "seasons and times, and months and years," – should people all these worlds with living beings of endless diversity of nature, providing for their support, their happiness, their mutual reliance, ordaining their constant decay and succession, not merely as individuals but as races, and adapting them in every minute particular to the conditions of their dwelling, – and should harmonise and blend together all the innumerable multitude of these actions, making their very perturbations sources of new powers; – when our knowledge is sufficiently advanced to comprehend these things, then shall we be led to a far higher and nobler conception of the Divine Mind than we have at present the means of forming. But, even then, how infinitely short of the reality will be any view that our limited comprehension can attain, seeing, as we ever must in this life, "as through a glass, darkly;" – how much will remain to be revealed to us in that glorious future, when the Light of Truth shall burst upon us in unclouded lustre, but when our mortal vision shall be purified and strengthened so as to sustain its dazzling brilliancy.

Notes

a Every sect had its own notion of the origin and nature of this entity; some regarding it as a kind of fire; others as a kind of air, or ether, or spirit; and others, again, as merely a kind of water. The fable of Prometheus embodies this doctrine in a mythological form, the artist being described as vivifying his clay statues by Fire stolen from the chariot of the Sun. And whatever was the idea entertained as to the character of this agent, all regarded it as universally pervading the World, and as actuating all its operations in the capacity of a Life or Soul, whilst a special division of it – a *divinæ particula auræ* – regulated all the concerns of each individual organism.

b Herschell's Preliminary Discourse, p. 37.

c Philos. Transactions, 1833.

d Linn. Trans. vol. xvii.

e This fact has also come under the author's own notice in the Durdham Down Nursery, near Bristol, two of the genera being the same as in the instance just quoted, but the third a different one, so that *four* may thus be regarded as of the same species.

f There are many instances in which foreign plants have been introduced into this country, and have received different specific names, but have been found capable of producing fertile hybrids in these cases a more accurate examination of the original locality has generally shown that the parents were nothing more than permanent varieties, or even hybrids naturally occurring between other varieties. This is particularly the case with many of the South American genera, such as that elegant garden flower, the Calceolaria; and this is probably the explanation of the almost indefinite number of splendid varieties, well known to horticulturists, which may be obtained from the South American Amaryllis.

g One or two instances have been mentioned in which a mule has, from union with a similar animal, produced offspring; but this is certainly the extreme limit, since no one has ever maintained that the race can be continued further than one generation, without admixture with one of the parent species.

h Several additional instances of this kind are related in Dr. Prichard's work, vol. i. p. 147 and in Mr. Combe's Constitution of Man, chapter v.

i Library of Useful Knowledge. Botany, p. 138.

j Gray, in Philos. Trans., 1833, p. 786.

k Phil. Trans. 1813.

l In a mongrel race of dogs employed by the inhabitants of the banks of the Magdalena almost exclusively in hunting the white-lipped Pecari, a peculiar instinct appears to have become hereditary, like that of the pointers and other dogs of this country. The address of these dogs consists in restraining their ardour, and attaching themselves to no animal in particular, but keeping the whole herd in check. Now among these dogs some are found which, the very first time they are taken to the woods, are acquainted with this mode of attack; whereas, a dog of another breed starts forward at once, is surrounded by the Pecari, and, whatever may be his strength, is destroyed in a moment. Mr. Lyell mentions that some Englishmen engaged in conducting the operations of the Real del Monte company in Mexico, carried out with them some greyhounds of the best breed, to hunt the hares which abound in that country. The great platform which is the scene of sport, is at an elevation of about 9000 feet above the level of the sea, and the mercury in the barometer stands habitually at the height of about 19 inches. It was found that the greyhounds could not support the fatigues of a long chase in this attenuated atmosphere; and before they could come up with their prey they lay down gasping for breath: but these same animals have produced whelps which have grown up, and are not in the least degree incommoded by the want of density in the air, but run down the hares with as much ease as the fleetest of their race in this country. Some curious instances of a similar propagation of acquired peculiarities connected with the natural habits of the race are given us by Mr. Knight, Phil. Trans. 1837.

m Montgomery on the Signs of Pregnancy, p. 16.

Editorial Headnote

21. "Carpenter's Principles of General and Comparative Physiology" 1840 *Edinburgh Medical and Surgical Journal* 53: 213–28.

In 1872, the Unitarian physiologist William Benjamin Carpenter (1813–1885) delivered the Presidential Address to the assembled British Association for the Advancement of Science.[1] Carpenter was thus at this time arguably the foremost representative of the British scientific establishment. He was well respected as a scientific practitioner and as a devout man of faith.[2] However, as the historian Alison Winter has pointed out, there was nothing inevitable about this outcome for Carpenter's career. Indeed, the reception of Carpenter's first published work, his *Principles of General and Comparative Anatomy* (1839), demonstrates just how different things might have been. In that work, Carpenter had made the claim that physiology should be recognized as subject to the same overarching laws as the physical sciences. In the context of the late 1830s and early 1840s, however, this was deeply controversial.[3] The science of physiology was still being read in light of the deep controversy that had followed the debate between the physiologist William Lawrence and the surgeon John Abernathy. The former had argued that life was an inherent quality of the organic world, and that this was as true for mental phenomena as for other aspects of physiology. In contrast, Abernathy had argued that life – and mind – were superadded to the material world.[4] Lawrence's position was thought by many critics to be monistic, reductive, materialist, and even atheistic, and Carpenter's work was initially interpreted in just this way. Indeed, there was perhaps good reason for this association. Carpenter had studied in Edinburgh in the 1830s,[5] and was also a supporter and publicist of the views of the French transmutationist and transcendental anatomist Étienne Geoffroy Saint-Hilaire.[6] Even if this had not been the case, though, the very notion of a physiology of mind would lead conservative commentators to suspect this kind of sympathy.[7]

Indeed, this is precisely the response that we see in the review of Carpenter's work that appeared in the *Edinburgh Medical and Surgical Journal*. The reviewer castigated Carpenter's works for perpetrating reductionist ideas that were "detrimental to the best interests of mankind," and which threatened to "lead the mind to the doctrines of materialism" (p. 228). In this event, and as Winter has shown, Carpenter was able to negotiate a response to such potentially career-threatening accusations only by appealing to established men of science for support.

Notes

1 William Benjamin Carpenter, "Inaugural Address of Dr. William Carpenter, F.R.S., President," *Nature*, 15th August 1872, pp. 206–12.

2 Piers J. Hale, "Between Intuition and Empiricism. William Benjamin Carpenter on Man, Mind, and Moral Responsibility," in Catherine Marshall, Bernard Lightman, & Richard England, (eds.), *The Metaphysical Society (1869–1880). Intellectual Life in Mid-Victorian England* (Oxford: Oxford University Press, 2019), pp. 205–27.

3 Alison Winter, "The Construction of Orthodoxies and Heterodoxies in the Early Victorian Life Sciences," in Bernard Lightman (ed.), *Victorian Science in Context* (Chicago: University of Chicago Press, 1997), pp. 24–50.

4 L.S. Jacyna, "Immanence or Transcendence: Theories of Life and Organization in Britain 1790–1835," *ISIS*, (1983), 74: 311–29.

5 Bill Jenkins, "Race Before Darwin: Variation, Adaptation, and the Natural History of Man in Post-Enlightenment Edinburgh, 1790–1835," *British Journal for the History of Science*, (2020), 53 (3): 333–350, on p. 350.

6 Adrian Desmond, *The Politics of Evolution. Medicine, Morphology, and Reform in Radical London* (Chicago: University of Chicago Press, 1989), pp. 213–20.

7 And in this case not without justification, for Carpenter later secretly helped revise the controversial evolutionary work, *The Vestiges of the Natural History of Creation* (1844), and wrote, later still, a favourable review of Darwin's *Origin of Species*. For more on Carpenter, and on *Vestiges* see the introductory essay to this volume, Caden C. Testa and Piers J. Hale, "Mapping evolutionary ideas in Britain in the first half of the nineteenth century," in Caden C. Testa and Piers J. Hale, (eds.), *Evolution Before Darwin* (London: Routledge, 2025), and the extract and headnotes on *Vestiges* in this collection.

21

"CARPENTER'S PRINCIPLES OF GENERAL AND COMPARATIVE PHYSIOLOGY"

Anon

Source: "Carpenter's Principles of General and Comparative Physiology", *Edinburgh Medical and Surgical Journal* 53 (1840), pp. 213–28

ART. V. – *Principles of General and Comparative Physiology, intended as an Introduction to the Study of Human Physiology, and as a Guide to the philosophical pursuit of Natural History.* By William B. Carpenter, Member of the Royal College of Surgeons, London, &c. London, 1839. 8vo. Pp. 478, six Plates.

LIVING bodies are distinguished from those which are inanimate, and have never presented the phenomena of life, by a peculiar species of structure; by consisting of various parts which, though analogous, are heterogeneous; by their parts possessing certain properties, and contributing to certain actions and functions; and by these properties, actions, and functions being regulated by laws of a peculiar character. The study of the phenomena evincing the existence of these properties, actions, and functions, and the investigation of the laws by which they are regulated, constitute a peculiar department of science to which the name of Physiology is applied. Physiology, according to this view, is a science of very great extent, since it combines the examination of the structures, properties, and functions of the whole organized world, both vegetable and animal.

Living bodies consist of two great divisions, those of the animal, and those of the vegetable world. Both of these divisions, however they differ in several of their minute component departments, make, nevertheless, a near approach to each other in many common features. While they differ from the objects of the dead or inorganic kingdom, they agree with each other in many circumstances of structure, of properties, and of functions. The study of the phenomena of this general class of bodies, the knowledge of the functions by which these are performed, and the investigation of the laws by which the existence of these properties, and the performance of these functions is regulated, constitutes a branch of physiology of a general nature, which directs the attention of the inquirer to the general

DOI: 10.4324/9781003490548-24

characters of living action, as distinguished from physical actions, and is, therefore, both a more exalted and more difficult department of physiological science, and is one more suited to the philosopher than the detailed physiological views, either of exclusive vegetable, or exclusive animal bodies.

On human physiology there are many treatises; on the general physiology of animals there are several; and on the physiology of the vegetable world there are a few. But on the general doctrines of the phenomena and laws of animal and vegetable life in common, there is scarcely a single treatise to which the physiological reader can look with confidence, except the first volume of Tiedemann's System, and several of the preliminary chapters of Treviranus's System, and his work on the Phenomena and Laws of Organic Life. The present may be regarded as an attempt intended to answer the object now specified.

A book that fulfils in moderate compass the promise held out in the title of Mr Carpenter's work, would, there can be little doubt, be of great service to those who, with limited time and means, enter upon the study of Physiology or Natural History. The information scattered through numberless memoirs, and in the language of different countries, when added to the ponderous treatises of successive ages, upon any branch of science, forms a barrier to the progress of knowledge which can only be successfully overcome by those who have wealth and leisure at command. It is therefore a useful and important duty to place in lucid order the scattered information gleaned from the researches of a thousand minds, and to show how their facts, their experiments, and their observations, bear upon the subject of their general inquiry. This analysis, concentration, and arrangement of the new facts daily established by labourers in the field of science, is now doubly necessary, from the wide-spread spirit of scientific inquiry, which is a characteristic feature of the present times. We are therefore disposed to look with favour upon every attempt to condense the existing information in any one branch of science; and in no department is this more requisite than in that of general physiology.

Mr Carpenter's volume is divided into three parts. The first of these, introductory to the others, is "on Organized Structures." This portion of the work treats of, 1. Organized structures in general; 2. The elementary structure of Vegetables; 3. The elementary structure of animals; 4. Of the transformation of tissues; 5. A general view of the Vegetable Kingdom; 6. A general view of the Animal Kingdom. – Then follow, in two books, I. The Principles of General Physiology; and II. The Principles of Special and Comparative Physiology.

In his account of the structure of the animal kingdom, arranged under the heads Vertebrata, Annulosa, Mollusca, and Radiata, Mr Carpenter notices the leading points of resemblance which they bear to corresponding groups in the vegetable kingdom. Thus, he conceives the Exogenous plants to correspond to the Vertebrata – the Endogenous plants to the Articulata – the Ferns to the Mollusca, and the Fungi to the Radiata; and from a comparative view of their structure and formations, he traces an analogy between the two great divisions of living bodies.

On subjects of such vast extent, and upon which so much has been written, it would be foreign to the object of this Journal to dilate at any length. And it would be equally out of place to attempt a condensed view of a work, professing to be itself an epitome of all that is known upon the subjects of which it treats. We shall, therefore, in the remarks which follow, confine ourselves to such general observations as the limits assigned to a notice like the present will admit.

After some preliminary remarks, Mr Carpenter treats in his second section "of Organized Structures in general;" and the object of this section is, he says, to "show the relations between Life or Vital Action and Organized Structures." These relations are stated in terms so new to us, and appear to lead to conclusions so dangerous, that, notwithstanding the modifications of expression in other parts of the book, or the confusion of ideas that may have suggested the use of such terms, we cannot pass them over without observation. In the analysis of the contents this passage occurs: – "All physical phenomena result from the excitement of the *physical properties* of matter by external agents; and when these are not in operation no change takes place. In like manner, vital actions result from the excitement of the *vital properties* of certain forms of matter by external agents or *stimuli;* and are not manifested or called into play without the influence of these." (P. ix.) – In another passage these external agents or vital stimuli are said to be heat, light, and electricity.

The same opinions are expressed in the following passages, which are given in the words of the author.

> "In the production of the changes which constitute the life of every animated being, we find an agent employed which is peculiar to the bodies that exhibit such changes, and which is entirely different from any thing we observe in the surrounding universe. This agent is the mechanism which is termed *organized structure.*"
>
> – P. 7.

> "It is scarely a sufficient argument for the existence of a set of vital affinities, distinct from those which hold inorganic substances in combination, to say that all organized tissues exhibit a tendency to spontaneous decomposition by the separation of their elements, or by their dissipation under simpler forms, immediately upon the loss of their vitality."
>
> – P. 17.

> "To refer all the operations of life which cannot be explained by physical laws to a 'vital principle,' is in reality to proceed just as philosophically as the ancients did in the cases just quoted, (the motions of the heavenly bodies;) since a strict examination into their character will show that, though not *identical with* physical phenomena, they are *analogous* to them; being the results of the operation of the vital properties which are

peculiar to organized matter, just as the physical changes of inorganic matter result from the operation of its mere physical properties."

– P. 130.

"The terms *vital principle* and *life*, are erroneously employed, almost synonymously, to imply the controlling agent by which the phenomena of living beings are directed, if not immediately produced." – "The doctrine of a vital principle is not only quite unnecessary to explain facts, but is totally unsupported by the analogies of nature, and by what we know of the Divine government in general."

– P. 131.

"They (the functions) are called into action by *stimuli* of various kinds, adapted to excite each of them to its own peculiar operations. Now, although the adaptation of the various functions to one another, and the manifest tendency of all the vital processes to a common end, would appear, at first sight, to favour the idea of a presiding power by which the whole is regulated, a little consideration will show that these *really* imply no more than an original adaptation of the structure and properties of the organs by which they are performed."

– P. 134.

"It is a question which has been vehemently discussed, but which is after all more one of words than of realities, whether vital properties are the *result* of organization, or that peculiar combination and arrangement in which the elementary particles of living beings are disposed, – or whether they are to be regarded as *superadded* to them."

– P. 136.

"It has been one object of the foregoing pages to show that *vital* properties are as essentially connected with certain forms of matter, as are those usually denominated *physical* with matter under its more common aspects."

– P. 463.

In these passages what are we to understand by the "vital properties of certain forms of matter," called into action by external stimuli? – or the agent of life described as "the mechanism which is termed organized structure?" Is life, either animal or vegetable, "the result of the operation of the vital properties which are peculiar to organized matter?" Is the very supposition of a vital principle unnecessary to explain the functions exercised by living beings? And is it possible that the voluntary exercise of these functions, with reference to a common end, "really imply no more than an original adaptation of the structure and properties of the organs by which they are performed?"

An answer to these questions would involve us in discussions unsuited to this Journal. It appears, however, from the extracts we have given, that Mr Carpenter wishes to impress the idea that combinations of matter may originate a living structure; and that life and vital action are the result of such combinations. If this be his true meaning, as it is the apparent one in the passages quoted above, it would require more than mere assertion to establish a proposition which has been repeatedly refuted. An organized body capable of being acted upon by a living principle, can in no case be said to be itself that principle; and though the connection between mind and matter, – between life and vitality, and the material body acted upon by this untangible, inexplicable agent, – may escape the scalpel of the anatomist, or the crucible of the chemist, yet is its existence not less palpable to our senses in every form of life around us. Without this vital principle to act upon organs suited to the nature of the individual, and of which it is perhaps the modifying agent, the tissues and organs of bodies would be inert substances. Has any chemist by a combination of the substances known to enter into the composition of the bodies of animals, been able to form muscular fibre, nerves, or bone? What concatenation of atoms has been observed to roll themselves into even the lowest or smallest of animal forms? The nerves cease to act, and the body to feel, when the living principle is suddenly extinguished, notwithstanding that all the apparatus for nutrition, circulation, and sensation, may be entire and uninjured. Let the vital spark – the breath of life – cease to act upon the most perfect assemblage of material organs, and these organs, the textures which compose these forms, are instantly subjected to chemical decomposition; form and texture are alike destroyed; and the matter which constituted them resolved into a few simple elements.

Life, or the vital principle, which when allied to matter controls its affinities and directs its forms, is, in fact, only known to us from its effects; in these alone is it recognized. These effects seem to consist in the faculty with which certain combinations of matter, when combined with vitality, are endowed, of existing for a certain period under a determinate form, and assimilating to their substance a part of the surrounding bodies; at the same time restoring to the elements part of their own substance. Baron Cuvier compares the mechanical action of life on matter to a vortex more or less rapid, more or less complicated, where the supply and waste of particles occasion a constant motion, and where the form is more essential than the matter. While this motion subsists the body lives; when it is stopt beyond recal the body dies. After death, the elements which composed it, left to the ordinary laws of chemistry, soon separate to form other and new combinations. It is by the vital movement or energy alone (according to Cuvier) that organized matter is held together in living forms. And these structures upon which the living principle acts, are so constructed and arranged by the Great Creator, as to be the medium of communication between the living agent which wills and feels, with the external and material world. The adaptation of these forms and structures to the nature and wants of the various tribes of living beings, affords ample space for research, without encumbering the mind with the investigation of unseen relations, which it is perhaps beyond the reach of the human faculties ever to unfold.

"Wherever we direct our inquiries (says Dugald Stewart[a]) whether to the anatomy and physiology of animals, to the growth of vegetables, to the chemical attractions and repulsions, or to the motions of the heavenly bodies, we perpetually perceive the effects of powers which cannot belong to mere matter. To a certain extent we are able to proceed; but in every research we meet with an impassable line – a line which is marked with sufficient distinctness, and which no man now thinks of passing who has just views of the nature and object of philosophy. It was indeed by pointing it out that Bacon did so much service to science."

Though we have considered it to be our duty to point out the tendency of such doctrines as are inculcated in the passages which have been quoted, yet it is but fair to the author to notice, that, in other passages of his book, he seems to allow that the vital principle which is found in connection with organized bodies is not a quality of the matter of which these bodies are composed. "Every organized body," says he, "is produced by a previously existing organism, no living being ever taking its origin from spontaneous combinations of existing matter." (P. 134.) And again, "organization and vital properties are simultaneously communicated to the germ by the structures of its parent." (P. 138.) And further, "physical properties themselves are dependent both for their existence and for their excitement to action, upon those vital processes which no mechanical contrivances, or chemical operations can produce or imitate." (P. 144.) And in one place he allows the whole of creation to be under the "continued superintendence of the Creator;" while in another passage, and in his concluding paragraph, he would seem to embrace the opinion, that all the visible creation was at first made so perfect that the machine of nature runs its allotted course, without requiring such continued superintendence.

Had these last been the only terms in which Mr Carpenter had enunciated his ideas regarding matter, we should have been spared the necessity of pointing out statements which we consider to be highly exceptionable, and totally destitute of foundation, from any facts known with regard to the qualities of mere matter. Scattered as are the first quoted statements over his volume, we are not sure that every reader will stumble upon the other passages, fewer in number, which seem to negative the passages first quoted, if it was really Mr Carpenter's wish that they should have this effect. Destined as this book is for young inquirers, any expressions tending to lead the mind to the doctrines of materialism ought to have been sedulously avoided. The passages we have noted as to this effect, are, we hope, not the real sentiments of the author, but merely unguarded expressions; for we hold that no investigator of Nature in any of her aspects can, for a moment, entertain such opinions. In a future edition we trust that these passages will be so modified or altered as to afford no ground for observation or challenge. We are far from wishing to limit the bounds of physical investigation; but there are questions, and the connection of mind and matter is one of them, which are never likely to be solved, and the pursuit of which is as aimless as unavailing. Life or vitality we do not and cannot know, but in its manifestations through organic bodies. Its effects we see daily in numerous forms of being, but the principle itself is otherwise concealed from our view.

Regarding the doctrine of the machinery of the material universe being at first so perfectly formed and arranged by its great Author under certain laws, as not to require continued superintendence, we do not mean to quarrel. Many eminent men have entertained the same idea; and even this opinion has been attributed to Aristotle. But the more common and more correct belief is, that not only was the universe at first created by Almighty Power, but that it is under the continued and immediate superintendence of its Divine Author. And this doctrine is supported by the highest authority – that of Revelation; for St Paul, quoting as is said from an ancient Greek poet, asserts, that "in Him we live, and move, and have our being:"

> "All are but parts of one stupendous whole,
> Whose body Nature is, and God the soul."

We pass over the introductory portions of Mr Carpenter's work, of which we have already given the contents, with the remark, that he seems to have exercised considerable industry in the selection of his materials. In addition to the list of books, which he has consulted for the purpose of his work, it would have been very desirable, and extremely useful to the student, to have had references at the foot of the pages to the specific books or portions of books referred to, as supporting or controverting the author's statements. This would not have added much either to the size or expense of the work; and room might have been advantageously made for these, by the omission of some details in the introductory part, not so essential.

For Mr Carpenter's epitome of General Physiology we must refer to the work itself. It consists of four chapters, – 1. On the nature and causes of vital actions. 2. Of vital stimuli. 3. General laws of organic developement. 4. General view of the functions.

The first chapter (Chap. v.) of the division entitled "Special and Comparative Physiology," is devoted to the consideration of the "Ingestion and Absorption of the Aliment." The various organs or apparatus by which alimentary matters are conveyed to the organized structures, both in the animal and vegetable kingdoms, are minutely described, and the analogy between the nutritive organs of these two departments of organized bodies is pointed out. All animals possess an internal organ for the reception of their food, and Mr Carpenter endeavours to establish an analogy between them and a few vegetables which he conceives to possess an organ something analogous to the structure of the animal stomach. The plants alluded to are the *Nepenthes, Dischidia*, and *Sarcenia*, which possess a kind of pitchers or bags for the reception of a fluid, which appears to contribute somewhat to the nourishment of the plants. The *Sarcenia* and *Dionæa* are in particular mentioned, as the flies which are entrapped in the pitchers of the former, or the leaves of the latter, seem in his opinion to furnish an aliment favourable to their growth. The lower classes of animals, as the Porifera, are also instanced as a connecting link between the animal and vegetable kingdoms, their porous surface being "apparently equally endowed with the power of appropriating to itself the nutritious materials contained in the water." Mr Carpenter is of opinion that there is a distinct chain of connection between all the various tribes of organized bodies, commencing at the lowest in the scale of the vegetable creation, and proceeding in an uninterrupted series up to man.

The "Circulation of a Nutritive Fluid," (Chap. vi.) is next noticed, as a process essential to the life of organized structures. The various means by which this process is effected, both in the vegetable and animal kingdoms, is minutely described. No vessels, it is observed, capable of carrying a circulating fluid, exist in those plants whose structure appears to be completely cellular; but in the *Lichenes* the cells present an elongated appearance, and possess the power of transmitting liquid to all parts of the plant. In the *Fungi*, in addition to the transmission of fluid by the cells, it is probable that the intercellular spaces also perform the same function. In the Ferns a proper vascular structure is observed, which is most perfect in the Phanerogamous plants. The sap ascends by the woody fibres to the leaves, where it acquires new properties, and descends by the cellular tissue and the intercellular passages of the bark. The ascent of the sap is explained on the principle of Endosmose and Exosmose; and Mr Carpenter thinks this fact is confirmed from the circumstance of its being observed that all vegetables throw out by their roots a peculiar secretion.

The author also describes minutely the circulating system in the various tribes of animals, from the simple movement of the nutritious fluid through the reticulated vessels of the Infusoria, "apparently similar to the circulation of the elaborated sap in the intercellular passages of plants," to the perfect vascular system of the Mammalia. Considerable weight is laid on the "influence of the capillary circulation in the general distribution of the blood." It is remarked that in the lower animals, and in plants, the circulation is entirely carried on by means of capillary attraction, and that even in the most perfect animals it is probable the circulation is very much modified by the same principle. The uterus during pregnancy is noticed as an example of this, as at that period a much larger quantity of blood is sent to that organ without affecting in any degree the general action of the heart. Other examples to the same effect are given, and amongst these the developement of the chick *in ovo*, where a capillary circulation is established before the formation of a heart. A few of the malformations of the vascular system, resulting from a want of organic developement, are also noticed.

"Interstitial Absorption" (Chap. vii.) is attributed almost entirely to the action of the lymphatic system of vessels; whilst too little notice is taken of the influence of the viens in producing the same effect.

The subject of the next chapter (Chap. vii.) is the "Nutrition and Formation of Tissues." It is remarked, that "it would seem to be a general law of organization that no solid textures can assimilate or convert into living structures like their own, matter which has not been previously formed into combinations, differing essentially from those which exist in the inorganic world. Thus we find in the blood a variety of ingredients, most of them peculiar to animal bodies, which have been produced subsequently to the first reception of its materials into the digestive cavity, and which are prepared for the reparation and maintenance of the several tissues; and, in like manner, the elaborated sap of vegetables contains other principles peculiar to the vegetable structure, and adapted to its maintenance. These are called, therefore, *organizable products*." – P. 268.

In the nutrition of vegetables the most prominent feature is the ascent of the sap, and its conversion in the leaves into a nutritious fluid, consisting in the great majority of instances of a gummy nature, in others, however, of a saccharine quality. The deposits of fecula and saccharine matters in different parts of plants is considered to be a store for their future nourishment. The mode in which the organizable products are converted into living tissues is endeavoured to be explained from analogical considerations, founded on what has been observed with regard to the cellular extension or growth observed in the Characeæ; but the subject is still involved in too great obscurity to permit of our arriving at any determinate conclusions.

In animal structures, again, the food is subjected to the process of digestion before being fitted for the nutrition of the tissues. It is remarked, that albumen appears to be the characteristic ingredient of the nutritious fluid in animals, as gum is in that of vegetables. It is not yet ascertained in what part of the system the nutritive fluid first takes on the form of globules, and acquires its vital characters; but Mr Carpenter thinks it probable that this process occurs in the lacteals, as globules are seen in the lacteals of the intestinal tube and mesentery. The whole of the vital properties do not appear to be communicated at once to the chyle, as it is not observed to possess the power of coagulation till it has passed through the mesenteric glands, and even then the coagulum is but imperfect. The properties of the blood and its constituent parts are next particularized, and a short table is given of the size of the blood globules in different animals.

In mentioning the constituents of the tissues, one substance, viz. gelatine, is noticed as being remarkably abundant in them, whilst its presence has not been detected in the blood. Mr Carpenter endeavours to explain this on the supposition, that "the conversion of the albumen of the blood into the gelatine of the tissues, (or into whatever principle occupies the place of that which is known as gelatine in a separate form,) may not improbably be the source of the carbon set free during the circulation of the blood through the system." (P. 289.) It is quite probable that a very small proportion of the carbon which is thrown off from the system may result from such a process as that described; but it would never account for the enormous quantity of carbon which is daily thrown off by human bodies. It is usually estimated that man throws off by the lungs alone about eleven ounces of solid carbon every twenty-four hours. If the conversion of albumen into gelatine were the only source of the carbon, what would become of the immense quantity of gelatine which would be daily formed? Or what purposes could gelatine serve in the animal economy?

"Respiration" (Chap. ix.) is the next function which is noticed; but, with a strange want of consistency, he attributes the changes effected on the nutritious fluid in plants by the air to a vital process, whilst in animals, he avers that the aeration of their nutritious fluid is "a change dependent on physical laws." He instances as this physical law, the experiments of Professor Graham on the diffusion of gases; and compares the interchange which takes place between gases when separated from each other by a membrane, to the elimination of carbonic acid from the lungs, and the absorption of oxygen gas. There is, however, no analogy between these processes; for if there were, instead of the oxygen being alone diminished, both the oxygen and the nitrogen would disappear; a circumstance which does not occur. If such were the case, oxygen would be found in a free state in the arterial blood, and carbonic acid, in much larger quantity than it has ever yet been met with, would be found in the venous blood; both strong arguments against such a supposition. Besides, such an assumption would never account for the fact that the oxygen consumed is, as nearly as can be estimated, in the exact ratio of the carbonic acid thrown out. Mr Carpenter, however, adduces, in proof of his theory, the experiments of Edwards and Müller, of enclosing frogs in hydrogen and nitrogen gases, when it was found that a certain quantity of carbonic acid was exhaled from their bodies. The fact, too, of a limb or the body of an animal throwing off a quantity of carbonic acid, when confined in a limited quantity of air, is also adduced in favour of the presence of carbonic acid in the blood. These experiments may be quite correct, and yet not prove the point in question. For if, as Mr Carpenter himself affirms, the conversion of albumen into gelatine sets free a proportion of carbon, its union with any oxygen which may exist in the blood will at once account for the appearance of carbonic acid in the confined air in these experiments, without having recourse to the supposition that it exists in a free state in any material quantity in the blood.

The respiration of plants is described as being effected chiefly by means of the green surfaces of the leaves and stem. They are described as deriving their chief source of carbon from the surrounding air, but that they can only receive this supply under the influence of light. The Fungi are, however, admitted to draw their supply of carbon almost entirely through their roots. The author appears, however, to overrate the functions of the leaves, for if their chief supply of nutritive carbon was extracted from the air, what would be the use of the roots? or why do potted plants fade as soon as they exhaust the soil in which they are placed, however well they may be supplied with water and air? It appears to be fully ascertained that the sensible properties of plants are developed in proportion to their free exposure to the influence of light; but it does not appear that their growth, or the addition of carbon to their substance, depends on their extracting that substance solely or even chiefly from the air. Blanching celery, rhubarb, or sea-kale, does not prevent their growth; it only makes them more tender, and less harsh-flavoured than in their natural state; and plants growing in confined cases, according to the plan of Mr Ward of London, continue to grow and flower, though they can receive no fresh supply of carbon but from the soil in which they grow or from their own exhalations. It is admitted on all hands, that it is only during sunshine that plants possess the power of decomposing carbonic acid, and, of course, it is only then they can have the power of extracting carbon from the air and uniting it to their system. But a fact is related by Mr Ellis, which goes far to overturn the theory, that their supply of carbon is chiefly derived from the atmosphere. For nearly twenty days in spring, the sun had not shone, and the vast forests of North America had meanwhile burst into nearly full leaf, but the leaves were of a pale or whitish colour. One forenoon, the sun broke forth in full brightness, and by the middle of the afternoon, the whole forest exhibited its usual summer dress. Here the growth had gone on as in the blanched pot-herbs, without the possibility of any part of it being attributable to the decomposition of the carbonic acid gas of the atmosphere. Some plants, however, appear to derive the greater portion of their nourishment from the atmosphere, but these are not, in general, furnished with the same active roots as those which draw their nutritive supplies chiefly from the soil.

Respiration in animals exerts a more immediate and more powerful influence over the system than in vegetables; and it is remarked that there is a constant correspondence between the activity of this function and the vital energy of the animal. The organs which are concerned in carrying on this function in the various classes of the animal kingdom are minutely described; the general purpose in all the varied contrivances being the exposure of the nutritive fluid to the agency of the atmospheric air.

The changes which air undergoes during the process of respiration are considered under four heads. "1. The disappearance of oxygen, which is absorbed. 2. The presence of carbonic acid, which has been exhaled. 3. The absorption of nitrogen. 4. The exhalation of nitrogen." The first two heads have been already noticed. With regard to the absorption and exhalation of nitrogen, there appears to be no doubt; and it would even seem that in most of the herbivorous animals the chief nitrogen of the system was derived from the atmosphere, as they live on food totally deprived of that principle; so that there is no occasion for supposing, as Mr Carpenter has done, that "no animal could exist long, if fed on purely unazotized principles."

The Exhalation of Aqueous Vapour (Chap. x.) is next accounted for, and the different structures which exist in both the vegetable and animal kingdoms, and are concerned in this process, explained. It is remarked, that, "though partly the effect of simple evaporation, it is principally dependent on a true process of secretion, by which it is liberated from the circulating fluid." Vegetables whose nutritious juices contain a very large quantity of aqueous fluid are provided with stomata, which are most abundant on the lower surface of the leaves, and also on the stems of herbaceous plants. It is by these apertures or pores that the aqueous exhalation is poured out. It is asserted that the vapour thrown off by the lungs of animals is probably the result of mere evaporation, whilst that from the skin is the result of a true secretion, particular organs being provided for this purpose.

Chapter xi. is devoted to the consideration of the Secretions. It is remarked that "the process of secretion is truly one of separation," as "the peculiar products contained in the secretions exist in the circulating fluid;" and that, though it is difficult to explain why each gland should secrete a fluid peculiar to itself, the fact seems substantiated by the occurrence of deposits of uric acid, cholesterine, &c. in various parts of the body. Mr Carpenter infers that the animal body is continually renewing by a process of interstitial absorption and deposition; a conclusion which, though often stated, has never been clearly proved. The only experiments upon which this assertion rests are those of Hunter and Monro, who found that, by feeding animals with madder, the bones became tinged of a red colour, but that if the madder were discontinued for a few weeks before the animal was killed, the colouring matter was all removed, and the original colour of the bone restored. The conclusion from this fact is, that the bones possess absorbents, and that colouring matters introduced into the circulation are deposited with facility in the bony structures, and as easily removed. Is it reasonable to suppose that, in the short space of a few weeks, every particle of the bony textures of the body should be taken up, and anew deposited? If such were the case, why is it that carious particles of bone are the cause of months, nay, years of suffering? Why is it that scars on any part of the body remain indelibly imprinted during the whole of life? The only answer which can be given is, that no such continued interstitial absorption and deposition takes place, at least during health. The analogy of the vegetable kingdom would lead to the same conclusion. Particles once deposited in the solid structures of plants remain unaltered during the life of the individual. Names cut on trees have been discovered hundreds of years afterwards, enclosed in the centre of the woody fibres, as fresh as the day they were first cut there. Mr Carpenter, however, thinks that the analogy, which he endeavours to prove in every other instance between vegetables and animals, does not exist here, on the ground that in animals "all the soft tissues possess a strong tendency to decomposition." This tendency to decomposition is, however, as effectually prevented in the one case as in the other by the presence of the vital principle.

The varied secretions which are met with in the vegetable kingdom are next noticed, and it is remarked generally that they are the result of the combinations of the elements of water with carbon, to which nitrogen is in some cases added; a few, however, are composed of hydrogen and carbon alone. The milkiness of the juices of many plants is attributed to their holding in solution or suspension various products besides those which contribute to the nutrition of the system. These juices are found in the leaves and bark alone. It is remarked with regard to the fixed oils, that like fecula they occupy the interior of common cells, and are probably transformed by germination into a material fit for the nutrition of the young plant. The excretion of gum, sugar, oils, &c. are regarded as being merely owing to an excess of these substances in the plant, or from their being found near the surface. The matter excreted from the root is described as being probably a noxious product, whose presence is deleterious to the plant. The various secretions in the animal body, and the secreting organs in the different classes of animals, are described with equal minuteness; but it appears unnecessary to notice them in detail.

The next chapter (xii.) contains a few short remarks on the Evolution of Light, Heat, and Electricity Phosphorescence is observed in many bodies both animal and vegetable when in a state of incipient decay; but it is also noticed in several Fungi during a state of active growth. That in the animal kingdom has attracted the greatest notice, a great number of the lower aquatic animals being possessed of phosphorescent properties. It is remarked, that in the Medusæ this phosphorescent property appears to reside in the mucus which is secreted around their tentacula; and it is not a little singular, that the mucous secretion of fuci, and also that of fishes, possesses the same property. Several insects present this phosphorescence.

The evolution of heat by vegetables, particularly during inflorescence, is noticed; but vegetables appear to be more dependent than animals on the temperature of the medium in which they live. A few facts are mentioned with regard to the evolution of heat by the invertebrated animals, which seem to prove that they possess the power, though to a limited extent, of generating heat. This power is most perfect in Birds and Mammalia, who can maintain a temperature quite independent of that of the surrounding medium. Mr Carpenter endeavours to explain the mode by which this increased temperature is maintained, by supposing that the oxygen, which he asserts is absorbed by the arterial blood from the atmosphere, is united to carbon in the capillaries, and that this union evolves heat. It is far more probable, however, from all we know of the process, that this union takes place at the lungs, by an immediate union of the oxygen with the carbon in the venous blood. If the process of union and consequent evolution of heat took place in the capillaries, the extremities ought to be warmer than the interior of the body; and exposure to cold would invariably cool the lungs before it acted on the extremities. Neither of these are known to occur, but, on the contrary, we are invariably told that the internal parts are several degrees higher in temperature than the surface of the body. These facts are well explained in a little work by Mr Grisenthwaite on food, where he proves that the blood in the lungs receives during ordinary respiration an increment in temperature of three-quarters of a degree, a quantity amply sufficient to maintain the wants of the system.

A few facts are stated with regard to the occurrence of electricity in plants and animals.

The next chapter (xiii.) contains many curious facts with regard to the Reproduction of Organized Beings, and the striking analogies which exist in this respect between plants and animals. Thus many of the lower animals, as polypi, annulose animals, &c. may be propagated by cutting them into pieces, each portion becoming a new and perfect animal. The usual mode of the reproduction of polypi is by gemmules or buds from the sides of the parent animal, which at length drop off perfect animals. In this respect they present a strong analogy with plants. Our space will not, however, admit of further details on this subject.

The Subordinate laws regulating the exercise of the Reproductive Function, – Distinction of Species, and Propagation of spontaneous or acquired Peculiarities, are treated of in the fourteenth chapter. Many interesting facts are recorded on all these heads, but they all tend to show how much is yet to be learned with regard to them. Many species of plants and also of animals, formerly thought to be distinct, are now found out to be derived from the same original source, and the modifying causes which have led to the appearances of distinct varieties are not yet known.

The fifteenth chapter treats of the "Sensible Motion of Living Beings." Sensible motions are noticed in several plants as well as in the animal creation; but it has never been proved that this motion in plants is the result of a nervous agency. The folding of the flowers and drooping of the leaves of plants have been usually attributed to a diminution of the activity of the vital processes, by which the turgescence of the softer parts of the structure is diminished, this appearing to depend much on the withdrawal of the stimulus of light and heat. In some cases, however, it appears to be in part of a periodic character. The spontaneous movement of the seed-vessel in many plants has been explained on the principle of endosmose. The outer cells being the largest, and possessing a denser fluid, that from the inner smaller cells passes to them, and by their distension causes the shutting of the valves of the seed-vessel, and the scattering of the seed. A special tissue, the muscular, is provided for the motive powers of animals.

The "Functions of the Nervous System" are treated of in the sixteenth chapter, and as vegetables possess no analogous structure, the whole chapter is devoted to the consideration of the principal forms which the nervous system presents in the animal kingdom, and the functions to which it ministers. The developement of this system is traced through the various classes of animals, from the simple nervous filaments of the Acrita, to its most perfect structure in the Mammalia. The division into motor and sensitive nerves is explained, as well as the functions which each of these performs in the animal economy.

The volume concludes with a short chapter on the "Evidences of Design presented by the structure of organized beings."

We regret that we cannot conscientiously recommend this book. Tainted, as it seems to be, with opinions detrimental to the best interests of mankind, we cannot hold it up in its present state as a safe guide, even in the study of those sciences, the principles of which it professes to place before the reader. The concluding chapter is but a feeble antidote to the previous statements.

Note

a The Philosophy of the Active and Moral Powers of Man. By Dugald Stewart, Vol. i. p. 212.

Editorial Headnote

22. [Charles Southwell & William Chilton], "Theory of Regular Gradation", *Oracle of Reason*, 2 vols. (1842–3), Vol. 1, pp. 5–6, Vol. 2, pp. 5–7, 21–3, 219–21, 229–30, 253–4, 279–80, 308–9, 318–19, 325, 332–4, 340–42, 347–8, 378–9.

Charles Southwell (1814–1860) and William Chilton (1815–1855), along with John Field, were founding editors of the *Oracle of Reason*, a radical and explicitly atheist penny publication. They began publication late in 1841 but were forced to close the paper in debt in December 1843. The *Oracle* was the most outspoken atheist publication in Britain, with, at its height, a weekly circulation of between 4,000 and 6,000 copies. It had a succession of different editors as a result of the imprisonment for blasphemy of each in turn. It was but one among many radical, socialist, Owenite, and atheist publications that came out of the 'pauper press.'[1] The historian Adrian Desmond has made the case that scholars need to appreciate the extent to which working-class radicals, socialists, and atheists appropriated contemporary science to their cause, and the science of evolution was no exception.[2] Often reaching back to Enlightenment authors such as Paul-Henri Thiry, Baron d'Holbach (1723–1789); Constantin François de Volney (1757–1820); Nicolas de Condorcet (1743–1794); and William Godwin (1756–1836), atheists used science and natural law to reject the idea of a divine or miraculous creation, but also in support of the importance of the environment in forming the mind, moral character, and physiology of mankind. In this context, the embrace of transmutation fit well with their political agenda, which centred on political agency as the mechanism for biological and social progress.[3] Further, a dynamic, upward, transformative change in nature undermined the establishment argument for immutable and natural social distinction. The existence of rich and poor was neither natural nor inevitable, they argued, but was the result of injustice.

The regular column "Theory of Regular Gradation" stretched to forty-eight articles. It borrowed from the sciences of chemistry, geology, palaeontology, anatomy, and physiology in order to make a case for transmutation by natural law. Begun by Southwell, it was continued by Chilton following Southwell's imprisonment for blasphemy. The series began with transmutationist ideas purloined from the French geologist and naturalist Pierre Boitard (1787–1859). Later articles referred to the controversial anatomist William Lawrence (1783–1867), whose anatomy lectures had been associated with materialism and atheism, and the French naturalist and transmutationist Jean-Baptiste Lamarck (1744–1829). Much of this information was lifted from *Chambers's Information* but also from Charles Lyell's critical account of Lamarck in the second volume of his *Principles of Geology* (1832).[4] Although an earlier generation of historians described the nineteenth century in terms of a clash between science and religion, it is by paying attention to these kinds of publications that we see that this narrative needs to be grounded in an appreciation of the national and local politics of religious belief. Chilton, Southwell, and others were successful in linking transmutation to atheism and socialism, and this motivated the rejection of evolutionary ideas among establishment men of science across the rest of the century by the politically conservative Anglican establishment.

Notes

1 Robert Owen (1771–1858) was a socialist and cooperator. He was a founder of the cooperative movement as well as being an industrialist and mill owner. He reformed the mill into a progressive community as a model to show that industry did not need to be a depredation upon workers but rather that it could be a source of physical, mental, and moral improvement. His *A New View of Society* (1813) was an influential work. For details on the *Oracle* and its editors, see Edward Royle, *Victorian Infidels. The Origins of the British Secularist Movement 1791–1866* (Manchester: Manchester University Press, 1972). For more on evolution and socialism, see Caden C. Testa and Piers J. Hale, (eds.), *Evolution and Socialism* (London: Routledge, 2025).

2 Adrian Desmond, "Artisan Resistance and Evolution in Britain 1819–1848," *Osiris* 3 (1987), pp. 77–110.

3 For more context on Enlightenment evolutionists, see Caden C. Testa and Piers J. Hale, "Mapping evolutionary ideas in Britain in the first half of the nineteenth century," in Caden C. Testa and Piers J. Hale, (eds.), *Evolution Before Darwin* (London: Routledge, 2025). Extracts from Godwin and Condorcet are provided in that collection.

4 Desmond, "Artisan Resistance and Evolution in Britain," p. 99. For Lyell's commentary on Lamarck see Caden C. Testa and Piers J. Hale, *Evolution Before Darwin* (London: Routledge, 2025).

22

"THEORY OF REGULAR GRADATION"

[Charles Southwell and William Chilton]

Source: [Charles Southwell & William Chilton], "Theory of Regular Gradation", *Oracle of Reason*, 2 vols. (1842–3), Vol. 1, pp. 5–6, Vol. 2, pp. 5–7, 21–3, 219–21, 229–30, 253–4, 279–80, 308–9, 318–9, 325, 332–4, 340–2, 347–8, 378–9

(Fossil Man.)

022 " 'Fossil Man,' 'Theory of Regular Gradation' from the *Oracle of Reason*.

Image in the public domain.

DOI: 10.4324/9781003490548-25

(*Fossil Man.*)

THEORY OF REGULAR GRADATION.

I.

> "I will show you how the earth has been peopled, how organic formation after organic formation has taken place, passing gradually from simple to compound bodies, and covering itself, as at this hour we find it, with plants and animals. In following matter through all its changes, noting its metamorphoses, from the most simple organization to the most complex, we shall find, without doubt, that point where man, brutish and savage, as at first he must have been, took rank among the creatures of the universe."
> – Free Translation from "L'Homme Fossile," (Fossil Man) of Boitard.

Nothing is more fatal to a common sense view of things, than the very prevalent habit of considering consequences before investigating and determining principles. It is not only illogical but absurd, and highly prejudicial to the cause of truth; for it fills timid minds with alarm, lest by a resolute and persevering search into the properties of matter and the nature of opinions, they may light upon results disastrous to their prejudices and perhaps fatal to their hopes of "singing hallelujah above the clouds." The rightly balanced, well ordered mind has the strongest assurance – an assurance drawn from the great fountain of human experience and the general analogies of things, that neither truth nor virtue are empty names, but real substantial blessings, whatever clerical obstructives may preach to the contrary. Take the question, the regular gradation of the human species, and note how philosophers have occupied themselves in the solution of it; and if you do not say that merry andrews would have gone a better way to work it will be odd indeed; for like those creatures that by a certain sort of instinct to avoid pursuit darken the medium through which they pass, so have our clerical swimmers in the sea of knowledge a happy knack of leaving a long tail of obscurity behind. Instead of carefully collecting facts, and opening their mental exchequer to receive something solid in the way of experience, they in general rest contented in most shameful ignorance, rather than their pride should be mortified by any discoveries in science, hostile to their cherished opinions. Solicitous, for their own advantage, to maintain what they call the honour of the human species, and its superiority over the brutes, all that could flatter and soothe the delightful idea is that alone which has been said and collected. Any attempt to establish a relationship, however remote, between man and the inferior animals, has always been scouted as impious, an insult to the creator, in whose image they tell us we are made, and little short of blasphemy against the holy ghost. Hundreds of sermons have been preached against the unlucky Bulliver, who insisted that in his day there was a Kentish family all tailed. Anathema after anathema was heaped upon Lochner, who in his Miscellanea Curiosa relates with great gravity and minuteness the case of a boy with a monstrous tail. As to Dr. Ferriar, who considered that the os coccygis must sometimes have an accidental elongation; and Dr. Grindant, who published many cases tending to give strength to the opinion that our forefathers were tailed animals also at first; both those gentlemen have been very roughly handled. The last-named doctor states, among other facts, as he styled them, that the islands of Moluccas, Formosa, and the Philippines, were at one time inhabited by whole races of men with tails. But to pass by all these, and the flagellations they received from our spiritual whips, for setting forth such *degrading* opinions, we need only refresh the reader's memory with the case of the famous Monboddo, who had such a predilection for monkey men that he has taken vast pains to prove that they formerly did wear tails; for insisting upon which anti-poetic notion he has been as soundly abused as any man that ever lived. That his theory was incomplete and erroneous in many important particulars, it is presumed these papers will show; but that there is *some* truth in it will be as clearly proved. Lord Monboddo was ignorant of many particulars, without a knowledge of which he could not but blunder, as the following, from the "Origin and Progress of Language," will show: "A whole nation, if I may call them so, have been found without the use of speech. This is the case of the ourang-outangs that are found in the kingdom of Angola, in Africa, and in several parts of Asia. They are exactly of the human form, walking erect,

WAS MAN ORIGINALLY SAVAGE?

not upon all fours; they use sticks for weapons, they live in society, they carry off negro girls, whom they make slaves of, and use both for work and pleasure." This, it must be confessed, is anything but "asserting the dignity of human nature," but it is of far more consequence than asserting human dignity, to know that when Lord Monboddo talks of whole nations of *monkey men*, without the use of speech, who walked erect, and not upon all fours; he gave us a legacy of nonsense, and bad men having possession of the precious relic use it most unscrupulously as a sharp weapon against those who maintain the substance and sense of Monboddo's theory, without its outrageous nonsense. To our minds one simple fact is sufficient to overturn the theory of Monboddo, a fact mentioned by White, which is, that the law of gradation, that philosophers of his school so strenuously maintain, would be destroyed if the hypothesis of tailed men were correct; for in descending through the species of apes, we meet with no tails till we reach the baboons, which are further removed from man than the apes are; but surely it is much better to state the fact or facts, and cure the delusion, than to persecute or abuse its author. It will hereafter be proved that Lord Monboddo is sufficiently refuted by a reference to the relative structure of men and monkeys; but enough has already been said to free us from the imputation of supposing that real *bona fide* men or women ever did wear so unsightly an appendage. The tendency of the multitude to be gratified with all that flatters their prejudices and puffs them up with the notion that they are in action like angels, in form, apprehension, &c. like gods, has strangely affected philosophers, especially modern ones, who, though at fitting opportunities will preach and moralize to admiration about corrupt, degraded, and fallen human nature, will yet insist that there is in that nature something divine, angelic, and the like; so that he had better been born a dog who, notwithstanding all this fuss, should affirm that man is but a two-legged, two-handed animal, less swift than the hare, less strong than the lion, and, though better organized for all the purposes of intelligence and happiness, has been perhaps less reasonable than either. But it is evident that in a state of society where it is treason to expose bad governments, however vile and corrupt, and flat blasphemy to propagate any opinions against a hireling priesthood and law-made religions; it is natural to suppose that the people should be systematically deceived, and reduced to that state of servile vassalage, both of mind and body, that wholesome truth is an abomination to them; and the public orator, who is loaded with popular applause, has good reason to demand, like Phocion, whether he have not given birth to something foolish. Between doctors for the body, and doctors for the soul, the credulous multitude have been physicked out of all plain sense. The divine has to administer to the soul, that is his profession; and whatever may be thought of its utility, it makes all that concerns the dignity of the soul the peculiar care of the divine, who would feel much scandalized were it proved that his flock were, after all, but the fiftieth cousins of sheep. While physicians, who in the practice of their honourable profession are as little interested in the entire cure of bodies as the divines are in the entire cure of souls, are shrewd enough to understand the real state of the question, which is this, that did correct and sober ideas prevail with regard to the nature of man, his rights, duties, and real position as a member of the great commonwealth, physic would be thrown to the dogs, for assuredly man would have none on't. So that these two brothers in the medicinal art have long made common cause together, most artfully raising such a cloud of prejudice that the patients can't see an inch in advance of their noses, but are led by that useful organ, to borrow an expression from the poet, "as asses are." If, however, which is not very common, a Taylor or a Lawrence steps forward to expose these peccadillos, why their mouths are stopped with a sop or a sword; they having the alternative, poverty and independence in prison, riches and degradation in a palace; for whatever may be said of individuals, it is certain that associated bodies act to the letter upon the scripture injunction, "if thine eye offend thee pluck it out" – for if any member offend them *they thrust him out*.

[...]

THEORY OF REGULAR GRADATION.
II.

> The man who does not find in animals younger brothers, and in plants cousins more or less removed, is unacquainted with his own nature, and is devoid of the elements of morality." – OLD TREATISE.

SOME one has defined man to be "an animal ashamed of his own body," which definition is not wide of the strict truth, for in all ages human beings have been the dupes of their own [Illegible Text] which hindered them from appreciating truth. The pride of rationality, which puffs [Illegible Text] up with the absurd notion that they are [Illegible Text] anomaly among animated existences, and in [Illegible Text] sense allied to inferior creatures, is specially strong in our day, when truth is too often sacrificed to false delicacy, and genuine philosophy made the sport of the most ridiculous conceits. In women, who, as Kotzebue justly remarked, are worse and better than we are, the feeling in favour of what are called the prerogatives of our species, is exceedingly strong; but women are more remarkable for warmth of feeling than solidity of judgment; and it is with regret we add, that both men and women are, in the gross, the creatures of sound rather than sense; and determined more by custom than reason.

The fatal effects of this spurious shamefacedness is a plentiful crop of folly, with an almost entire ignorance on the part of both sexes as to those subjects which they should be most concerned to know. A distaste for the naked truth, whatever such nakedness may discover to us, shows a depraved & unwholesome state of mind. Besides, false delicacy is not merely the antipodes but direst antagonist of genuine modesty, which revels, but does not glory, in its own perfections. That genuine modesty and such spurious delicacy, aptly called by our French neighbours *mauvaise honte*, are wide as the poles in nature, is abundantly proved by every day experience, which shows your more *nice-than-wise-people*, who would be shocked at the bare supposition that man is but a link in the great chain of animated existences, yet do not hesitate to perform the most vicious and unjust actions. Philosophers, who are obliged to feel the public pulse before they venture to teach, are, in consequence of the sickly condition of the public mind, tempted to prescribe all sorts of quack medicines; for philosophers, like the man who stole the loaf, must live; not to mention that it is a maxim now in much repute, that a man who is one grain more honest than the times he lives in is little better than a fool. This is one of those obvious facts which need no particular illustration, every man by his own experience illustrating them; so we cannot feel surprised that Mr. White, in his admirable Treatise on "The Regular Gradation in Man," found it necessary, or at least thought it prudent, to preface the work by a kind of apology for its contents, to the effect, that he had no desire to elevate the brute creation to the rank of humanity, nor to reduce the human species to a level with the brutes, all of which fine stuff about elevating and reducing, &c., is, or ought to be, quite foreign to the matter contained in the book, which should be judged according to its merits, and its teachings, if true, abided by, whether it elevate or degrade human conceptions of itself. This is what Dr. Samuel Clark calls appealing from reason to vulgar opinion, which he justly contends philosophers should not do, because it is not the rule of truth. Philosophers have no business to confound consequences with principles, such conduct being only admissible in starved men of science, who "turn country parsons and go about deceiving the vulgar." It is evident that neither the wishes of Mr. White, nor any other theory propounder, can possibly affect the truths of science, which are no more disturbed or affected by human wishes than the waves of the sea or the man in the moon. The object of Mr. White's treatise was to show, that from the simplest being up to man, there is a complete and regular gradation; or, as it were, a growing out of creature from creature; that there is a linking together of all the kinds and species; the exact idea of the facetious author, an extract from whose work stands at the head of this article, that men are the elder brothers of inferior animals, and cousins many times removed of plants and flowers. Mr. White remarks in his first page, that "Every one who has made natural history his study, must have been led occasionally to contemplate the beautiful gradation that subsists among created beings from the highest to the lowest. From man down to the smallest reptile, whose existence can be discovered only by the microscope, nature exhibits to our view an immense chain of beings, endued with various degrees of intelligence and active powers, suited to their stations in the general system.

"In the animal kingdom the different classes into which nature seems to have divided her productions, are so blended by creatures apparently anomalous to all system, that it is often difficult, and sometimes impossible, to draw lines of distinction. The inhabitants of the earth and air encroach upon each other; for there are flying *macaucoes*, flying *mice*, flying *squirrels* and *bats*, which, though quadrupeds, have wings to buoy themselves up in the air; and on the other hand, there are some birds that cannot fly at all, as the *ostrich*, the *cassouwary*, the *touyou*, and the *dodo*. The porcupine, though a quadruped, has quills. Nor are even the inhabitants of the sea and those of the air much better discriminated, for we meet with the flying fish, and birds that inhabit the waters. Amphibious animals link the terrestrial & aquatic; we find some fish with, and others without, lungs. The distinction of animals into viviparous (brought forth alive) and oviparous (produced from eggs) is not more definite; for vipers and cartilaginous animals produce their eggs within their bodies previous to the expulsion of the live animal; and some animals are like vegetables, propagated by germs, neither viviparous nor oviparous. Bipeds, quadrupeds, and quadrumanuses equally encroach on each other. There are some apes that walk only on their hind legs, and others that walk on all four, and the greatest part of them use their hind fect in the manner of hands. The genus lacerta, from the alligator of twenty feet to the lizard of three inches, forms a class of animals connecting the race of quadrupeds with that of reptiles. To those useful facts may be added, one of at least equal importance, that the distinction usually drawn between what are called the mineral, vegetable, and animal kingdoms, however useful, is purely arbitrary. There is no distinction save that which exists in the minds of men; nor is the equally common distinction between what is called the dead and living matter of our globe one whit better founded, or any thing more than one of the many chimeras paraded as science. These fictions of philosophy, when received as facts, are the causes of incredible mischief; for men are led to the conclusion that such distinctions, which are but arbitrary aids to the weakness of human understanding, really exist, a something actually in matter; hence the confusion of ideas that prevails. All will readily admit that men disagree so much, only because they know so little, and that it is the paucity of facts and emptiness of our heads, with regard to the relationship of things, which causes us to arrive at such strange and opposing conclusions.

The reader should therefore be upon his guard and carefully separate the truths of things, that is relations or descriptions of things as they actually exist, from the arbitrary division made by men of science, which, though intended as helps to the understanding, often clog and perplex it.

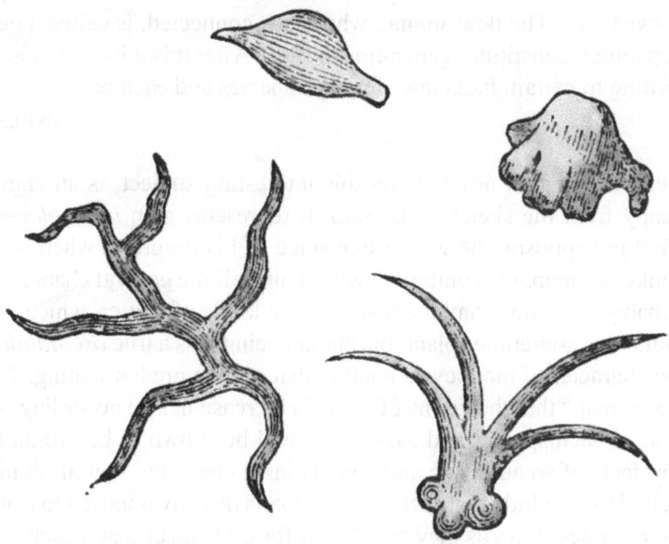

The above engravings show what is called by naturalists the first type of organization, an infusoria or animalcule. It has been drawn from the original of Boitard, who calls it le protèe changeant (proteus diffluens), literally, changing proteus; the term proteus denoting with the ancients, matter; that they held to be oldest or first. The word proteus properly signifying oldest or first. This kind of changeful matter is seen above as through a microscope and prodigiously enlarged. Boitard observes – "It is exceedingly difficult to describe it; for its body changes form every instant and assumes an infinite variety of magnitudes. Sometimes it is of a round or oval figure; presently you may see it, as it were, carried upon four or five tubercles or pimples; an instant after, it divides and subdivides into small and irregular thongs; then it takes the form of a flower with four petals or leaves, perfectly regular and pointed, finally; before I have time to observe its motions it has already assumed a thousand successive and varied forms." In answer to those who consider that matter capable of such an infinite variety of transformations must be animal matter, living and organized, it is urged, that naturalists have long been of the same opinion, but that the observations and researches of Scopoli Regnier, Otto de Munchausen, Buttner and others, have shown that at a period very remote, when bodies of water were stagnant upon certain parts of the globe, the humidity, heat, &c., caused certain combinations, and matter became organized, in harmony with the modes known by the name of chemical affinities.

[...]

THEORY OF REGULAR GRADATION.
III.

> "All living beings have, in the first place, formed part of a body like their own; have been attached to a parent before the period of their independent existence. The new animal, while thus connected, is called a germ; its separation constitutes generation or birth. After this it increases in size according to certain fixed laws for each species and each part."
>
> LAWRENCE.

At the head of our first article upon this interesting subject, is an engraving, a faithful copy from the sketch of Boitard. It represents man *undeveloped*, as we are justified in supposing he was at that stage of his progress, when not exactly either monkey or man, yet combining within himself the general characteristics of each. Probably, like those ambiguous, doubtful kinds of matter, which are sometimes animal and sometimes plant, the human being was a true *oscillatoire*, at one period the character of monkey, at another that of man predominating. The assertion of Lawrence, "that the germs of animals increase in size according to certain fixed laws for each species and each part," will be shown to be groundless, and absurd, by facts of so authentic and convincing a character, that all thinkers will be compelled to conclude that such an assertion is directly opposed to truth. There are no fixed modes (laws as they are styled) for each species and each part, there being nothing *fixed* in the parts of nature, which are in a state of continual flux or change. We also enter our protest against the notion of life that the paragraph at the head of this paper tends to perpetuate. It is quite unphilosophical; a spurious idea which has crept into the heads of scientific men, in consequence of their very loose and unscientific method of pursuing truth. It is the common error of certain extremely logical minds, when pretending to interpret nature in harmony with the grand principles of philosophy, to talk of living beings; for the use of these terms presuppose dead beings or dead substances. Now, no one is prepared to show that there are such things as dead substances; indeed, the notion of dead matter has long since been discarded by the most eminent physiologists. Life, we are told, is "a property of organization," which will do very well for people who are amused by sounds, thinking they have a new idea, when it is only a new word; but cannot satisfy any reflecting mind, which will require to *understand* words. Life is a property of organization. Well! What is organization? The only reply that can be given is, that organization is an assemblage of organs, made up of organized or living matter: so first we are told, with great gravity, that living matter is organized matter, and then, to clarify the explanation, we are told that organized matter is living matter, life being a property of organization, and organization itself, known by its display of vital properties. The Bishop of Llandaff has some admirable remarks in his "Essays on Chemistry," respecting this fallacious mode of reasoning. "If it be asked," he observes, "what are the discriminative characteristics of mineral, vegetable, and animal life, as opposed to one another, I plainly answer, that I don't know any, either from natural history or chemistry, which can wholly be relied upon." Nor, it may be safely affirmed, does any one else. It would, indeed, be strange, if he did, seeing that there are no such "discriminative characteristics:" the distinctions which are the plague of science, usually drawn between dead and living, earthy, mineral, vegetable, and animal matter, have, as the learned bishop well observes, "no real foundation in nature, and should not be depended upon too far, as they often perplex and impede the progress of a free inquirer."

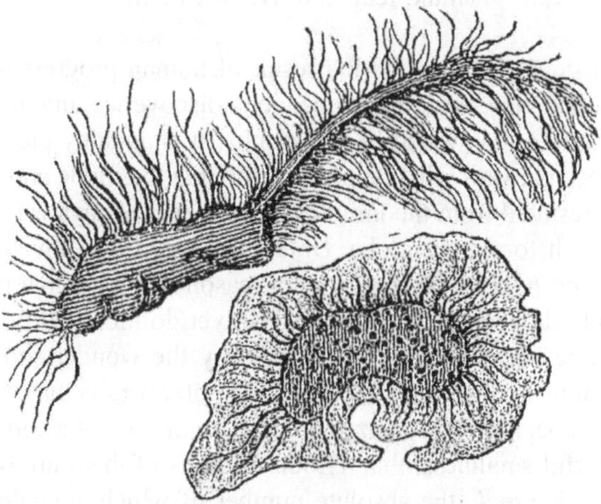

(Oscillatoire des murailles.)

In the last paper was shown the first type or form of what naturalists have thought proper to call living matter. The present engraving displays the hesitating, or living matter, already alluded to. It is considered a plant by some naturalists, and an animal by others, while according to Bory, Saint Vincent, Ingenhouz, and Agardh, its character is very doubtful, being sometimes plant at others animal. It is called by Boitard *(oscillatoire des murailles)* literally balancer of the walls; our word oscillatory, which answers to *oscillatoire*, signifying a something that moves backwards and forwards like a pendulum. "When seen through a microscope," says Boitard, "it displays a multitude of streaked or strongly marked filaments, of a deep green colour, inclosed in small transparent sheaths, bearing some resemblance to a glove, and may be seen to blow, in a brush-like shape, at their openings." It may be found during the rainy periods of spring and autumn, along the walls and terraces, or even upon the trees among the moss. If we suppose a chain of beings, at the one end of which is man, and at the other true monads, that is, creatures without any discernible length, breadth, or thickness, we must be content in our researches to begin with the most minute species of organism, that the microscope makes sensible to sight; at the same time carefully guarding ourselves against the radically erroneous supposition, that the smallest living creatures we behold, are, in fact, *the smallest;* for improvements in every department of science, are now so rapid, that without being deemed hair-brained visionaries, we may hope to see so vast a field of vision opened by improved optical instruments, that substances hitherto deemed dead, as our chalky cliffs, even granite itself, will be seen teeming with life, and its now indiscernible particles, actual moving, breathing bodies, which living things may, according to the beautifully poetic idea of Shelley –

"Think, feel, and live like man."

We are led to believe such will be the result of human progression, by analogy; analogy, that guided Newton to his stupendous discoveries, making the fall of an apple the prelude to a new world of philosophy. It is analogy proclaims that man could not have been always what he now is; and that the mode and character of all existences have resulted from the infinitely varied, and always varying, conditions of the matter which forms our globe. Dr. Johnson said, "he loved a good hater," and we love a good *hoper;* and may we not reasonably hope that the microscope, which has already done so much for man, will yet do more, and that all past discoveries in science may hereafter be eclipsed by the wonder showings of some extraordinary genius now in embryo. Professor Ehrenberg actually *saw*, through a powerful microscope, highly organized animals composed of a sort of living earth, of such a wonderful smallness, that 41,000 millions of them are often within the volume of *one cubic inch*, the absolute number of which animalculi is certainly greater than that of all other living creatures taken together. Yet, who shall be rash enough to declare that improvement has gone to the end of its tether, and that, as our ancestors would have been astonished could they have looked through Professor Ehrenberg's glasses, ages to come, could we rise from our graves, our children's children might not show *us* even more subject for wonderment? All that relates to the world of intellect is progressive or retrogressive, like the waves of the sea incessantly in motion, never fixed one single instant. Each generation having its wonders, which are spoiled by familiarity; familiarity as certainly destroying wonder as fact destroys fiction. The first form of life, or organized substance, is the *changing proteus*, an engraving of which was given in the last paper, called first, not because there are no forms of life yet more minute, but because we do not see them, so that they are accepted as the first at present perceived link in the organic chain; the second, and rather more complex organic apparatus, is the *oscillatoire* described above.

The *changing proteus* and the *oscillatoire* are the two simplest known kinds of organism. All other forms of life, supposing them to exist, the most powerful microscope cannot discover to us. Proceeding thus, we shall illustrate the whole series of animated existences, from the insect, to which "the blade of grass is an unbounded world," up to man, the microcosm, or little world within himself, confessedly the most compound and re-compounded of all animals. The principle that any distinction between living and dead matter, as that usually drawn between the animal and vegetable kingdoms, is purely arbitrary, should be kept steadily in view. This point is thus strenuously insisted upon, because it is vital, the true pivot of sound philosophy. That rocks, stones, sand, &c., are sensitive or living bodies, cannot be proved in our present state of knowledge; but it has been roundly asserted, that stones *vegetate, grow*, and contain within them germs of other stones. De Boot thought that crystals, amethysts, and rocks, grew like mushrooms, to which opinion the Bishop of Llandaff seems very much inclined, for he says, "rock, crystals, precious stones, and amethysts, often contain in them several heterogeneous particles, a circumstance which proves them to have been once in a fluid state, and induces a suspicion that in their formation they may resemble the germs and resins extravasated (forced out) from various species of vegetables. The vegetation of stones has been admitted by many; and some have contended *that minerals, as well as animals and vegetables, spring from seed; the greatest being nothing but the expansion of the parts of a minute grain of sand*." Such an authority as this in favour of living, growing stones, will startle many Christian readers, who think every thing dead that seems to be so. It is careful reasoning by analogy, that is so profitable to man; for it teaches how to judge of things unseen by things seen, enabling us to say with perfect confidence, like the harlequin in the Emperor of the Moon, *as it is here so it is there*.

[...]

THEORY OF REGULAR GRADATION.
XXXVIII.

MANKIND have been so accustomed to meet with theories based upon vague generalities, instead of well-determined facts and experiments – have been so accustomed to the synthetical, instead of the analytical, mode of reasoning – that an individual, wanting a name, who does not possess the tact or facilities for placing in a popular or entertaining light important facts, which he thinks deserving consideration, stands but slight chance of being read. If the evil ended in the rejection of the unpopular author, it might readily be tolerated, but unfortunately men tire in the pursuit of truth when they find the road to it rugged and uninteresting, instead of smooth and pleasant, as they anticipated – forgetful that the gem when obtained is beyond all price, and will amply repay them for the trouble which its acquisition has cost.

The talented editor of the *Investigator* has said, in reference to the multitude of rubbishing works which have been written upon metaphysics, that if the celebrated Alexandrian Library consisted of such books, he did not regret its destruction. But though he might *now* regret the bushels of chaff which he has turned over, only to obtain a few grains of corn, still he has the satisfaction of knowing that he has not heedlessly rejected any man's opinions, but is able to give a satisfactory reason for the faith which is within him. This he could not have done, had he left unexamined authors considered by the world worthy of credit. Hume, in his beautiful and masterly essay on "Liberty and Necessity," says, "I own that this dispute has been so much canvassed on all hands, and has led philosophers into such a labyrinth of obscure sophistry, that it is no wonder if a sensible reader indulge his ease so far as to turn a deaf ear to the proposal of such a question, from which he can expect neither instruction nor entertainment. But the state of the argument here proposed may, perhaps, serve to renew his attention; as it has more novelty, promises at least some decision of the controversy, and will not much disturb his ease by any intricate or obscure reasoning."

When an apology was deemed necessary by such an accomplished and able writer as Hume to introduce a subject of such immense importance as the one upon which he was treating – how much more necessary is an apology from me for the *manner* in which I have treated the subject of my labours? With all sincerity, then, I offer the most ample apology to the readers of the *Oracle* for the uninteresting and unpopular manner in which I have treated the important and really interesting question, when properly handled, of animal gradation. The fault was in the operator and not in the subject. I was induced to enter upon the investigation from the very favourable reception which the articles of my friend Southwell had met with – forgetful that therein lay the reason which should check, rather than incite, me to the attempt. Having, though the medium of geology and comparative anatomy, been led to suspect the truth of the generally received opinion of man's origin – I thought I could not do better than place in a consecutive form the facts which had first led me to think upon, and ultimately to dispute the truth of, that opinion. This course in other hands might have been fraught with beneficial results, but in my case I fear it has failed. A much valued friend, and one who has had ample opportunities of arriving at a correct conclusion, writes me that my articles upon the Theory of Regular Gradation "are wished to be discontinued." In justice to myself, and without egotism, I may say, that contrary opinions have come to my knowledge. In deference, then, to the general wishes of the readers of the *Oracle*, and agreeably to my own inclinations, I shall conclude the subject as quickly as possible. But, inasmuch as I have some interesting matter, which may not be in the possession of many parties, I trust I may be excused for occupying their attention for a few more weeks.

I have brought to a close my digest of the comparative anatomy of the animal kingdom. In some twenty numbers, comprising at the most forty pages, I have given the chief points of analogy which the researches of modern investigators have proved to exist between the various beings of the animate world. Professor Evers, to whose "Compendium of Comparative Anatomy," I have principally been indebted, says that he has consulted the works of Baron Cuvier, Linnæus, Grant, Lamark, Owen, Lawrence, Sir Charles Bell, Blumenbach, Sir E. Home, Hunter, and many others, a guarantee that the statements made are based upon well-ascertained facts, and not upon mere conjecture.

Before concluding, I purpose giving a few extracts from Mr. Charles White's work, entitled "An account of the Regular Gradation in man, and in different animals and vegetables; and from the former to the latter." A great portion, in fact *all*, of this interesting book would be worth reprinting, but the smallness of the space which can be devoted to the subject in these pages renders it impossible, even if it were advisable, and I shall, therefore, confine myself principally to his concluding remarks. Afterwards, I shall give the philosophy of the celebrated M. Lamark, as explained by Mr. Lyell, which, with some few observations of my own will, I believe, bring my labours to a close.

Mr. White, in the work above named, says, "The inferences to be drawn from the facts given, are these: – That there is a general gradation from man through the animal race; from animals to vegetables, and through the whole vegetable system. By gradation, I mean the various degrees in the powers, faculties, and organisation. The gradation from man to animals is not by one way; the person and actions descend to the ourang-outang, but the voice to birds, as has been observed.

"That there are many quadrupeds, insects, birds, and fishes, which appear to be created for particular climates, and cannot live in any other.

"That many animals and vegetables exist in the old world, which were not found in the new one, when discovered by Columbus; and that there are many animals and vegetables found in the new world, which were never known in the old.

"Lastly, that those animals which were common to both worlds, were only to be met with in the northern hemisphere, in which the new and old world had probably communications near the north pole. These animals were about twenty-six in number.

"The gradation that exists in the animal kingdom, extends to the organs of sensation, as these differ in different subjects with respect to sensibility, dimensions, figure, &c. It also extends to the various dispositions in the animal economy. It may be traced from the brain of the human European, which is the largest of any animal we are acquainted with, be its size ever so enormous, to the polypus, who has none at all – from the keen eye of an eagle and a hawk, and the great number of eyes of a fly, to those animals that have none, as the blind worm – from the acute olfactory nerves of a dog, to those animals who have not the faculty of smelling – from the large ears of an ass, to those that are void of that organ – from animals whose feelings is exquisite; to those that have comparatively little sensation – from the catamenia of an European female, to those animals that have no such discharge – from animals that have the greatest evacuation by sweat, to such as have not that discharge, as the dog – from the quills of the porcupine, the spines of the hedge-hog, and the long coarse hair of an arctic bear, to the soft sleek hair of the mole and the dormouse – from the thin skin of the human European face, to the thick coat of mail of the elephant and the rhinoceros – from the milk white to the jet black colour of many animals and vegetables, both birds, quadrupeds, fishes, insects, reptiles, and the flowers of plants – from the high prominent nose of the human European, to those that are perfectly flat – from the projecting chin of man, to those animals who have no chin – from the perpendicular face of the human European, to the horizontal one of the woodcock – from the short jawbone of man to that of the whale, which is the longest known – from the double row of sharp teeth of the shark, the grinders and tusks of an elephant, and the teeth of a lobster, which are placed in the stomach, to those which have none – from the gizzard of a bird, which is composed of two strong muscles placed opposite to and acting upon each other as two grindstones (the two flat lateral surfaces of the grinding cavity being lined with a thick horny substance) to the most tender membranous stomach, which digests the food by its own gastric juice, the former being more peculiarly adapted for vegetable and the latter for animal food – from those animals which bring forth but one at a birth, to those who are seen to produce a thousand, as the spider and the beetle.

"Lavater is of opinion, that the same gradation holds good in physiognomy; he says, 'From the weakest of winged insects up to the towering eagle, from the worm which crawls under our feet, up to the elephant, up to the formidable lion, you everywhere discover a gradation of physiognomical expression.' "

[...]

THEORY OF REGULAR GRADATION.
XXXIX.

"M. Bonnet observes, that if we survey the principal productions of nature, we shall perceive, that betwixt those of a different class, and even those of a different species, there will always be found some which will apparently link the classes or species together. The polypus forms the most striking instance, which combines the properties of an animal with those of a vegetable. He has given us a scale of beings on the principle of gradation, and pointed out those particular subjects which seem to connect the different classes, so as to form but one group of organised bodies. Whatever exceptions may be made to the scale, it is certainly ingenious; we shall give it a place. Idea of a scale of natural beings by M. Bonnet:

Man.
Orang-outang
Monkey
Quadrupeds.
Flying-squirrel
Bat
Ostrich
Birds.
Aquatic bird
Amphibious bird
Flying-fish
Fish.
Creeping-fish
Eel
Water-serpents
Reptiles.
Slug
Snail

Snails (with shells).
Pipe-worms
Moth
Insects.
Gall-insects
Tape-worm
Polypus
Sea-nettle
Sensitive-plant
Plants.
Lichens
Mouldiness
Mushrooms
Truffles
Coral
Lithophytes
Amianthes
Tales, Gypsums, Selenites
Slate
Stones.
Figured stones
Chrystallised stones
Salts.
Vitriols
Metals.
Semi-metals
Brimstone.
Bitumens
Earths.
Pure Earth
Water.
Air.
Fire.
More Subtile Matter.

"Where, for the illustration of any subject, a variety of facts are collected from every part of the globe, many of them must of course depend upon the relation of persons whose impartiality, or whose accuracy of observation, may in some instances be questioned. Enough, however, it is hoped, is ascertained to disprove the theories by which naturalists have attempted to account for what they denominate varieties of the human species; and to establish that of a gradation, as well of the human race, as of the animal and vegetable kingdoms in general.

"A gradation in the human race, supposing all to have descended from one pair, could only be the temporary result of accidental causes, and would scarcely merit a minute investigation. But as a contemplation of the facts produced, leads to the conclusion, that various species of men were originally created and separated, by marks sufficiently discriminative, it becomes an important object, in general physiology, to trace the lines of distinction. Previously to discussing the question of species, it seems necessary to consider the signification of the term as used by naturalists.

"It has been found convenient for the purposes of science, to divide the three kingdoms into classes, orders, genera, and species, each superior denomination comprehending one or more of the inferior. With respect to the three first divisions, nature herself does not seem to define, or even to recognise them, but leaves it for the sagacity of the naturalist to seize the leading characteristics, and to arrange her productions accordingly. Of the last, however, there must be some determinate and invariable number, otherwise the productions of nature would be liable to change every generation; and the species of animals, vegetables, and minerals, at this day, might be expected to be very different from what they were one or two thousand years ago. We find the fact to be, that were nature is left to herself, and not interrupted by the artifice of man, as in wild animals, all kinds maintain their respective specific distinctions, through a series of generations; and that anomalous productions are rarely met with. This leaves us to infer, as most naturalists have done, that species were originally so created and constituted, as to be kept apart from each other, with certain characteristic distinctions, which form a proper subject for investigation. These distinctions, notwithstanding, have not, as it should seem, been ascertained in all cases – a diversity of opinion respecting them still prevailing.

"The most generally received characteristic of species is derived from generation. When animals, however unlike, can breed together, and their offspring is prolific, it has been deemed sufficient to warrant the conclusion that they are of the same species, the diversity of the parent animals being attributed to accidental circumstances. On this principle, not only the various kinds of dogs, but even foxes, wolves, and jackalls, must be considered as of one species. It is allowed, however, that some animals, of different species, will breed together, as the horse and ass; but the circumstance of their offspring being barren, effectually precludes the possibility of their being of the same species. Some proofs have indeed been adduced, and upon good authority, of mules generating, although it must be acknowledged that such instances are extremely rare.

"This hypothetic characteristic of species, derived from generation, will, I am afraid, not be found to agree with facts well ascertained. It is known to every one, notwithstanding individuals of the same species are all discriminated one from another, yet that like animals generally produce their like, within certain degrees – even when the parent animals differ much from each other, the offspring, under the influence of this law, is of an intermediate nature, partaking of the properties of both. This being admitted, it is next to impossible to conceive in what manner the species of dogs, for instance, could have branched out into that diversity of kinds above alluded to, supposing that they are descended from one pair. But, even waving the inquiry how they came to be so diversified, taking them as they really are at this moment, and allowing them a free intercourse, they ought all to be reduced, in a few generations, to one common mould, resulting from a mixture of all the varieties now subsisting. This certainly ought to follow, on the supposition that they are all of one species, and equally prolific.

"It may be said, the supposition of a free intercourse is not admissible, because the varieties once subsisting are carefully preserved by attention in the breeding. This is perhaps in some degree true; but still an inquiry will be suggested – why does not nature, though under these restrictions, occasionally produce those varieties which she is supposed to have been the author of at some former period? Why does not the greyhound occasionally produce a mastiff, the spaniel a bull-dog and the lap-dog a wolf? These are difficulties which the advocates for the hypothesis ought, if possible, to obviate; since, whilst they exist, they seem to render it altogether untenable.

"I should rather suppose that the different kinds of dogs, which, from time immemorial, have preserved their distinctive qualities, are in reality separate species of animals; and that all others are only varieties, or mongrels, produced by the intermixture of those species, and which, like the mule, in one, two, or more generations, in the mongrel line, lose their prolific quality, and consequently become extinct. Neither the truth nor falsehood of this opinion can, I apprehend, be proved absolutely from any facts known at present; but thus much must be admitted, that the present and former states of the dog kind, are perfectly in agreement with the hypothesis just advanced. Most certainly, the state of this domestic animal, at present, in regard to kinds and varieties, compared with what it was at any former period, does not lead us to conclude that the varieties are upon the point of swallowing up the different kinds, which are as numerous and excellent as in any former period – yet this conclusion is always presented to us by the other theory."

[...]

THEORY OF REGULAR GRADATION.
XL.

"Mr. Long, in his history of Jamacia, affirms that he never knew two mulattoes have any offspring; and he seems suspicious, that in the few instances where a mulatto-woman, married to a mulatto-man, may have had offspring, it is very probably derived from another quarter. I should, however, suppose, that numerous instances might be found where two mulattoes have had offspring; yet, certainly, the fact related by Mr. Long is sufficient to warrant the conjecture, that mulattoes, confined to themselves, are not so prolific as others. The small proportionate number of mulattoes in the West Indies and in America, compared with what they should have been if equally prolific with other people, is a striking circumstance.

"If this doctrine be admitted, it will be asked, how are we to distinguish species from varieties? In answer to this, it may be observed, that varieties are reducible to the common stock again. Thus, amongst men, Albinos are varieties. which do not fail in succeeding generations to return to the common stock. In vegetables, the variegated holly, for instance, will return to the common green holly, when propagated by seed, and can only be preserved as a variety by grafting. With respect to family, provincial, or national varieties of feature and complexion, it may be observed, that the more confined and circumscribed the intercourse of any people may be, the more they will assume and retain a sameness of appearance; by reason that every anomaly of feature is worn out, through continual intermixture with those more nearly resembling the standard. The people of every country, therefore, which has been long inhabited, and from which foreigners have been in a great measure excluded, will have the characteristics of its first settlers, who, if few in number, might entail a peculiar cast of features on their posterity. In England, where the intercourse with foreigners has been great, and consequently a great variety of features is found, we could select individuals who would entail, if circumstanced as above, the Scotch, the Irish, or other European national characteristics, on their posterity; whereas, were we to range over Europe, we should not find men likely to propagate a race of negroes or copper-coloured Americans.

"From the numerous facts which have been adduced, it must appear evident, that various differences exist in the human race; some of which are generally known – but others, it is presumed, have never been before pointed out. In the bony system, it has been shown that the head, the arms, and the feet, differ materially; characteristic differences have also been pointed out in the hair, the colour of the skin, the complexion, the being adapted to a particular climate, and the being subject to different diseases in the same situation.

"There are but two ways of accounting for this great diversity in the human frame and constitution. 1. To suppose that the diversity, great as it is, might be produced from one pair, by the slow operation of natural causes. 2. Or to suppose that different species were originally created with those distinctive marks which they still retain.

"The advocates for the first opinion, have endeavoured to account for the colour of man from the effects of the climate in which he resides. Indeed, if climate can account for any particular diversity, it must be that of colour; for it is difficult to conceive in what manner it can affect the other differences, one way or another. Some have endeavoured to maintain, that in the course of about two centuries, during which white people from Europe have resided in the torrid zone, and negroes from Africa in the temperate zone, there is a small but visible change in colour, the whites approaching a little towards black, and the blacks towards white; but they have by no means made out the fact in either case. Indeed, as has been already observed, the fact seems to be, that no general and permanent affection of colour is produced by climate. The temporary discolouration of the skin, called tanning, seems to have no relation to the permanent colour of the skin: it arrives at its maximum a very short time after it begins, and is as soon worn off again; whereas the permanent black colour (supposing, for argument sake, it could be effected in time) must require many centuries to effect it. Thus the father, it is supposed, transmits his degree of colour to the son, and the climate still keeps increasing it; and hence, ultimately, from the climate alone, or at least from external circumstances, we are taught to expect the complete change from black to white, or the contrary, in the course of perhaps fifty or a hundred generations.

"As to the opinion, that the constitution of man may be adapted to any climate by long residence, it is not only unwarranted by facts, but is in direct opposition to all analogy, drawn from the animal and vegetable kingdoms at large.

"On the other hypothesis we can easily account for these and other diversities in the human race; or rather indeed the hypothesis itself presumes upon such diversities. Besides, we find that, in those animals which most nearly resemble man in their bodily conformation, there are a great number of species differing but in small degrees one from another. The same observation, indeed, may be extended to the animal kingdom in general. Why then should we seek to infringe this apparent law of nature in regard to man, unless to serve an hypothesis?

"The opinion here maintained, so far from degrading, tends much more to dignify the human race than the opposite one. For if, according to the latter, we admit that such great varieties can be produced in the same species as we find to exist in man, it would be easy to maintain the probability that several species of simiæ are but varieties of the species man; since they differ as little in their organisation from some individuals of the species, as these do from men in general. And if the argument be still further extended, almost all the animal kingdom might be deduced from one pair, and be considered as one family; than which a more degrading notion certainly cannot be entertained. But the opinion advanced above, effectually precludes any such consequences, as it places each species upon its own proper basis, and debars them from intermixing with other species, unless nearly resembling themselves, and even that in a limited manner.

"Different species of men being once admitted, it will become a proper object of physiological inquiry to determine their number and distinction, with the merits, excellencies, and defects of each. In pursuing this inquiry there is no doubt but gradation will afford the proper clue to direct us. What the number of species may be, is not perhaps easy to determine. The four quarters of the globe will each, probably, furnish us with at least one. In Africa, however, there seems to be more than one species; and perhaps the lowest degree of the human race resides there. I am inclined to think that hair, rather than colour, ought to guide us in that quarter; and that it is not the blackest inhabitants, but those with extremely short hair, and a most ungracious appearance, as the Hottentots, who may be reckoned the lowest on the scale of humanity. The Negro, the American, some of the Asiatic tribes, and the European, seem evidently to be different species.

"Ascending the line of gradation, we come at last to the white European; who being most removed from the brute creation, may, on that account, be considered as the most beautiful of the human race. No one will doubt his superiority in intellectual powers; and I believe it will be found that his capacity is naturally superior also to that of every other man. Where shall we find, unless in the European, that nobly arched head, containing such a quantity of brain, and supported by a hollow conical pillar, entering its centre? Where the perpendicular face, the prominent nose, and round projecting chin? Where that variety of features, and fulness of expression; those long, flowing, graceful ringlets; that majestic beard, those rosy cheeks, and coral lips? Where that erect posture of the body and noble gait? In what other quarter of the globe shall we find the blush that overspreads the soft features of the beautiful women of Europe, that emblem of modesty, of delicate feelings, and of sense? Where that nice expression of the amiable and so ter passions in the countenance; and that general elegance of features and complexion? Where, except on the bosom of the European woman, two such plump and snowy white hemispheres, tipped with vermillion?"

[...]

THEORY OF REGULAR GRADATION.
XLI.

I CONCLUDED in my last my extracts from Mr. White, and I shall now proceed, in conformity with my plan, to give Mr. Lyell's digest of M. Lamarck's theory of transmutation of species. Mr. Lyell says:

"Before we can advance a step, we must be able to define precisely the meaning which we attach to the term species. This is even more necessary in geology than in the ordinary studies of the naturalist; for they who deny that such a thing as a species exists, concede nevertheless that a botanist or zoologist may reason as if the specific character were constant, because they confine their observations to a brief period of time. Just as the geographer, in constructing his maps from century to century, may proceed as if the apparent places of the fixed stars remained absolutely the same, and as if no alteration was brought about by the precession of the equinoxes, so it is said in the organic world, the stability of a species may be taken as absolute, if we do not extend our views beyond the narrow period of human history; but let a sufficient number of centuries elapse, to allow of important revolutions in climate, physical geography, and other circumstances, and the characters, say they, of the descendants of common parents may deviate indefinitely from their original type.

"Now, if these doctrines be tenable, we are at once presented with a principle of incessant change in the organic world, and no degree of dissimilarity in the plants and animals which may formerly have existed, and are found fossil, would entitle us to conclude that they may not have been the prototypes and progenitors of the species now living. Accordingly, M. Geoffroy St. Hilaire has declared his opinion, that there has been an uninterrupted succession in the animal kingdom effected by means of generation, from the earliest ages of the world up to the present day; and that the ancient animals whose remains have been preserved in the strata, however different, may nevertheless have been the ancestors of those now in being. Although this notion is not generally received, we feel that we are not warranted in assuming the contrary, without fully explaining the data and reasoning by which we conceive it may be refuted.

"We shall begin by stating as concisely as possible all the facts and ingenious arguments by which the theory has been supported, and for this purpose we cannot do better than offer the reader a rapid sketch of Lamarck's statement of the proofs which he regards as confirmatory of the doctrine, and which he has derived partly from the works of his predecessors, and in part from original investigations.

"We shall consider his proofs and inferences in the order in which they appear to have influenced his mind, and point out some of the results to which he was led while boldly following out his principles to their legitimate consequences.

"The name of species, observes Lamarck, has been usually applied to 'every collection of similar individuals, produced by other individuals like themselves.' This definition, he admits, is correct, because every living individual bears a very close resemblance to those from which it springs. But this is not all which is usually implied by the term species, for the majority of naturalists agree with Linnæus in supposing that all the individuals propagated from one stock have certain distinguishing characters in common which will never vary, and which have remained the same since the creation of each species.

"In order to shake this opinion, Lamarck enters upon the following line of argument. The more we advance in the knowledge of the different organised bodies which cover the surface of the globe, the more our embarrassment increases, to determine what ought to be regarded as a species, and still more how to limit and distinguish genera. In proportion as our collections are enriched, we see almost every void filled up, and all our lines of separation effaced; we are reduced to arbitrary determinations, and are sometimes fain to seize upon the slight differences of mere varieties, in order to form characters for what we choose to call a species, and sometimes we are induced to pronounce individuals but slightly differing, and which others regard as true species, to be varieties."

[...]

THEORY OF REGULAR GRADATION.
XLII.

(Lyell's Digest of Lamark's Theory.)

"The greater the abundance of natural objects assembled together, the more do we discover proofs that everything passes, by insensible shades, into something else; that even the more remarkable differences are evanescent, and that nature has, for the most part, left us nothing at our disposal for establishing distinctions, save trifling, and, in some respects, puerile particularities.

"We find that many genera amongst animals and plants are of such an extent, in consequence of the number of species referred to them, that the study and determination of these last has become almost impracticable. When the species are arranged in a series, and placed near to each other, with due regard to their natural affinities, they each differ in so minute a degree from those next adjoining, that they almost melt into each other, and are in a manner confounded together. If we see isolated species, we may presume the absence of some more closely connected, and which have not yet been discovered. Already are there genera, and even entire orders – nay, whole classes, which present an approximation to the state of things here indicated.

"If, when species have been thus placed in a regular series, we select one, and then, making a leap over several intermediate ones, we take a second, at some distance from the first, these two will, on comparison, be seen to be very dissimilar; and it is in this manner that every naturalist begins to study the objects which are at his own door. He then finds it an easy task to establish generic and specific distinctions; and it is only when his experience is enlarged, and when he has made himself master of the intermediate links, that his difficulties and ambiguities begin. But while we are thus compelled to resort to trifling and minute characters, in our attempt to separate species, we find a striking disparity between individuals which we know to have descended from a common stock, and these newly-acquired peculiarities are regularly transmitted from one generation to another, constituting what are called *races*.

"From a great number of facts, continues the author, we learn that, in proportion as the individuals of one of our species change their situation, climate, and manner of living, they change also, by little and little, the consistence and proportions of their parts, their form, their faculties, and even their organisation, in such a manner, that everything in them comes at last to participate in the mutations to which they have been exposed. Even in the same climate a great difference of situation and exposure causes individuals to vary; but if these individuals continue to live and to be reproduced under the same difference of circumstances, distinctions are brought about in them which become in some degree essential to their existence. In a word, at the end of many successive generations, these individuals, which originally belonged to another species, are transformed into a new and distinct species. Thus, for example, if the seeds of a grass, or any other plant which grows naturally in a moist meadow, be accidentally transported, first to the slope of some neighbouring hill, where the soil, although at a greater elevation, is damp enough to allow the plant to live; and if, after having lived there, and having been several times regenerated, it reaches by degrees the drier and almost arid soil of a mountain declivity, it will then, if it succeeds in growing and perpetuates itself for a series of generations, be so changed that botanists who meet with it will regard it as a particular species. The unfavourable climate in this case, deficiency of nourishment, exposure to the winds, and other causes, give rise to a stunted and dwarfish race, with some organs more developed than others, and having proportions often quite peculiar.

"What nature brings about in a great lapse of time, we occasion suddenly by changing the circumstances in which a species has been accustomed to live. All are aware that vegetables taken from their birth-place and cultivated in gardens, undergo changes which render them no longer recognisable as the same plants. Many which were naturally hairy become smooth or nearly so; a great number of such as were creepers and trailed along the ground, rear their stalks and grow erect. Others lose their thorns or asperities; others, again, from the ligneous state which their stem possessed in hot climates, where they were indigenous, pass to the herbaceous, and, among them, some which were perennials become mere annuals. So well do botanists know the effects of such changes of circumstances, that they are averse to describe species from garden specimens, unless they are sure that they have been cultivated for a very short period.

" 'Is not the cultivated wheat' (*triticum sativum*), asks Lamarck, 'a vegetable brought by man into the state in which we now see it? Let any one tell me in what country a similar plant grows wild, unless where it has escaped from cultivated fields? Where do we find in nature our cabbages, lettuces, and other culinary vegetables, in the state in which they appear in our gardens? Is it not the same in regard to a great quantity of animals which domesticity has changed or considerably modified?' Our domestic fowls and pigeons are unlike any wild birds. Our domestic ducks and geese have lost the faculty of raising themselves into the higher regions of the air, and crossing extensive countries in their flight, like the wild ducks and wild geese from which they were originally derived. A bird which we breed in a cage cannot, when restored to liberty, fly like others of the same species which have been always free. This small alteration of circumstances, however, has only diminished the power of flight, without modifying the form of any part of the wings. But when individuals of the same race are retained in captivity during a considerable length of time, the form even of their parts is gradually made to differ, especially if climate, nourishment, and other circumstances, be also altered.

"The numerous races of dogs which we have produced by domesticity, are nowhere to be found in a wild state. In nature we should seek in vain for mastiffs, harriers, spaniels, greyhounds, and other races, between which the differences are sometimes so great, that they would be readily admitted as specific between wild animals; 'Yet all these have sprung originally from a single race, at first approaching very near to a wolf, if, indeed, the wolf be not the true type which at some period or other was domesticated by man.'

"Although important changes in the nature of the places which they inhabit modify the organisation of animals as well as vegetables, yet the former, says Lamarck, require more time to complete a considerable degree of transmutation, and, consequently, we are less sensible of such occurrences. Next to a diversity of the medium in which animals or plants may live, the circumstances which have most influence in modifying their organs are differences in exposure, climate, the nature of the soil, and other local particulars. These *circumstances* are as varied as are the characters of species, and, like them, pass by insensible shades into each other, there being every intermediate gradation between the opposite extremes. But each locality remains for a very long time the same, and is altered so slowly that we can only become conscious of the reality of the change, by consulting geological monuments, by which we learn that the order of things which now reigns in each place has not always prevailed, and by inference anticipate that it will not always continue the same.

"Every considerable alteration in the local circumstances in which each race of animals exists, causes a change in their wants, and these new wants excite them to new actions and habits. These actions require the more frequent employment of some parts before but slightly exercised, and then greater development follows as a consequence of their more frequent use. Other organs no longer in use are impoverished and diminished in size, nay, are sometimes entirely annihilated, while in their place new parts are insensibly produced for the discharge of new functions."

[...]

THEORY OF REGULAR GRADATION.
XLIII.

(Lyell's Digest of Lamarck's Theory.)

"We must here (says Mr. Lyell) interrupt the author's argument, by observing that no positive fact is cited to exemplify the substitution of some *entirely new* sense, faculty, or organ, in the room of some other suppressed as useless. All the instances adduced go only to prove that the dimensions and strength of members and the perfection of certain attributes may, in a long succession of generations, be lessened and enfeebled by disuse; or, on the contrary, be matured and augmented by active exertion, just as we know that the power of scent is feeble in the greyhound, while its swiftness of pace and its acuteness of sight are remarkable – that the harrier and stag-hound, on the contrary, are comparatively slow in their movements, but excel in the sense of smelling.

"We point out to the reader this important chasm in the chain of the evidence, because he might otherwise imagine that we had merely omitted the illustrations for the sake of brevity, but the plain truth is, that there were no examples to be found; and when Lamarck talks 'of the efforts of internal sentiment,' 'the influence of subtle fluids,' and the 'acts of organisation,' as causes whereby animals and plants may acquire *new organs*, he gives us names for things, and with a disregard to the strict rules of induction, resorts to fictions, as ideal as the 'plastic virtue,' and other phantoms of the middle ages.

"It is evident, that if some well authenticated facts could have been adduced to establish one complete step in the process of transformation, such as the appearance, in individuals descending from a common stock, of a sense or organ entirely new, and a complete disappearance of some other enjoyed by their progenitors, that time alone might then be supposed sufficient to bring about any amount of metamorphosis. The gratuitous assumption, therefore, of a point so vital to the theory of transmutation, was unpardonable on the part of its advocate.

"But to proceed with the system; it being assumed as an undoubted fact, that a change of external circumstances may cause one organ to become entirely obsolete, and a new one to be developed such as never before belonged to the species, the following proposition is announced, which, however staggering and absurd it may seem, is logically deduced from the assumed premises. 'It is not the organs, or, in other words, the nature and form of the parts of the body of an animal which have given rise to its habits, and its particular faculties, but on the contrary, its habits, its manner of living, and those of its progenitors have in the course of time determined the form of its body, the number and condition of its organs, in short, the faculties which it enjoys. Thus otters, beavers, water-fowl, turtles, and frogs, were not made web-footed in order that they might swim; but their wants having attracted them to the water in search of prey, they stretched out the toes of their feet to strike the water and move rapidly along its surface. By the repeated stretching of their toes, the skin which united them at the base acquired a habit of extension, until in the course of time the broad membranes which now connect their extremities were formed.

" 'In like manner the antelope and the gazelle were not endowed with light agile forms, in order that they might escape by flight from carnivorous animals; but having been exposed to the danger of being devoured by lions, tigers, and other beasts of prey, they were compelled to exert themselves in running with great celerity, a habit which, in the course of many generations, gave rise to the peculiar slenderness of their legs, and the agility and elegance of their forms.

" 'The cameleopard was not gifted with a long flexible neck because it was destined to live in the interior of Africa, where the soil was arid and devoid of herbage, but being reduced by the nature of that country to support itself on the foliage of lofty trees, it contracted a habit of stretching itself up to reach the high boughs, until its fore-legs became longer than the hinder, and its neck so elongated, that it could raise its head to the height of twenty feet above the ground.'

"Another line of argument is then entered upon, in farther corroboration of the instability of species. In order it is said that individuals should perpetuate themselves unaltered by generation, those belonging to one species ought never to ally themselves to those of another; but such sexual unions do take place, both among plants and animals; and although the offspring of such irregular connexions are usually sterile, yet such is not always the case. Hybrids have sometimes proved prolific where the disparity between the species was not too great; and by this means alone, says Lamarck, varieties may gradually be created by near alliances, which would become races, and in the course of time would constitute what we term species.

"But if the soundness of all these arguments and inferences be admitted, we are next to inquire, what were the original types of form, organisation, and instinct, from which the diversities of character, as now exhibited by animals and plants, have been derived? We know that individuals which are mere varieties of the same species, would, if their pedigree could be traced back far enough, terminate in a single stock; so according to the train of reasoning before described, the species of a genus, and even the genera of a great family, must have had a common point of departure. What then was the single stem from which so many varieties of form have ramified? Were there many of these, or are we to refer the origin of the whole animate creation, as the Egyptian priests did that of the universe, to a single egg?

"In the absence of any positive data for framing a theory on so obscure a subject, the following considerations were deemed of importance to guide conjecture.

"In the first place, if we examine the whole series of known animals, from one extremity to the other, when they are arranged in the order of their natural relations, we find that we may pass progressively, or at least with very few interruptions, from beings of more simple to those of a more compound structure; and in proportion as the complexity of their organisation increases, the number and dignity of their faculties increase also. Among plants a similar approximation to a graduated scale of being is apparent. Secondly, it appears from geological observations, that plants and animals of more simple organisation existed on the globe before the appearance of those of more compound structure, and the latter were successively formed at later periods: each new race being more fully developed than the most perfect of the preceding era.

"Of the truth of the last-mentioned geological theory, Lamarck seems to have been fully persuaded; and he also shows that he was deeply impressed with a belief prevalent amongst the older naturalists, that the primeval ocean invested the whole planet long after it became the habitation of living beings, and thus he was inclined to assert the priority of the types of marine animals to those of the terrestrial, and to fancy, for example, that the testacea of the ocean existed first, until some of them, by gradual evolution, were *improved* into those inhabiting the land.

"These speculative views had already been, in a great degree, anticipated by Delamétherie in his Teliamed, and by several modern writers, so that the tables were completely turned on the philosophers of antiquity, with whom it was a received maxim, that created things were always most perfect when they came first from the hands of their maker, and that there was a tendency to progressive deterioration in sublunary things when left to themselves –

– – omnia fatis
In pejus ruere, ac retrò sublapsa referri.

"So deeply was the faith of the ancient schools of philosophy imbued with this doctrine, that to check this universal proneness to degeneracy, nothing less than the re-intervention of the deity was thought adequate; and it was held, that thereby the order, excellence, and pristine energy of the moral and physical world had been repeatedly restored."

[...]

THEORY OF REGULAR GRADATION.
XLIV.

(Lyell's Digest of Lamarck's Theory.)

"But when the possibility of the indefinite modification of individuals descending from common parents was once assumed, as also the geological generalisation respecting the progressive development of organic life, it was natural that the ancient dogma should be rejected, or rather reversed; and that the most simple and imperfect forms and faculties should be conceived to have been the originals whence all others were developed. Accordingly, in conformity to these views, inert matter was supposed to have been first endowed with life; until, in the course of ages, sensation was superadded to mere vitality: sight, hearing, and other senses, were afterwards acquired; and then instinct and the mental faculties; until, finally, by virtue of the tendency of things to *progressive improvement*, the irrational was developed into the rational.

"The reader, however, will immediately perceive, that if all the higher orders of plants and animals were thus supposed to be comparatively modern, and to have been derived in a long series of generations from those of more simple conformation, some further hypothesis became indispensable, in order to explain why, after an indefinite lapse of ages, there were still so many beings of the simplest structure. Why have the majority of existing creatures remained stationary throughout this long succession of epochs, while others have made such prodigious advances? Why are there still such multitudes of infusoria and polypes, or of confervæ and other cryptogamic plants? Why, moreover, has the process of development acted with such unequal and irregular force on those classes of beings which have been greatly perfected, so that there are wide chasms in the series; gaps so enormous, that Lamarck fairly admits we can never expect to fill them up by future discoveries?

"The following hypothesis was provided to meet these objections. Nature, we are told, is not an intelligence, nor the deity, but a delegated power – a mere instrument – a piece of mechanism acting by necessity – an order of things constituted by the supreme being, and subject to laws which are the expressions of his will. This nature is *obliged* to proceed gradually in all her operations; she cannot produce animals and plants of all classes at once, but must always begin by the formation of the most simple kinds; and out of them elaborate the more compound, adding to them successively, different systems of organs, and multiplying more and more their number and energy.

"This nature is daily engaged in the formation of the elementary rudiments of animal and vegetable existence, which correspond to what the ancients termed *spontaneous generations*. She is always beginning anew, day by day, the work of creation, by forming monads, or 'rough draughts' (ébauches), which are the only living things she ever gives birth to *directly*.

"There are distinct primary rudiments of plants and animals, and *probably* of each of the great divisions of the animal and vegetable kingdoms. These are gradually developed into the higher and more perfect classes by the slow, but unceasing agency of two influential principles: first, *the tendency to progressive advancement* in organisation, accompanied by greater dignity in instinct, intelligence, &c.; secondly, *the force of external circumstances*, or of variations in the physical condition of the earth, or the mutual relations of plants and animals. For as species spread themselves gradually over the globe, they are exposed from time to time to variations in climate, and to changes in the quantity and quality of their food; they meet with new plants and animals which assist or retard their development, by supplying them with nutriment, or destroying their foes. The nature also of each locality is in itself fluctuating, so that even if the relation of other animals and plants were invariable, the habits and organisation of species would be modified by the influence of local revolutions.

"Now, if the first of these principles, *the tendency to progressive development*, were left to exert itself with perfect freedom, it would give rise, says Lamarck, in the course of ages, to a graduated scale of being, were the most insensible transition might be traced from the simplest to the most compound structure, from the humblest to the most exalted degree of intelligence. But in consequence of the perpetual interference of the *external causes* before mentioned, this regular order is greatly interfered with, and an approximation only to such a state of things is exhibited by the animate creation, the progress of some races being retarded by unfavourable, and that of others accelerated by favourable, combinations of circumstances. Hence, all kinds of anomalies interrupt the continuity of the plan, and chasms into which whole genera or families might be inserted, are seen to separate the nearest existing portions of the series."

[...]

THEORY OF REGULAR GRADATION.
XLV.

(Lyell's Digest of Lamarck's Theory concluded.)

"SUCH is the machinery of the lamarckian system; but our readers will hardly, perhaps, be able to form a perfect conception of so complicated a piece of mechanism, unless we exhibit it in motion, and show in what manner it can work out, under the author's guidance, all the extraordinary effects which we behold in the present state of the animate creation. We have only space for exhibiting a small part of the entire process by which a complete metamorphosis is achieved, and shall, therefore, omit the mode whereby, after a countless succession of generations, a small gelatinous body is transformed into an oak or an ape. We pass on at once to the last grand step in the progressive scheme, whereby the orang-outang, having been already evolved out of a monad, is made slowly to attain the attributes and dignity of man.

"One of the races of quadrumanous animals which had reached the highest state of perfection, lost, by constraint of circumstances (concerning the exact nature of which tradition is unfortunately silent), the habit of climbing trees, and of hanging on by grasping the boughs with their feet, as with hands. The individuals of this race being obliged for a long series of generations to use their feet exclusively for walking, and ceasing to employ their hands as feet, were transformed into bimanous animals, and what before were thumbs became mere toes, no separation being required when their feet were used solely for walking. Having acquired a habit of holding themselves upright, their legs and feet assumed insensibly a conformation fitted to support them in an erect attitude, till at last these animals could no longer go on all fours without much inconvenience.

"The Angola orang, *simia troglodytes*, Linn., is the most perfect of animals, much more so than the Indian orang, *simia satyrus*, which has been called the orang-outang, although *both* are *very inferior* to man in corporeal powers and intelligence. These animals frequently hold themselves upright, but their organisation has *not yet* been sufficiently modified to sustain them habitually in this attitude, so that the standing posture is very uneasy to them. When the Indian orang is compelled to take flight from pressing danger, he immediately falls down upon all fours, showing clearly that this was the original position of the animal. Even in man, whose organisation, in the course of a long series of generations, has advanced so much farther, the upright posture is fatiguing, and can only be supported for a limited time, and by aid of the contraction of many muscles. If the vertebral column formed the axis of the human body, and supported the head and all the other parts in equilibrium, then might the upright position be a state of repose; but as the human head does not articulate in the centre of gravity; as the chest, belly, and other parts, press almost entirely forward with their whole weight, and as the vertebral column reposes upon an oblique base, a watchful activity is required to prevent the body from falling. Children which have large heads and prominent bellies can hardly walk at the end even of two years, and their frequent tumbles indicate the natural tendency in man to resume the quadrupedal state.

"Now, when so much progress had been made by the quadrumanous animals before mentioned, that they could hold themselves habitually in an erect attitude, and were accustomed to a wide range of vision, and ceased to use their jaws for fighting, and tearing, or for clipping herbs for food, their snout became gradually shorter, their incisor teeth became vertical, and the facial angle grew more open.

"Among other ideas which the natural *tendency to perfection* engendered, the desire of ruling suggested itself, and this race succeeded at length in getting the better of the other animals, and made themselves masters of all those spots on the surface of the globe which best suited them. They drove out the animals which approached nearest to them in organisation and intelligence, and which were in a condition to dispute with them the good things of this world, forcing them to take refuge in deserts, woods, and wildernesses, where their multiplication was checked, and the progressive development of their faculties retarded, while in the mean time the dominant race spread itself in every direction, and lived in large companies where new wants were successively created, exciting them to industry, and gradually perfecting their means and faculties.

"In the supremacy and increased intelligence acquired by the ruling race, we see an illustration of the natural tendency of the organic world to grow more perfect, and in their influence in repressing the advance of others, an example of one of those disturbing causes before enumerated, that *force of external circumstances*, which causes such wide chasms in the regular series of animated beings.

"When the individuals of the dominant race became very numerous, their ideas greatly increased in number, and they felt the necessity of communicating them to each other, and of augmenting and varying the signs proper for the communication of ideas. Meanwhile the inferior quadrumanous animals, although most of them were gregarious, acquired no new ideas, being persecuted and restless in the deserts, and obliged to fly and conceal themselves, so that they conceived no new wants. Such ideas as they already had remained unaltered, and they could dispense with the communication of the greater part of these. To make themselves, therefore, understood by their fellows, required merely a few movements of the body or limbs – whistling, and the uttering of certain cries, varied by the inflexions of the voice.

"On the contrary, the individuals of the ascendant race, animated with a desire of interchanging their ideas, which became more and more numerous, were prompted to multiply the means of communication, and were no longer satisfied with mere pantomimic signs, nor even with all the possible inflexions of the voice, but made continual efforts to acquire the power of uttering articulate sounds, employing a few at first, but afterwards varying and perfecting them according to the increase of their wants. The habitual exercise of their throat, tongue and lips, insensibly modified the conformation of these organs, until they became fitted for the faculty of speech.

"In affecting this mighty change, 'The exigencies of the individuals were the sole agents, they gave rise to efforts, and the organs proper for articulating sounds were developed by their habitual employment.' Hence, in this peculiar race, the origin of the admirable faculty of speech; hence also the diversity of languages, since the distance of places where the individuals composing the race established themselves, soon favoured the corruption of conventional signs."

[...]

THEORY OF REGULAR GRADATION.
XLVI.

SUCH a length of time has elapsed since I first commenced this series of articles, and so many accidents have broken the thread of my discourse, making it difficult for the reader, who, for the first time, perhaps, has given his attention to the subject, to follow me, and adapt the many particular illustrations I have given of the general principle – I shall now, in conclusion, give a summary of the evidence I have adduced, show its bearing upon the question, and re-assert the superior *probability* of the theory of regular gradation, or of the transmutation of species, over the asserted *absolute creation* of all material substances and forms by an intelligent god – which is contended for by theologians and religionists all over the world.

Those of my readers who have long since thrown off the fetters of superstitious prejudice, and who upon the dawn of geology saw the consequences which would result from the establishment of the truth of its revelations, must not suppose that all the world, nor even all the thinking and intellectual world, are of the same opinion as themselves – for if they do imbibe so flattering a notion, they will be most grievously deceived. Far from christians generally of the present day, giving up the mosaic cosmogony, some of the ablest writers of the day contend for its truth over geology. In the *Morning Herald* of September last, is a long extract from the "Church of England Quarterly Review," on Scripture Geology. Upon the subject of creation is the following:

"They (geologists) talk loosely and familiarly of creation and extinction, as if these were two of the laws of matter, like attraction and repulsion – as if these were laws now in operation, the progress of which we can watch, and measure, and weigh – when they mean no such thing – and when to both words they attach *two meanings*, and, between the two, shirk the exact and rigid application of either. For creation and extinction are not co-relatives – the proper opposite to creation is annihilation – but this would not serve the same turn. They mean creation proper – the bringing into being of new forms of life: they do not mean uncreation proper – the making those beings as though they had never been – but extinction of the life, without destroying the form – and the evil of improperly assorting such terms as creation and extinction, is sure to extend to both terms. For though creation proper is the only meaning which truly agrees with their argument, since their extinction at each stage does mean the disposal of all the then existing creation in the stratum then deposited, and consequently to bring in the new inhabitants of the next stratum, there is absolutely nothing for it, in their theory, but new creation – yet they shrink from the idea, and would fain imagine that the germs of the new forms of being had been latent somewhere, in the air, or the sea, or the bowels of the earth, and were brought to light by chance, rather than by the power of god. Nay, even in the acknowledged *beginning* of all things, they will scarcely allow the proper meaning of creation, and often flatly deny it – as is the case of Dr. Buckland, quoting with approbation from Dr. Chalmers, as follows: – 'Does Moses ever say that, when god created the heavens and the earth, he did no more at that time alluded to, than transform them out of previously-existing materials?' – (Bridgewater Treatise.) So then, to create is to transform! But what is gained by this? The previously existing materials *must* have, some time or other, *been brought out of nothing!* – *must* have been *created proper* when they began to be! – *and what other time is this but the beginning, the time when god created the heavens and the earth?*"

Here, then, we have an absolute creation of something (matter) out of nothing (no matter) resolutely contended for, upon the authority of the bible, this will be further seen from the following, a part of the same article:

"Now there is one volume *which we rank higher than the book of nature*, or the volumes of science, *and to the standard of which we would bring the book of nature, assured that where they differ we have not read the book of nature aright.* We are sure that the bible speaks of only *one* creation, and at the beginning – that it speaks of only *one* revolution or catastrophe, and at the deluge – and we, therefore, scrutinise narrowly that interpretation of the book of nature which professes to discover *many* creations, and *many* catastrophes – and we assert that on scrutiny, we find it to be manifestly erroneous – inconsistent with itself and with science – and, therefore, even leaving the bible out of the question, this interpretation cannot stand."

It is unnecessary that I should endeavour to controvert these absurd dogmas, for I have already given my readers a sufficient number of geological *facts*, from the highest authorities, in disproof of the mosaic cosmogony. The two extracts following, from our two greatest geologists and natural philosophers, will be illustrative of my previous arguments, and render immediate reference to former articles unnecessary. Dr. Buckland says:

"The study of these remains will form our most interesting and instructive subject of inquiry, since it is in them that we shall find the great master-key, whereby we may unlock the secret history of the earth. They are documents which contain the evidences of revolutions and catastrophes *long antecedent* to the creation of the human race – they open the book of nature, and swell the volumes of science with the records of *many successive series of animal and vegetable creations*, of which the creation and extinction would have been equally unknown to us, but for recent discoveries in the science of geology."

Baron Cuvier, upon the same subject, says:

"It is certain that we are at present at least in the midst of a fourth succession of terrestrial animals – and that after the age of reptiles, after that of palæotheria, after that of mammoths, mastodonta, and megatheria, the age arrived in which the human species, together with some domestic animals, governs and fertilises the earth peaceably – and it is only in formations subsequent to this period, in alluvial deposits, in turf-bogs, in the recent concretions, that those bones are found in a fossil state which all belong to animals known and now existing."

Although I said, but a few minutes since, that I would not trouble you with a refutation of the Church of England Review, yet the monstrous folly of such arguments, and the unblushing effrontery of the men who can submit them to the light of day, and to the scrutiny of reason is so great, that I will, for the satisfaction of the tyro, offer one single fact as damnatory of the doctrine contended for. They say, "We are sure that *the bible* speaks of only *one* creation, *and at the beginning* – that it speaks of only *one* revolution or catastrophe, *and at the deluge*," and they thence insist that all the phenomena belonging to the science of geology, was the result of the noachian deluge. The bible declares that the noachian deluge was a consequence of man's iniquity, and that all men who were upon the face of the earth, with the exception of Noah and his family, were drowned. The different strata of the earth is crowded with the petrified remains of all other animals, excepting men and monkeys – how comes it then that no remains of these are found, if *one* convulsion was sufficient to entomb all the other varieties of animated nature? Surely, if man's wickedness induced the deity to entirely annihilate *him*, leaving not a wreck behind – the monkeys had not similarly sinned, and their remains might have been preserved, as a monument of their previous existence – but no, they also are not to be found. Cuvier says:

"It is wonderful that, among all these mammifera, of which, at the present day, the greater part have a congerminate species in the warm climates, *there has not been one quadrumanous animal, not a single bone or a single tooth of a monkey* (found), not even a bone or a tooth of an extinct species of this animal. *Neither are there any remains of man.*"

Thus much for the foundation of my theory – our next step is an examination of the forms of extinct and existing animals, and here we perceive a regular gradation of organs, beginning with the almost imperceptible evidences of animation, or animated substance, and proceeding upwards to the most complicated and positive evidence of such existence.

[...]

THEORY OF REGULAR GRADATION.
XLVII.

I HAVE traced in my various articles, the transmutation of species, or the *blending* of one animal into another, the growing out of, or changing of, one form into another. I commenced with the fossils found in the different geological strata of the earth, and showed, that not only were simple forms associated with more complex ones, but that the simplest forms of animals were contemporaneous with the lowest, or earliest strata, and that, as we proceed upwards from the granites to the tertiary, the number of complex forms increase, until at last, upon the tertiary strata, we find man, associated, however, with forms as simple as those found in the lowest strata. The first evidences of animal life, the sponges and zoophytes, would scarcely be discoverable but for chemical analysation. From the sponges and zoophytes, we next proceed to either the articulata or mollusca, both of which exhibit some points of resemblance to the class above and below them – in both articulata and mollusca are animals much superior to the *lowest* vertebrata, and in both, also, we find species which are in many respects below the *highest* radiata. Here we see one of the facts for which Lamarck contends – the blending of animals one into the other, so as to defy the discovery of *distinct* species. In the mollusca we find a near approach to the cartilaginous fishes, or the lowest order of fishes. In the lowest class of the vertebrata (fishes) we find the first type of the shoulder blade, bone of the upper arm, bones of the fore or lower arm, bones of the wrist, the oschyoides, this latter reaching a high degree of development, and many other approximations to higher orders, as well as assimilations to lower ones. After fishes come the amphibia – but here we have scarcely any advance upon fishes – they connect fishes with reptiles. In the reptilia, we meet with a still higher grade of development than in the amphibia, both in the skeleton, the viscera, and other parts of the system. Next in order come birds, in which the approaches towards the human type are very numerous. At last we arrive at the mammalia, or the class to which man belongs. This class contains thirteen orders, and the approximation of each to the human type is proportioned to the distance naturalists have assigned it from that standard. The dolphin and whale, although inhabitants of the great deep, present many points of resemblance to man – not only from the peculiar circumstance of their suckling their young, by means of mammæ, or teats, but also in respect of their osseous system.

That there is a regularity in the succession of the strata of the earth, is not disputed, and the lower or earliest formations are considered to be more crude and simple than the later or highest deposits. In the vegetable world the plants are classed according to the simplicity or complexity of their structure – so also, in the animal kingdom, men are guided by the same rule in determining the order of arrangement. But although whilst thus employed, men begin with those phenomena which are of so doubtful a character as to divide opinion upon their real nature – although they proceed step by step, by almost imperceptible degrees from one form of being to another – although they perceive throughout all their investigations no other agent in operation but matter – notwithstanding that their most minute search after a principle independent of matter, has ended in disappointment – still, do they pertinaciously and dogmatically insist that such a principle *is* in existence, and that the various operations of nature could not be carried on without it. To *disprove* this assertion is impossible, but all facts are opposed to its *probability*. Immaterialists, aware of the weakness of their position, cunningly demand of materialists a disproof of *their* assumption – such a challenge should never be heeded, it is sufficient to show its utter absurdity and untenableness. To attempt positively to demonstrate the non-existence of an immaterial principle, that "directs the atom and controls the aggregate of nature," is waste of time – but a few facts in illustration of the folly of such a belief may not be ill-timed.

The foundation of atheism is materialism, that is, the doctrine of the infinite extension, and eternal duration of matter. The atheist rejects as puerile the idea of the *creation*, either of matter or its modes. Accepting reason alone as his guide, he follows wherever she leads, and halts where she halts. Reason teaches that that which cannot be destroyed, can never have been created, and as matter cannot be destroyed, it can never have been created – hence all the modes or forms of matter which we see, must have been derived from matter's inherent, unassisted properties. The same principle which produces the cohesion of particles in a stone, causes the moon to revolve round the earth, and the earth round the sun – causes the sap to ascend the plant, and the blood to circulate through the veins and arteries of an animal – why the results should be so different we cannot tell, but that they are so we know.

Religionists, on the other hand, are compelled, to be consistent, to contend for a *bona-fide* absolute creation of matter out of nothing, for the obvious reason, that if matter was *not* created, its properties were not given to it, for its properties cannot be detached from it – we know matter's properties, but not matter itself. If matter was not created, the various forms and modes of its existence which we see around us, may reasonably be supposed to result from matter's action. Motion is a property of matter, matter has eternally existed, matter then must have been eternally in motion – must have been eternally changing its modes and forms of existence, exhibiting an infinite variety of shapes and phenomena.

Materialism being the foundation of atheism, the idea of a superior being or power, who directs and controls, manufactures and keeps in order, the infinite variety of forms and phenomena displayed in the universe is necessarily rejected, and the atheist has to seek for reasons for what he sees, in what he believes can alone be the cause, namely, the properties of matter. Satisfied with explaining visible effects, he never dreams of solving final causes – hence, in this series of articles, an attempt has been made to show the reasonableness of the belief that animal organisms have derived their existence from matter's action, and not from the manipulation or conjuration of an immaterial spirit, or being, called god.

[...]

THEORY OF REGULAR GRADATION. XLVIII.

THE fundamental principle of all religions is the belief in the existence of an immaterial being, whose attributes are infinite power and wisdom, and sometimes goodness. But all nature cries aloud against this assumption, for the earth is crowded with imperfections and folly – there is no thing perfect, no not one. Man is the victim of numberless diseases, and the earth is continually convulsed by the jarring elements. Neither animals nor plants are secure from decay, and the merest trifles often produce disease and death. These obstacles to the belief in a wise, powerful, and good god interfere not with the natural theologian or the revelationist – they *will not* see them unless they are pointed out, and when they are forced upon their attention they declare they are necessary, to enable us to appreciate the good which exists, or, that god, in his inscrutable wisdom, has undoubtedly a reason, good and sufficient, for creating such circumstances, or, rather, in their jargon, for permitting sin and death to continue – which reasons, however, his inscrutable wisdom does not see fit to make known to us. This worse than ridiculous – this vilely pernicious teaching, the atheist rejects with contempt and disgust – contempt for those who would enforce such principles, and disgust for principles thus sought to be enforced. But so long as men remain in ignorance such ideas will obtain – not only because the untutored mind finds in them a ready solution of the greatest of all difficulties – but, also, because wherever there is ignorance there is sure to be cunning, turning that ignorance to its own account, and no article in the fraudulent market has sold so rapidly, so generally, and produced such high prices, as the god article. Iron is, *par excellence*, the staple of English manufacture, and has produced, from first to last, an immense, unaccountable amount of wealth – but, in comparison with the profits which have resulted from god-mongery, the profits upon ironmongery are as a grain of sand to a mountain.

When the atheist rejects, as puerile and demoralising, the fables of the goddist, he must show a reason for the faith which is within him, or submit to be calumniated, ridiculed, and to stand alone without sympathy or support. If he disbelieves the statements of priests, it is asked *why?* if he rejects a manufacturer of natural phenomena, he is asked *why?* He should be able to reply – he should have satisfied himself *why* he differs from the majority, and if he has done so, the odds are but he will be able to explain himself to another.

The first query with which a scepticism of god belief is met, is – how can you account for natural phenomena without a god? The sceptic replies that he does not attempt to explain, or pretend to understand, the causes of recognised effects. But, persists the theist, if you reject the solution of the difficulty which I offer you, and which to me appears reasonable and conclusive, you have, doubtless, some motive for it – give me the reasons for your contrary belief that I may test their value, and become of your opinion if I find them better than my own. Although you frankly declare that you do not attempt to explain, or solve final causes – you surely do not sit apathetically down, an indifferent spectator of passing events, never caring to think about the hidden springs which produce such regular and perpetual motion. Things exist. Supposing you can never *demonstrate* why they exist – can you prevent your brain endeavouring to frame a reasonable hypothesis for their existence? Where there is mystery there is always curiosity – though men know they cannot solve the universe, still will they endeavour to satisfy themselves for its existence. Men may *talk* of the pivot of doubt, but it is a phantom, a self-delusion, a weakness of which honest men should be ashamed. Men may believe without being bigotted – those who are guided by reason have nothing to fear from embracing error, tis only necessary that they see their folly to relinquish it. To me god-belief appears reasonable, and by it I explain *all* difficulties – how do you *attempt*, without a god, to account for what you cannot understand?

This would be my language were I a theist, and I should be disappointed if the atheist replied that he only knew that things did exist, but why they existed as they did rather than in any other form, he cared not to inquire, because he was certain he should never succeed in solving the enigma. Men must have reasons for the faith which is within them – they may frequently be bad reasons, but they are, nevertheless, *the* reasons of the holders, and if you require them to be surrendered you must offer better ones in their place. The believer in a god will not give up his belief for the barren satisfaction of knowing that all is a mystery, which can never be explained – for he can, with his god, explain it to his own satisfaction – and that is far preferable to doubt and uncertainty, if a man could endure in such a state, but he cannot – and failing reasons in materialism, or atheism, which will relieve his mind, he relapses into theism, and rests satisfied.

Many theists believe in the eternity of matter, but contend for a god as necessary to fashion it into the infinite variety of shapes and conditions observable around us. The properties of matter, they say, are not sufficient alone to produce the phenomena of animal and vegetable life – it requires intelligence to produce life and to perpetuate it, and this intelligence is god. This, though a mere assumption, must be met, and an attempt made to show that the interference of a superior being is quite unnecessary, a work of supererogation – for an attentive examination of natural phenomena satisfies the reasonable man that matter is not only all that exists, but that it is all-sufficient to produce the varied phenomena observable in the universe.

Editorial Headnote

23. [Robert Chambers] *Vestiges of the Natural History of Creation* (London: John Churchill, 1844); (London: George Routledge & Sons, 1887), pp. 145–65, 174, 176–7.

The Vestiges of the Natural History of Creation was arguably the most popular evolutionary work of the nineteenth century. Indeed, it was, as the historian of science James Secord has shown, a Victorian sensation.[1] Darwin's name and ideas may have subsequently had a greater and lasting impact, but *Vestiges* far outsold *On the Origin of Species* across the nineteenth century. As a result, we need to appreciate *Vestiges* not only as indicative of the extent of debate about evolution or 'transmutation' in the decades before *Origin* but also because, when people read Darwin's work, they would have done so very much in light of *Vestiges*.[2]

Written by the Edinburgh publisher Robert Chambers, *Vestiges* was published anonymously by John Churchill, a respected publisher of scientific and medical works. The secret of who actually wrote the book was carefully guarded for a number of years, a fact that only added to the excitement the book generated. In *Vestiges,* Chambers not only presented an argument for the transmutation of one species from another, but did so in a manner that placed humans within the same natural category as other primate species.[3] Further, he did so in the context of an evolving universe.[4] Progressive change subject to natural law was the dominant theme. Chambers was careful to couch this as a law established by an omnipotent and beneficent creator – if He had chosen to create humanity through the operation of the natural laws he had created, just as He had chosen to ordain the creation, growth, and development of each human being through the natural laws pertaining to reproduction and parturition, who were we to find such fact degrading?[5] Chambers' efforts in this direction did little to dampen the religious hostility his book provoked among prominent Anglicans. William Whewell and Adam Sedgwick, for example, were both ordained churchmen, but also among the most prominent men of science of their day, and they rejected *Vestiges* and its message as profoundly irreligious.[6] The anatomist Thomas Huxley also attacked the work as that of a rank amateur.[7] Huxley was certainly no friend of the Anglican establishment and later championed Darwin and his work as a "veritable Whitworth gun in the armoury of liberalism."[8] The unitarian physiologist William Benjamin Carpenter, however, did appreciate Chambers' efforts and actually worked secretly with the publisher to revise and improve some of the scientific claims for the second edition.[9] The debate that surrounded *Vestiges* in the 1840s in many ways prepared the way for Darwin's later publication of *Origin* in 1859.

The first edition of *Vestiges* was published by Churchill in 1844; it subsequently went through many often-revised editions across the rest of the century. The text reproduced here is from the 1887 edition published by George Routledge and Sons. It was published as a part of "Morley's Library," a set of works intended to appeal to a middle-class readership. As the historian James Secord has noted, *Vestiges* was published in many different forms and formats, each intended to appeal to a different readership.[10]

Notes

1 James Secord, *Victorian Sensation. The Extraordinary Publication, Reception, and Secret Authorship of The Vestiges of the Natural History of Creation* (Chicago: University of Chicago Press, 2000).

2 Indeed, many readers of *Origin* would have been aware of the vast extent of the debate over transmutation represented by the other texts in this volume: Caden C. Testa and Piers J. Hale, *Evolution Before Darwin* (London: Routledge, 2025).

3 *Vestiges of the Natural History of Creation* (London: George Routledge & Sons, 1887), p. 204.

4 *Vestiges of the Natural History of Creation*, pp. 11–15.

5 *Vestiges of the Natural History of Creation*, p. 176–7.

6 Adam Sedgwick, "Review of Vestiges of the Natural History of Creation," *Edinburgh Review*, (July 1845), pp1–85. On Whewell, see Secord, *Victorian Sensation*, p. 136.

7 Secord, *Victorian Sensation*, p. 500–506.

8 Thomas Henry Huxley, "On the Origin of Species," *Westminster Review*, (April,1860).

9 Secord, *Victorian Sensation*, p. 143.

10 Secord, *Victorian Sensation*, pp. 124–152.

23

VESTIGES OF THE NATURAL HISTORY OF CREATION

[Robert Chambers]

Source: [Robert Chambers] *Vestiges of the Natural History of Creation* (London: John Churchill,1844); (London: George Routledge & Sons, 1887), pp. 145–65, 174, 176–7

HYPOTHESIS OF THE DEVELOPMENT OF THE VEGETABLE AND ANIMAL KINGDOMS.

It has been already intimated, as a general fact, that there is an obvious gradation amongst the families of both the vegetable and animal kingdoms, from the simple lichen and animalcule respectively, up to the highest order of dicotyledonous trees and the mammalia. Confining our attention, in the meantime, to the animal kingdom – it does not appear that this gradation passes along one line, on which every form of animal life can be, as it were, strung. There seems to be a plurality of lines; how many there are it is not necessary at present to decide. There may be two or more lines; or the lines may be branching; or the whole may be in a circle composed of minor circles, as has been recently suggested. But still it is incontestable that there are general appearances of a scale beginning with the simple and advancing to the complicated. The animal kingdom was divided by Cuvier into four sub-kingdoms, or divisions, and these exhibit an unequivocal gradation in the order in which they are here enumerated: –

Radiata (polypes, &c.).
Mollusca (pulpy animals). Articulata (jointed animals).
Vertebrata (animals with internal skeleton);

the mollusca and articulata being, as it were, two distinct parallel lines by which we pass from the radiata to the vertebrata. The gradation can, in like manner, be clearly traced in the *classes* into which the sub-kingdoms are subdivided, as, for instance, when we take those of the vertebrata in this order – fishes, reptiles, birds, mammals.

While the external forms of all these various animals are so different, it is very remarkable that the whole are, after all, variations of a fundamental plan, which can be traced as a basis throughout the whole, the variations being merely

DOI: 10.4324/9781003490548-26

modifications of that plan to suit the particular conditions in which each particular animal has been designed to live. Starting from the primeval germ which, as we have seen, is the representative of a particular order of full-grown animals, we find all others to be merely advances from that type, with the extension of endowments and modification of forms which are required in each particular case; each form, also, retaining a strong affinity to that which precedes it, and tending to impress its own features on that which succeeds. This unity of structure, as it is called, becomes the more remarkable when we observe that the organs, while preserving a resemblance, are often put to different uses. For example: the ribs become, in the serpent, organs of locomotion, and the snout is extended, in the elephant, into a prehensile instrument.

It is equally remarkable that analogous purposes are served in different animals by organs essentially different. Thus, the mammalia breathe by lungs; the fishes, by gills. These are not modifications of one organ, but distinct organs. In mammifers. the gills exist and act at an early stage of the fœtal state, but afterwards go back and appear no more; while the lungs are developed. In fishes, again, the gills only are fully developed; while the lung structure either makes no advance at all, or only appears in the rudimentary form of an air-bladder. So, also, the baleen of the whale and the teeth of the land mammalia are different organs. The whale, in embryo, shows the rudiments of teeth; but these, not being wanted, are not developed, and the baleen is brought forward instead. The land animals, we may also be sure, have the rudiments of baleen in their organisation. In many instances, a particular structure is found advanced to a certain point in a particular set of animals (for instance, feet in the serpent tribe), although it is not there required in any degree; but the peculiarity, being carried a little farther forward, is perhaps useful in the next set of animals in the scale. Such are called rudimentary organs. With this class of phenomena are to be ranked the useless mammæ of the male human being, and the unrequired process of bone in the male opossum, which is needed in the female for supporting her pouch. Such curious features are most conspicuous in animals which form links between various classes.

As formerly stated, the marsupials, standing at the bottom of the mammalia, show their affinity to the oviparous vertebrata, by the rudiments of two canals passing from near the anus to the external surfaces of the viscera, which are fully developed in fishes, being required by them for the respiration of aërated waters, but which are not needed by the atmosphere-breathing marsupials. We have also the peculiar form of the sternum and rib-bones of the lizards *represented* in the mammalia in certain white cartilaginous lines traceable among their abdominal muscles. The struthionidæ (birds of the ostrich type) form a link between birds and mammalia, and in them we find the wings imperfectly or not at all developed, a diaphragm and urinary sac (organs wanting in other birds), and feathers approaching the nature of hair. Again, the ornithorhynchus belongs to a class at the bottom of the mammalia, and approximating to birds, and in it behold the bill and web-feet of that order!

For further illustration, it is obvious that, various as may be the lengths of the upper part of the vertebral column in the mammalia, it always consists of the same parts. The giraffe has in its tall neck the same number of bones with the pig, which scarcely appears to have a neck at all.[a] Man, again, has no tail; but the notion of a much-ridiculed philosopher of the last century is not altogether, as it happens, without foundation, for the bones of a caudal extremity exist in an undeveloped state in the *os coccygis* of the human subject. The limbs of all the vertebrate animals are, in like manner, on one plan, however various they may appear. In the hind-leg of a horse, for example, the angle called the hock is the same part which in us forms the heel; and the horse, and all other quadrupeds, with the almost solitary exception of the bear, walk, in reality, upon what answers to the toes of a human being. In this and many other quadrupeds the fore-part of the extremities is shrunk up in a hoof, as the tail of the human being is shrunk up in the bony mass at the bottom of the back. The bat, on the other hand, has these parts largely developed. The membrane, commonly called its wing, is framed chiefly upon bones answering precisely to those of the human hand; its extinct congener, the pterodactyle, had the same membrane extended upon the fore-finger only, which in that animal was prolonged to an extraordinary extent. In the paddles of the whale and other animals of its order, we see the same bones as in the more highly developed extremities of the land mammifers; and even the serpent tribes, which present no external appearance of such extremities, possess them in reality, but in an undeveloped or rudimental state.

The same law of development presides over the vegetable kingdom. Amongst phanerogamous plants, a certain number of organs appear to be always present, either in a developed or rudimentary state; and those which are rudimentary can be developed by cultivation. The flowers which bear stamens on one stalk and pistils on another, can be caused to produce both, or to become perfect flowers, by having a sufficiency of nourishment supplied to them. So, also, where a special function is required for particular circumstances, nature has provided for it, not by a new organ, but by a modification of a common one, which she has effected in development. Thus, for instance, some plants destined to live in arid situations, require to have a store of water which they may slowly absorb. The need is arranged for by a cup-like expansion round the stalk, in which water remains after a shower. Now the *pitcher*, as this is called, is not a new organ, but simply the metamorphosis of a leaf.

These facts clearly show how all the various organic forms of our world are bound up in one – how a fundamental unity pervades and embraces them all, collecting them, from the humblest lichen up to the highest mammifer, in one system, the whole creation of which must have depended upon one law or decree of the Almighty, though it did not all come forth at one time. After what we have seen the idea of a separate exertion for each must appear totally inadmissible. The single fact of abortive or rudimentary organs condemns it; for these, on such a supposition, could be regarded in no other light than as blemishes or blunders – the thing of all others most irreconcilable with that idea of Almighty Perfection which a general view of nature so irresistibly conveys. On the other hand, when the organic creation is admitted to have been effected by a general law, we see nothing in these abortive parts but harmless peculiarities of development, and interesting evidences of the manner in which the Divine Author has been pleased to work.

We have yet to advert to the most interesting class of facts connected with organic development. It is only in recent times that physiologists have observed that each animal passes, in the course of its germinal history, through a series of changes resembling the *permanent forms*, first of the various orders inferior to it in the entire scale, and then of its own order. This is a department of natural history in which only a few facts have been collected; but these are of such a nature, that we cannot doubt of their being the indications of some great general law. Thus, for instance, the comatula, a free-swimming star-fish, is, at one stage of its early progress, a crinoid – that is, a star-fish fixed upon a stalk to the bottom of the sea. It advances from the form of one of the lower to that of one of the higher echinodermata. The animals of its first form were, as we have seen, among the most abundant in the earliest fossili-ferous rocks: they began to decline in the new red sandstone era, and they were succeeded in the oolitic-age by animals *of the form of the mature comatula*. Thus, too, the insect, standing at the head of the articulated animals, is, in the larva state, an annelid or worm, the annelida being the lowest in the same class. The higher crustacea, as the crab or lobster, at their escape from the ovum, resemble the perfect animal of the inferior order entomostraca, and pass through all the forms of transition which characterise the intermediate tribes of crustacea. The salmon, a highly organised fish, exhibits, in its early stages, as has been remarked, the gelatinous dorsal cord, the heterocercal tail, and inferior position of the mouth, which mark the mature example of the lower tribes of fishes, the placoids and ganoids. The frog, again, for some time after its birth, is a fish with external gills, and other organs fitting it for an aquatic life, all of which are changed as it advances to maturity, and becomes a land animal. The mammifer only passes through still more stages, according to its higher place in the scale. Nor is man himself exempt from this law. His first form is that which is permanent in the animal-cule. His organisation gradually passes through conditions generally resembling a fish, a reptile, a bird, and the lower mammalia, before it attains its specific maturity. At one of the last stages of his fœtal career, he exhibits an intermaxillary bone, which is characteristic of the perfect ape; this is suppressed, and he may then be said to take leave of the simial type, and become a true human creature. Even, as we shall see, the varieties of his race are represented in the progressive development of an individual of the highest, before we see the adult Caucasian, the highest point yet attained in the animal scale.

To come to particular points of the organisation. The brain of man, which exceeds that of all other animals in complexity of organisation and fulness of development, is, at one early period, only "a simple fold of nervous matter, with difficulty distinguishable into three parts, while a little tail-like prolongation towards the hinder parts, and which had been the first to appear, is the only representation of a spinal marrow. Now in this state it perfectly resembles the brain of an adult fish, thus assuming *in transitu* the form that in the fish is permanent. In a short time, however, the structure is become more complex, the parts more distinct, the spinal marrow better marked; it is now the brain of a reptile. The change continues; by a singular motion, certain parts (*corpora quadrigemina*) which had hitherto appeared on the upper surface, now pass towards the lower; the former is their permanent situation in fishes and reptiles, the latter in birds and mammalia. This is another advance in the scale, but more remains yet to be done. The complication of the organ increases; cavities termed *ventricles* are formed, which do not exist in fishes, reptiles, or birds; curiously organised parts, such as the corpora striata, are added; it is now the brain of the mammalia. Its last and final change alone seems wanting, that which shall render it the brain of MAN."[b] And this change in time takes place.

So also with the heart. This organ, in the mammalia, consists of four cavities, but in the reptiles of only three, and in fishes of two only, while in the articulated animals it is merely a prolonged tube. Now in the mammal fœtus, at a certain early stage, the organ has the form of a prolonged tube; and a human being may be said to have then the heart of an insect. Subsequently it is shortened and widened, and becomes divided by a contraction into two parts, a ventricle and an auricle; it is now the heart of a fish. A subdivision of the auricle afterwards makes a triple-chambered form, as in the heart of the reptiles tribes; lastly, the ventricle being also subdivided, it becomes a full mammal heart.

It is certainly very remarkable that, corresponding generally to these progressive forms in the development of individuals, has been the succession of animal forms in the course of time. Our earth, as we have seen, bore crinoidea before it bore the higher echinodermata. It presented crustacea before it bore fishes, and when fishes came, the first forms were those ganoidal and placoidal types which correspond with the early fœtal condition of higher orders. Afterwards there were reptiles, then mammifers, and finally, as we know, came man. The tendency of all these illustrations is to make us look to *development* as the principle which has been immediately concerned in the peopling of this globe, a process extending over a vast space of time, but which is nevertheless connected in character with the briefer process by which an individual being is evoked from a simple germ. What mystery is there here – and how shall I proceed to enunciate the conception which I have ventured to form of what may prove to be its proper solution! It is an idea by no means calculated to impress by its greatness, or to puzzle by its profoundness. It is an idea more marked by simplicity than perhaps any other of those which have explained the great secrets of nature. But in this lies, perhaps, one of its strongest claims to our faith.

The whole train of animated beings, from the simplest and oldest, up to the highest and most recent, are, then, to be regarded as a series of *advances of the principle of development*, which have depended upon external physical circumstances, to which the resulting animals are appropriate. I contemplate the whole phenomena as having been in the first place arranged in the counsels of Divine Wisdom, to take place, not only upon this sphere, but upon all the others in space under necessary modifications, and as being carried on, from first to last, here and elsewhere, under immediate favour of the creative will or energy.[c] The nucleated vesicle, the fundamental form of all organisation, we must regard as the meeting-point between the inorganic and the organic – the end of the mineral and beginning of the vegetable and animal kingdoms, which thence start in different directions, but in perfect parallelism and analogy. We have already seen that this nucleated vesicle is itself a type of mature and independent being in the infusory animalcules, as well as the starting-point of the fœtal progress of every higher individual in creation, both animal and vegetable. We have seen that it is a form of being which there is some reason to believe electric agency will produce – though not perhaps usher into full life – in albumen, one of those compound elements of animal bodies, of which another (urea) has been made by artificial means. Remembering these things, we are drawn on to the supposition that the first step in the creation of life upon this planet was *a chemico-electric operation, by which simple germinal vesicles were produced.* This is so much, but what were the next steps? Let a common vegetable infusion help us to an answer. There, as we have seen, simple forms are produced at first, but afterwards they become more complicated, until at length the life-producing powers of the infusion are exhausted. Are we to presume that, in this case, the simple engender the complicated? Undoubtedly, this would not be more wonderful as a natural process than one which we never think of wondering at, because familiar to us – namely, that in the gestation of the mammals, the animalcule-like ovum of a few days is the parent, in a sense, of the chick-like form of a few weeks, and that in all the subsequent stages – fish, reptile, &c. – the one may, with scarcely a metaphor, be said to be the progenitor of the other. I suggest, then, as an hypothesis already countenanced by much that is ascertained, and likely to be further sanctioned by much that remains to be known, that the first step was *an advance under favour of peculiar conditions, from the simplest forms of being, to the next more complicated, and this through the medium of the ordinary process of generation.*

Unquestionably, what we ordinarily see of nature is calculated to impress a conviction that each species invariably produces its like. But I would here call attention to a remarkable illustration of natural law which has been brought forward by Mr. Babbage, in his "Ninth Bridgewater Treatise." The reader is requested to suppose himself seated before the calculating machine, and observing it. It is moved by a weight, and there is a wheel which revolves through a small angle round its axis, at short intervals, presenting to his eye successively, a series of numbers engraved on its divided circumference.

Let the figures thus seen be the series, 1, 2, 3, 4, 5, &c., of natural numbers, each of which exceeds its immediate antecedent by unity.

"Now, reader," says Mr. Babbage, "let me ask you how long you will have counted before you are firmly convinced that the engine has been so adjusted, that it will continue, while its motion is maintained, to produce the same series of natural numbers? Some minds are so constituted, that after passing the first hundred terms, they will be satisfied that they are acquainted with the law. After seeing five hundred terms few will doubt, and after the fifty thousandth term the propensity to believe that the succeeding term will be fifty thousand and one will be almost irresistible. That term *will* be fifty thousand and one; and the same regular succession will continue; the five millionth and the fifty millionth term will still appear in their expected order, and one unbroken chain of natural numbers will pass before your eyes, from *one* up to *one hundred million.*

"True to the vast induction which has been made, the next succeeding term will be one hundred million and one; but the next number presented by the rim of the wheel, instead of being one hundred million and two, is one hundred million *ten*

	1	
	2	
	3	
	4	
	5	
	...	
	....	
	.....	
	......	
	99,999,999	
	100,000,000	
regularly as far as	100,000,001	
	100,010,002	the law changes.
	100,030,003	
	100,060,004	
	100,100,005	
	100,150,006	
	100,210,007	
	100,280,008	
	....	
	....	

thousand and two. The whole series from the commencement being thus –

"The law which seemed at first to govern this series failed at the hundred million and second term. This term is larger than we expected by 10,000. The next term is larger than was anticipated by 30,000, and the excess of each term above what we had expected forms the following table: –

10,000
30,000
60,000
100,000
150,000
. . . . . .
. . . . . .

being, in fact, the series of *triangular numbers*,[d] each multiplied by 10,000.

"If we now continue to observe the numbers presented by the wheel, we shall find, that for a hundred, or even for a thousand terms, they continue to follow the new law relating to the triangular numbers; but after watching them for 2761 terms we find that this law fails in the case of the 2762nd term.

"If we continue to observe, we shall discover another law then coming into action, which also is dependent, but in a different manner, on triangular numbers. This will continue through about 1430 terms, when a new law is again introduced which extends over about 950 terms, and this, too, like all its predecessors, fails, and gives place to other laws, which appear at different intervals.

"Now it must be observed that *the law that each number presented by the engine is greater by unity than the preceding number*, which law the observer had deduced from an induction of a hundred million instances, *was not the true law that regulated its action*, and that the occurrence of the number 100,010,002 at the 100,000,002nd term was *as necessary a consequence of the original adjustment, and might have been as fully foreknown at the commencement, as was the regular succession of any one of the intermediate numbers to its immediate antecedent.* The same remark applies to the next apparent deviation from the new law, which was founded on an induction of 2761 terms and also to the succeeding law, with this limitation only – that, whilst their consecutive introduction at various definite intervals, is a necessary consequence of the mechanical structure of the engine, our knowledge of analysis does not enable us to predict the periods themselves at which the more distant laws will be introduced."

It is not difficult to apply the philosophy of this passage to the question under consideration. It must be borne in mind that the gestation of a single organism is the work of but a few days, weeks, or months; but the gestation (so to speak) of a whole creation is a matter probably involving enormous spaces of time. Suppose that an ephemeron, hovering over a pool for its one April day of life, were capable of observing the fry of the frog in the water below. In its aged afternoon, having seen no change upon them for such a long time, it would be little qualified to conceive that the external branchiæ of these creatures were to decay, and be replaced by internal lungs, that feet were to be developed, the tail erased, and the animal then to become a denizen of the land. Precisely such may be our difficulty in conceiving that any of the species which people our earth is capable of advancing by generation to a higher type of being. During the whole time which we call the historical era, the limits of species have been, to ordinary observation, rigidly adhered to. But the historical era is, we know, only a small portion of the entire age of our globe. We do not know what may have happened during the ages which preceded its commencement, as we do not know what may happen in ages yet in the distant future. All, therefore, that we can properly infer from the apparently invariable production of like by like is, that such is the ordinary procedure of nature in the time immediately passing before our eyes. Mr. Babbage's illustration powerfully suggests that this ordinary procedure may be subordinate to a higher law which only *permits* it for a time, and in proper season interrupts and changes it. We shall soon see some philosophical evidence for this very conclusion.

It has been seen that, in the reproduction of the higher animals, the new being passes through stages in which it is successively fish-like and reptile-like. But the resemblance is not to the adult fish or the adult reptile, but to the fish and reptile at a certain point in their fœtal progress; this holds true with regard to the vascular, nervous, and other systems alike. It seems as if gestation consisted of two distinct and independent stages – one devoted to the development of the new being through the conditions of the inferior types, or, rather, through the corresponding *first stages of their development;* another perfecting and bringing the new being to a healthy maturity, on the basis of the point of development reached. This may be illustrated by a simple diagram. The fœtus of all the four classes may be supposed to advance in an identical condition to the point A. The fish there diverges and passes along a line apart, and peculiar to itself, to its mature state at F. The reptile, bird, and mammal, go on together to C, where the reptile diverges in like manner, and advances by itself to R. The bird diverges at D, and goes on to B. The mammal then goes forward in a straight line to the highest point of organisation at M. This diagram shows only the main ramifications; but the reader must suppose minor ones, representing the subordinate differences of orders, tribes, families, genera, &c., if he wishes to extend his views to the whole varieties of being in the animal kingdom. Limiting ourselves at present to the outline afforded by this diagram, it is apparent that the only thing required for an advance from one type to another in the generative process is that, for example, the fish embryo should not diverge at A, but go on to C before it diverges, in which case the progeny will be, not a fish, but a reptile. To protract the *straightforward part of the gestation over a small space* – and from species to species the space would be small indeed – is all that is necessary.

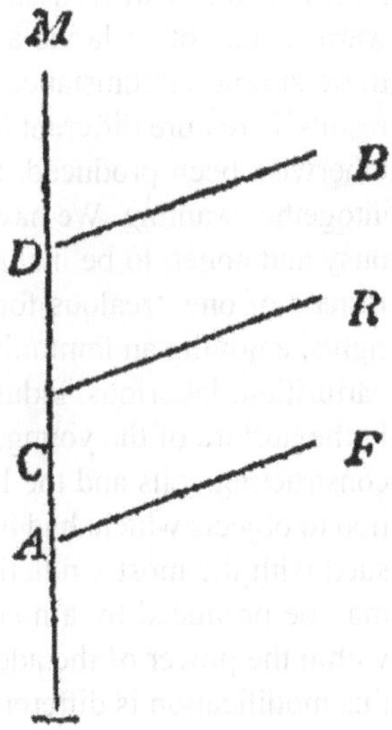

This might be done by the force of certain external conditions operating upon the parturient system. The nature of the conditions we can only conjecture, for their operation, which in the geological eras was so powerful, has in its main strength been long interrupted, and is now perhaps only allowed to work in some of the lowest departments of the organic world, or under extraordinary casualties in some of the higher, and to these points the attention of science has as yet been little directed. But though this knowledge were never to be clearly attained, it need not much affect the present argument, provided it be satisfactorily shown that there must be some such influence within the range of natural things.

To this conclusion it must be greatly conducive that the law of organic development is still daily seen at work to certain effects, only somewhat short of a transition from species to species. Sex we have seen to be a matter of development. There is an instance, in a humble department of the animal world, of arrangements being made by the animals themselves for adjusting this law to the production of a particular sex. Amongst bees, as amongst several other insect tribes, there is in each community but one true female, the queen bee, the workers being false females or neuters; that is to say, sex is carried on in them to a point where it is attended by sterility. The preparatory states of the queen bee occupy sixteen days; those of the neuters, twenty; and those of males, twenty-four. Now it is a fact, settled by innumerable observations and experiments, that the bees can so modify a worker in the larva state, that when it emerges from the pupa, it is found to be a queen or true female. For this purpose they enlarge its cell, make a pyramidal hollow to allow of its assuming a vertical instead of a horizontal position, keep it warmer than other larvæ are kept, and feed it with a peculiar kind of food. From these simple circumstances, leading to a shortening of the embryotic condition, results a creature different in form, and also in dispositions, from what would have otherwise been produced. Some of the organs possessed by the worker are here altogether wanting. We have a creature "destined to enjoy love, to burn with jealousy and anger, to be incited to vengeance, and to pass her time without labour," instead of one "zealous for the good of the community, a defender of the public rights, enjoying an immunity from the stimulus of sexual appetite and the pains of parturition; laborious, industrious, patient, ingenious, skilful; incessantly engaged in the nurture of the young, in collecting honey and pollen, in elaborating wax, in constructing cells and the like! – paying the most respectful and assiduous attention to objects which, had its ovaries been developed, it would have hated and pursued with the most vindictive fury till it had destroyed them!"[e] All these changes may be produced by a mere modification of the embryotic progress, which it is within the power of the adult animals to effect. But it is important to observe that this modification is different from working a direct change upon the embryo. It is not the different food which effects a metamorphosis. All that is done is merely to accelerate the period of the insect's perfection. By the arrangements made and the food given, the embryo becomes sooner fit for being ushered forth in its imago or perfect state. Development may be said to be thus arrested at a particular stage – that early one at which the female sex is complete. In the other circumstances it is allowed to go on four days longer, and a stage is then reached between the two sexes, which in this species is designed to be the perfect condition of a large portion of the community. Four days more make it a perfect male. It is at the same time to be observed that there is, from the period of oviposition, a destined distinction between the sexes of the young bees. The queen lays the whole of the eggs which are designed to become workers, before she begins to lay those which become males. But probably the condition of her reproductive system governs the matter of sex, for it is remarked that when her impregnation is delayed beyond the twenty-eighth day of her entire existence, she lays only eggs which become males.

We have here, it will be admitted, a most remarkable illustration of the principle of development, although in an operation limited to the production of sex only. Let it not be said that the phenomena concerned in the generation of bees may be very different from those concerned in the reproduction of the higher animals. There is a unity throughout nature which makes the one case an instructive reflection of the other.

We shall now see an instance of development operating within the production of what approaches to the character of variety of species. It is fully established that a human family, tribe, or nation is liable, in the course of generations, to be either advanced from a mean form to a higher one, or degraded from a higher to a lower, by the influence of the physical conditions in which it lives. The coarse features, and other structural peculiarities of the negro race only continue while these people live amidst the circumstances usually associated with barbarism. In a more temperate clime, and higher social state, the face and figure become greatly refined. The few African nations which possess any civilisation also exhibit forms approaching the European; and when the same people in the United States of America have enjoyed a within-door life for several generations, they assimilate to the whites amongst whom they live. On the other hand, there are authentic instances of a people originally well-formed and good-looking, being brought, by imperfect diet and a variety of physical hardships, to a meaner form. It is remarkable that prominence of the jaws, a recession and diminution of the cranium, and an elongation and attenuation of the limbs, are peculiarities always produced by these miserable conditions, for they indicate an unequivocal retrogression towards the type of the lower animals. Thus we see nature alike willing to go back and to go forward. Both effects are simply the result of the operation of the law of development in the generative system. Give good conditions, it advances; bad ones, it recedes. Now, perhaps, it is only because there is no longer a possibility, in the higher types of being, of giving sufficiently favourable conditions to carry on species to species, that we see the operation of the law so far limited.

Let us trace this law also in the production of certain classes of monstrosities. A human fœtus is often left with one of the most important parts of its frame imperfectly developed: the heart, for instance, goes no farther than the three-chambered form, so that it is the heart of a reptile. There are even instances of this organ being left in the two-chambered or fish-form. Such defects are the result of nothing more than a failure of the power of development in the system of the mother, occasioned by weak health or misery, and bearing with force upon that sub-stage of the gestation at which the perfecting of the heart to its right form ought properly to have taken place. Here we have apparently a realisation of the converse of those conditions which carry on species to species, so far, at least, as one organ is concerned. Seeing a complete specific retrogression in this one point, how easy it is to suppose an access of favourable conditions sufficient to reverse the phenomenon, and make a fish mother develop a reptile heart, or a reptile mother develop a mammal one. It is no great boldness to surmise that a super-adequacy in the measure of this under-adequacy (and the one thing seems as natural an occurrence as the other) would suffice in a goose to give its progeny the body of a rat, and produce the ornithorhynchus, or might give the progeny of an ornithorhynchus the mouth and feet of a true rodent, and thus complete at two stages the passage from the aves to the mammalia.

[...]

Early in this century, M. Lamarck, a naturalist of the highest character, suggested an hypothesis of organic progress which deservedly incurred much ridicule, although it contained a glimmer of the truth. He surmised, and endeavoured, with a great deal of ingenuity, to prove, that one being advanced in the course of generations to another, in consequence merely of its experience of wants calling for the exercise of its faculties in a particular direction, by which exercise new developments of organs took place, ending in variations sufficient to constitute a new species. Thus he thought that a bird would be driven by necessity to seek its food in the water, and that, in its efforts to swim, the outstretching of its claws would lead to the expansion of the intermediate membranes, and it would thus become web-footed. Now it is possible that wants and the exercise of faculties have entered in some manner into the production of the phenomena which we have been considering; but certainly not in the way suggested by Lamarck, whose whole notion is obviously so inadequate to account for the rise of the organic kingdoms, that we only can place it with pity among the follies of the wise. Had the laws of organic development been known in his time, his theory might have been of a more imposing kind. It is upon these that the present hypothesis is mainly founded. I take existing natural means, and show them to have been capable of producing all the existing organisms, with the simple and easily conceivable aid of a higher generative law, which we perhaps still see operating upon a limited scale. I also go beyond the French philosopher to a very important point, the original Divine conception of all the forms of being which these natural laws were only instruments in working out and realising.

[...]

But the idea that any of the lower animals have been concerned in any way with the origin of man – is not this degrading? Degrading is a term, expressive of a notion of the human mind, and the human mind is liable to prejudices which prevent its notions from being invariably correct. Were we acquainted for the first time with the circumstances attending the production of an individual of our race, we might equally think them degrading, and be eager to deny them, and exclude them from the admitted truths of nature. Knowing this fact familiarly and beyond contradiction, a healthy and natural mind finds no difficulty in regarding it complacently. Creative Providence has been pleased to order that it should be so, and it must therefore be submitted to. Now the idea as to the progress of organic creation, if we become satisfied of its truth, ought to be received precisely in this spirit. If it has pleased Providence to arrange that one species should give birth to another, until the second highest gave birth to man, who is the very highest; be it so, it is our part to admire and to submit. The very faintest notion of there being anything ridiculous or degrading in the theory – how absurd does it appear when we remember that every individual amongst us actually passes through the characters of the insect, the fish, and reptile (to speak nothing of others), before he is permitted to breathe the breath of life! But such notions are mere emanations of false pride and ignorant prejudice. He who conceives them little reflects that they, in reality, involve a contempt for the works and ways of God. For it may be asked, if he, as appears, has chosen to employ inferior organisms as a generative medium for the production of higher ones, even including ourselves, what right have we, his humble creatures, to find fault? There is, also, in this prejudice, an element of unkindliness towards the lower animals, which is utterly out of place. These creatures are all of them part products of the Almighty Conception, as well as ourselves. All of them display wondrous evidences of his wisdom and benevolence. All of them have had assigned to them by their Great Father a part in the drama of the organic world, as well as ourselves. Why should they be held in such contempt? Let us regard them in a proper spirit, as parts of the grand plan, instead of contemplating them in the light of frivolous prejudices, and we shall be altogether at a loss to see how there should be any degradation in the idea of our race having been genealogically connected with them.

Notes

a D'Aubenton established the rule, that all the viviparous quadrupeds have seven vertebræ in the neck.

b Lord's "Popular Physiology." It is to Tiedemann that we chiefly owe these curious observations; but ground was first broken in this branch of physiological science by John Hunter.

c When I formed this idea, I was not aware of one which seems faintly to foreshadow it – namely, Socrates' doctrine, afterwards dilated on by Plato, that "previous to the existence of the world, and beyond its present limits, there existed certain archetypes, the embodiment (if we may use such a word) of general ideas; and that these archetypes were models, in imitation of which all particular beings were created."

d The numbers, 1, 3, 6, 10, 15, 21, 28, &c. are formed by adding the successive terms of the series of natural numbers thus:

$$1 = 1$$
$$1+2 = 3$$
$$1+2+3 = 6$$
$$1+2+3+4 = 10, \&c.$$

They are called triangular numbers, because a number of points corresponding to any term can always be placed in the form of a triangle; for instance –

1 3 6 10

e Kirby and Spence.

Editorial Headnotes

24. Herbert Spencer. "The Developmental Hypothesis," *The Leader*, (March 1852), pp. 280–81.

25. Spencer, Herbert. "A Theory of Population, Deduced from the General Law of Animal Fertility" *Westminster Review*, Vol. 57, (1852), pp. 9–10, 12–3, 22, 23–7, 30–5.

26. Herbert Spencer: "Progress: Its Law and Cause", *The Westminster Review*, Vol. 67 (April 1857), pp 445–85.

Herbert Spencer has long been characterized and caricatured as the archetypal "social Darwinist," largely as a result of the immense success of Richard Hofstadter's book *Social Darwinism in American Thought*.[1] As a number of scholars have argued, however, this was by no means an accurate representation of Spencer's views.[2] Indeed, for all that he supported Darwin's ideas and coined the term "the survival of the fittest" to describe Darwin's theory of natural selection, Spencer's own ideas about evolution were always much closer to those of the French naturalist Jean-Baptiste Lamarck than they were to Darwin's. Like Lamarck, Spencer believed that nature was subject to a largely progressive development in which organisms adapted to their environments as a result of the use or disuse of their various organs and the characteristics that life under a given set of conditions demanded. Spencer believed that as individuals and thus species became better adapted to nature, so a point of equilibrium would be reached and, in the case of humanity, social harmony would be achieved.[3] It was for this reason that he was opposed to state aid to help those who struggled to support themselves and their families. Such actions would only interfere with nature's work and would only result in those individuals adapting to a state of dependence. The label 'social Darwinist,' which is usually applied to Spencer's later work, also draws attention away from the fact that Spencer's first public statement on evolution predates that made by Darwin by several years. His views on the 'developmental hypothesis' were representative of the views he discussed with Mary Ann Evans, George Henry Lewes, and others at the salon hosted by the radical liberal editor of the *Westminster Review*, John Chapman.[4] Unlike Darwin, Spencer felt no need for caution regarding the application of his ideas to the human species. Indeed, this was his main concern. He wrote a number of essays and books in which he explored the evolution of human society, and of human moral and social sentiments in particular, a number of which are extracted here.[5] Notably, while Darwin and Wallace embraced the idea of competition and struggle that they saw to be the consequence of the incommensurable ratio between population growth and available resources at the heart of Malthus's *Essay on the Principle of Population*, Spencer remained opposed to Malthus's contentions. Spencer argued that although Malthus had been right to point out that all life was subject to a struggle for existence, he had missed the point that as human intellect and reason developed, sociability and morality would increase and fertility levels would drop. Spencer was of the opinion that reproduction was necessarily at odds with the progressive developmental force of 'individuation.' In humans the force of individuation had resulted in the development of mind, intellect, and reason. This process had used, and would continue to use up, the available 'neurine' – the nutriment involved in the production of sperm cells. As humans continued to adapt and improve as a species, Spencer suggested that we would increase in intellect and moral capacity. As the species became better adapted to social life, the excess production of offspring would thus be lessened as the powers of mind and morality developed. The ratio between population and resources was thus, Spencer believed, very much amenable to change.

When Darwin's *On the Origin of Species* was published in 1859, Spencer welcomed it as a significant contribution to evolutionary thought and later incorporated natural selection into his own evolutionary framework. However, selection always remained very much a secondary mechanism for Spencer, acting only to cull those individuals that were unable to adapt to the prevailing circumstances and thus aiding in the broad theory of adaptation to circumstance that he had already laid out in detail across the previous decade. Alfred Russel Wallace admired Spencer and his work immensely and named his son Herbert in honour of him.

Notes

1 Richard Hofstadter, *Social Darwinism in American Thought* (Boston: Beacon, 1955). For further context see Caden C. Testa and Piers J. Hale, "Mapping evolutionary ideas in Britain in the first half of the nineteenth century," in Caden C. Testa and Piers J. Hale, (eds.), *Evolution Before Darwin* (London: Routledge, 2025).

2 Robert J. Richards, *Darwin and the Development of Evolutionary Theories of Mind and Behavior* (Chicago: University of Chicago Press, 1987), pp. 243–330; Piers J. Hale, *Political Descent. Malthus, Mutualism and the Politics of Evolution in Victorian England* (Chicago: University of Chicago Press, 2014), pp. 66–105.

3 Spencer explored this more fully in his 1851 book *Social Statics*. An extract from that work is included in this volume, Caden C. Testa and Piers J. Hale, *Evolution Before Darwin* (London: Routledge, 2025); for further context see their essay in that volume, Caden C. Testa and Piers J. Hale, "Mapping evolutionary ideas in Britain in the first half of the nineteenth century," in Caden C. Testa and Piers J. Hale, (eds.), *Evolution Before Darwin* (London: Routledge, 2025).

4 Hale, *Political Descent,* pp. 93–4.

5 In addition to the essays extracted here, see Herbert Spencer, *Social Statics: or, The Conditions Essential to Human Happiness Specified, and the First of Them Developed* (London: John Chapman, 1851); and Hebert Spencer, *Principles of Psychology* (London: Longman, Brown, Green & Longman, 1855). For further context and commentary see Caden C. Testa and Piers J. Hale, (eds.), *Evolution Before Darwin* (London: Routledge, 2025).

24

"THE DEVELOPMENTAL HYPOTHESIS"

Herbert Spencer

Source: Herbert Spencer, "The Developmental Hypothesis", *The Leader*, (March 1852), pp. 280–1

Portfolio.

We should do our utmost to encourage the Beautiful, for the Useful encourages itself. – GOETHE.

THE HAYHORNE PAPERS.

No. II. – The Development Hypothests.

024 "Herbert Spencer" – image in the public domain

Image in the public domain.

DOI: 10.4324/9781003490548-27

In a debate upon the development hypothesis, lately narrated to me by a friend, one of the disputants was described as arguing that, as in all our experience we know of no such phenomenon as the transmutation of species, it is unphilosophical to assume that transmutation of species ever takes place. Had I been present, I think that, passing over his assertion, which is open to criticism, I should have replied that, as in all our experience we have never known a species *created*, it was, by his own showing, unphilosophical to assume that any species ever had been created.

Those who cavalierly reject the theory of Lamarck and his followers, as not adequately supported by facts, seem quite to forget that their own theory is supported by no facts at all. Like the majority of men who are born to a given belief, they demand the most rigorous proof of any adverse doctrine, but assume that their own doctrine needs none. Here we find scattered over the globe vegetable and animal organisms numbering, of the one kind (according to Humboldt), some 320,000 species, and of the other, if we include insects, some *two millions* of species (see Carpenter); and if to these we add the numbers of animal and vegetable species that have become extinct (bearing in mind how geological records prove that, from the earliest appearance of life down to the present time, different species have been successively replacing each other, so that the world's Flora and Fauna have completely changed many times over), we may safely estimate the number of species that have existed, and are existing on the earth, at not less than ten millions. Well, which is the most rational theory about these ten millions of species? Is it most likely that there have been ten millions of special creations? or is it most likely that by continual modifications, due to change of circumstances, ten millions of varieties may have been produced, as varieties are being produced still? One of the two theories must be adopted. Which is most countenanced by facts?

Doubtless many will reply that they can more easily conceive ten millions of special creations to have taken place, than they can conceive that ten millions of varieties have been produced by the process of perpetual modification. All such, however, will find, on candid inquiry, that they are under an illusion. This is one of the many cases in which men do not really believe, but rather *believe they believe*. It is not that they can truly conceive ten millions of special creations to have taken place, but that they *think they can do so*. A little careful introspection will show them that they have never yet realized to themselves the creation of even *one* species. If they have formed a definite conception of the process, they will be able to answer such questions as – How is a new species constructed? and How does it make its appearance? Is it thrown down from the clouds? or must we hold to the notion that it struggles up out of the ground? Do its limbs and viscera rush together from all the points of the compass? or must we receive some such old Hebrew notion as, that God goes into a forest-cavern, and there takes clay and moulds a new creature? If they say that a new creature is produced in none of these modes, which are too absurd to be believed, then they are required to describe the mode in which a new creature *may* be produced – a mode which does *not* seem absurd; and such a mode they will find that they neither have conceived nor can conceive.

Should the believers in special creations consider it unfair thus to call upon them to describe how special creations take place, I reply, that this is far less than they demand from the supporters of the development hypothesis. They are merely asked to point out a *conceivable* mode; on the other hand, they ask, not simply for a *conceivable* mode, but for the *actual* mode. They do not say – Show us how this *may* take place; but they say – Show us how this *does* take place. So far from its being unreasonable to ask so much of them, it would be reasonable to ask not only for a possible mode of special creation, but for an ascertained mode; seeing that this is no greater a demand than they make upon their opponents.

And here we may perceive how much more defensible the new doctrine is than the old one. Even could the supporters of the development hypothesis merely show that the production of species by the process of modification is conceivable, they would be in a better position than their opponents. But they can do much more than this. They can show that the process of modification has effected and is effecting great changes in all organisms subject to modifying influences. Though, from the impossibility of getting at a sufficiency of facts, they are unable to trace the many phases through which any existing species has passed in arriving at its present form, or to identify the influences which caused the successive modifications, yet they can show that any existing species – animal or vegetable – when placed under conditions different from its previous ones, *immediately begins to undergo certain changes of structure fitting it for the new conditions*. They can show that in successive generations these changes continue until ultimately the new conditions become the natural ones. They can show that in cultivated plants, in domesticated animals, and in the several races of men, these changes have uniformly taken place. They can show that the degrees of difference so produced are often, as in dogs, greater than those on which distinctions of species are in other cases founded. They can show that it is a matter of dispute whether some of these modified forms *are* varieties or separate species. They can show, too, that the changes daily taking place in ourselves – the facility that attends long practice, and the loss of aptitude that begins when practice ceases – the strengthening of passions habitually gratified, and the weakening of those habitually curbed – the development of every faculty, bodily, moral, or intellectual, according to the use made of it – are all explicable on this same principle. And thus they can show that throughout all organic nature there *is* at work a modifying influence of the kind they assign as the cause of these specific differences – an influence which, though slow in its action, does, in time, if the circumstances demand it, produce marked changes – an influence which, to all appearance, would produce in the millions of years, and under the great varieties of condition which geological records imply, any amount of change.

Which, then, is the most rational hypothesis; that of special creations which has neither a fact to support it nor is even definitely conceivable; or that of modification, which is not only definitely conceivable, but is countenanced by the habitudes of every existing organism?

That by any series of changes a zoophyte should ever become a mammal, seems to those who are not familiar with zoology, and who have not seen how clear becomes the relationship between the simplest and the most complex forms, when all intermediate forms are examined, a very grotesque notion. Habitually looking at things rather in their statical than in their dynamical aspect, they never realize the fact that, by small increments of modification, any amount of modification may in time be generated. That surprise which they feel on finding one whom they last saw as a boy, grown into a man, becomes incredulity when the degree of change is greater. Nevertheless, abundant instances are at hand of the mode in which we may pass to the most diverse forms by insensible gradations. Arguing the matter some time since with a learned professor, I illustrated my position thus: – You admit that there is no apparent relationship between a circle and an hyperbola. The one is a finite curve; the other is an infinite one. All parts of the one are alike; of the other no two parts are alike. The one incloses a space; the other will not inclose a space, though produced for ever. Yet opposite as are these curves in all their properties, they may be connected together by a series of intermediate curves, no one of which differs from the adjacent ones in any appreciable degree. Thus, if a cone be cut by a plane at right angles to its axis we get a circle. If, instead of being perfectly at right angles, the plane subtends with the axis an angle of 89° 59′, we have an ellipse which no human eye, even when aided by an accurate pair of compasses can distinguish from a circle. Decreasing the angle minute by minute the ellipse becomes first perceptibly eccentric, then manifestly so, and by and by acquires so immensely elongated a form, as to bear no recognisable resemblance to a circle. By continuing this process the ellipse passes insensibly into a parabola; and ultimately, by still further diminishing the angle, into an hyperbola. Now here we have four different species of curve – circle, ellipse, parabola, and hyperbola – each having its peculiar properties, and its separate equation, and the first and last of which are quite opposite in nature, connected together as members of one series, all producible by a single process of insensible modification.

But the blindness of those who think it absurd to suppose that complex organic forms may have arisen by successive modifications out of simple ones, becomes astonishing when we remember that complex organic forms are daily being thus produced. A tree differs from a seed immeasurably in every respect – in bulk, in structure, in colour, in form, in specific gravity, in chemical composition; differs so greatly that no visible resemblance of any kind can be pointed out between them. Yet is the one changed in the course of a few years into the other – changed so gradually, that at no moment can it be said – Now the seed ceases to be, and the tree exists. What can be more widely contrasted than a newly-born child and the small, semi-transparent, gelatinous spherule constituting the human ovum? The infant is so complex in structure that a cyclopædia is needed to describe its constituent parts. The germinal vesicle is so simple that a line will contain all that can be said of it. Nevertheless a few months suffices to develop the one out of the other, and that, too, by a series of modifications so small that were the embryo examined at successive minutes not even a microscope would disclose any sensible changes. That the uneducated and the ill-educated should think the hypothesis that all races of beings, man inclusive, may in process of time have been evolved from the simplest monad, a ludicrous one, is not to be wondered at. But for the physiologist, who knows that every individual being *is* so evolved, – who knows further, that in their earliest condition the germs of all plants and animals whatever are so similar, "that there is no appreciable distinction amongst them which would enable it to be determined whether a particular molecule is the germ of a conferva or of an oak, of a zoophyte or of a man"[a] – for him to make a difficulty of the matter is inexcusable. Surely, if a single structureless cell may, when subjected to certain influences, become a man in the space of twenty years, there is nothing absurd in the hypothesis that under certain other influences, a cell may in the course of millions of years give origin to the human race. The two processes are generically the same, and differ only in length and complexity.

We have, indeed, in the part taken by many scientific men in this controversy of "Law *versus* Miracle," a good illustration of the tenacious vitality of superstitions. Ask one of our leading geologists or physiologists whether he believes in the Mosaic account of the creation, and he will take the question as next to an insult. Either he rejects the narrative entirely, or understands it in some vague non-natural sense. Yet one part of it he unconsciously adopts; and that, too, literally. For, whence has he got this notion of "special creations," which he thinks so reasonable, and fights for so vigorously. Evidently he can trace it back to no other source than this myth which he repudiates. He has not a single fact in nature to quote in proof of it; nor is he prepared with any chain of abstract reasoning by which it may be established. Catechise him, and he will be forced to confess that the notion was put into his mind in childhood as part of a story which he now thinks absurd. And why, after rejecting all the rest of this story, he should strenuously defend this last remnant of it as though he had received it on valid authority, he would be puzzled to say.

Note

a Carpenter's *Principles of Physiology*. 3rd ed. p. 867.

25

"A THEORY OF POPULATION, DEDUCED FROM THE GENERAL LAW OF ANIMAL FERTILITY"

Herbert Spencer

Source: Herbert Spencer, "A Theory of Population, Deduced from the General Law of Animal Fertility", *Westminster Review*, 57 (1852), pp. 9–10, 12–13, 22, 23–7, 30–5

A
THEORY OF POPULATION,
ETC.

§1. On contemplating its general circumstances, we perceive that any race of organisms is subject to two sets of conflicting influences. On the one hand by natural death, by enemies, by lack of food, by atmospheric changes, &c., it is constantly being destroyed. On the other hand, partly by the strength, swiftness, and sagacity of its members, and partly by their fertility, it is constantly being maintained. These conflicting sets of influences may be conveniently generalized as – the forces destructive of race, and the forces preservative of race.

§2. Whilst any race continues to exist, the forces destructive of it and the forces preservative of it must perpetually tend towards equilibrium. If the forces destructive of it decrease, the race must gradually become more numerous, until, either from lack of food or from increase of enemies, the destroying forces again balance the preserving forces. If, reversely, the forces destructive of it increase, then the race must diminish, until, either from its food becoming relatively more abundant, or from its enemies dying of hunger, the destroying forces sink to the level of the preserving forces. Should the destroying forces be of a kind that cannot be thus met (as great change of climate), the race, by becoming extinct, is removed out of the category. Hence this is necessarily the *law of maintenance* of all races; seeing that when they cease to conform to it they cease to be.

Now the forces preservative of race are two – ability in each member of the race to preserve itself, and ability to produce other members – power to maintain individual life, and power to propagate the species. These must vary inversely. When, from lowness of organization, the ability to contend with external dangers

 DOI: 10.4324/9781003490548-28

is small, there must be great fertility to compensate for the consequent mortality; otherwise the race must die out. When, on the contrary, high endowments give much capacity of self-preservation, there needs a correspondingly low degree of fertility. Given the dangers to be met as a constant quantity; then, as the ability of any species to meet them must be a constant quantity too, and as this is made up of the two factors – power to maintain individual life and power to multiply – these cannot do other than vary inversely.

[...]

Reproduction, under all its forms, may be described as the separation of portions of a parent plant or animal for the purpose of forming other plants or animals. Whether it be by spontaneous fission, by gemmation, or by gemmules; whether the detached products be bulbels, spores or seeds, ovisacs, ova or spermatozoa; or however the process of multiplication be modified, it essentially consists in the throwing off of parts of adult organisms for the purpose of making new organisms. On the other hand, self-preservation is fundamentally a maintenance of the organism in undiminished bulk. Amongst the lowest forms of life, aggregation of tissue is the only mode in which the self-preserving power is shown. Even in the highest, sustaining the body in its integrity is that in which self-preservation most truly consists – is the end which the widest intelligence is indirectly made to subserve. Whilst, on the one side, it cannot be denied that the increase of tissue constituting growth is self-preservation both in essence and in result; neither can it, on the other side, be denied that a diminution of tissue, either from injury, disease, or old age, is in both essence and result the reverse.

Hence the maintenance of the individual and the propagation of the race, being respectively aggregative and separative, *necessarily* vary inversely. Every generative product is a deduction from the parental life; and, as already pointed out, to diminish life is to diminish the ability to preserve life. The portion thrown off is organised matter; vital force has been expended in the organisation of it, and in the assimilation of its component elements; which vital force, had no such portion been made and thrown off, *would have been available for the preservation of the parent.*

Neither of these forces, therefore, can increase, save at the expense of the other. The one draws in and incorporates new material; the other throws off material previously incorporated. The one adds to; the other takes from. Using a convenient expression for describing the facts (though one that must not be construed into an hypothesis), we may say that the force which builds up and repairs the individual is an attractive force, whilst that which throws off germs is a repulsive force. But whatever may turn out to be the true nature of the two processes, it is clear that they are mutually destructive; or, stating the proposition in its briefest form – Individuation and Reproduction are antagonistic.

[...]

§ 11. Throughout both the animal and vegetable kingdoms, generation is effected "by the union of the contents of a 'sperm-cell' with those of a 'germ-cell;' the latter being that from within which the embryo is evolved, whilst the former supplies some material or influence necessary to its evolution."[a]

[...]

If, then, there be truth in the assumption that the two elements, which, by their union, give origin to a new organism, typify the two essential constituents of such new organism, we must infer that the sperm-cell and germ-cell respectively consist of co-ordinating matter and matter to be co-ordinated – neurine and nutriment. That apparent identity of sperm-cell and germ-cell seen in the lowest forms of life may thus be understood as significant of the fact that no extended co-ordination of actions exists in the generative product – each cell being a separate individual; and the dissimilarity seen in higher organic types may, conversely, be understood as expressive of, and consequent upon, the increasing degree of co-ordination exhibited[b].

That the sperm-cell and germ-cell are thus contrasted in nature and function may further be suspected on considering the distinctive characteristics of the sexes. Of the two elements they respectively contribute to the formation of a fertile germ, it may be reasonably supposed that each furnishes that which it possesses in greatest abundance and can best spare. Well, in the greater size of the nervous centres in the male, as well as in the fact that during famines men succumb sooner than women, we see that in the male the co-ordinating system is relatively predominant. From the same evidence, as well as from the greater abundance of the cellular and adipose tissues in women, we may infer that the nutritive system predominates in the female[c]. Here, then, is additional support for the hypothesis that the sperm-cell, which is supplied by the male, contains co-ordinating matter, and the germ-cell, which is supplied by the female, contains matter to be co-ordinated.

The same inference may, again, be drawn from a general view of the maternal function. For if, as we see, it is the office of the mother to afford milk to the infant, and during a previous period to afford blood to the fœtus, it becomes probable that during a yet earlier stage it is still the function to supply nutriment, though in another form. Indeed, when, ascending gradually the scale of animal life, we perceive that this supplying of milk, and before that of blood, is simply a continuation of the previous process, we may be sure that, with Nature's usual consistency, this process is essentially one from the beginning.

Quite in harmony with this hypothesis concerning the respective natures of the sperm-cell and germ-cell is a remark of Carpenter's on the same point: –

> "Looking," he says, "to the very equal mode in which the characters of the two parents are mingled in *hybrid* offspring, and to the certainty that the *material* conditions which determine the development of the germ are almost exclusively female, it would seem probable that the *dynamical* conditions are, in great part, furnished by the male."[d]

§ 12. Could nothing but the foregoing indirect evidence be adduced in proof of the proposition that the spermatozoon is essentially a neural element, and the ovum essentially a hæmal element, we should scarcely claim for it anything more than plausibility. On finding, however, that this indirect evidence is merely introductory to evidence of a quite direct nature, its significance will become apparent. Adding to their weight taken separately the force of their mutual confirmation, these two series of proofs will be seen to give the hypothesis a high degree of probability. The direct evidence now to be considered is of several kinds.

On referring to the description of the process of multiplication in monads, quoted some pages back (§ 5), from Professor Owen, the reader will perceive that it is by the pellucid nucleus that the growth and reproduction of these single-celled creatures are regulated. The nucleus controls the circulation of the plasmatic fluid; the fission of the nucleus is the first step towards the formation of another cell; each half of the divided nucleus establishes round itself an independent current; and, apparently, it is by the repulsion of the nuclei that the separation into two individuals is finally effected. All which facts, when generalised, imply that the nucleus is the governing or *co-ordinating* part. Now, Professor Owen subsequently points out that the matter of the sperm-cell performs in the fertilised germ-cell just this same function which the nucleus performs in a single-celled animal. We find the absorption by a germ-cell of the contents of a sperm-cell "followed by the appearance of a pellucid nucleus in the centre of the opaque and altered germ-cell; we further see its successive fissions governed by the preliminary division of the pellucid centre;" and, led by these and other facts, Professor Owen thinks that "one cannot reasonably suppose that the nature and properties of the nucleus of the impregnated germ-cell and that of the monad can be different."[e] And hence he further infers that "the nucleus of the monad is of a nature similar to, if not identical with," the matter of the spermatozoon. But we have seen that in the monad the nucleus is the co-ordinating part; and hence to say that the sperm-cell is, in nature, identical with it, is to say that the sperm-cell consists of co-ordinating matter.

Chemical analysis affords further evidence, though, from the imperfect data at present obtained, less conclusive evidence than could be wished. Partly from the white and gray nervous substances having been analysed together instead of separately, and partly from the difficulty of isolating the efficient contents of the sperm-cells, a satisfactory comparison cannot be made. Nevertheless, possessing in common, as they do, one element, by which they are specially characterised, the analysis, as far as it goes, supports our argument. The table on the following page, made up from data given in the *Cyclopædia of Anatomy and Physiology, Art.* Nervous System, gives the proportion of this element in the brain in different conditions, and shows how important is its presence.

This connection between the quantity of phosphorus present and the degree of mental power exhibited, is sufficiently significant; and the fact that in the same individual the varying degrees of cerebral activity are indicated by the varying quantities of alkaline phosphates excreted by the kidneys[f]. still more clearly shows the essentialness of phosphorus as a constituent of nervous matter. Respecting the constitution of sperm-cells chemists do not altogether agree. One thing, however, is certain – that they contain unoxidized phosphorus; and also a fatty acid, that is

	In Infants.	*In Youth.*	*In Adults.*	*In Old Men.*	*In Idiots.*
Solid constituents in a hundred parts of brain	17·21	25·74	27·49	26·15	29·07
Of these solid constituents the phosphorus amounts to	0·8	1·65	1·80	1·00	0·85
Which gives a percentage of phosphorus in the solid constituents of	4·65	6·41	6·54	3·82	2·92

not improbably similar to the fatty acid contained in neurine[g]. In fact, there would seem to be present the constituents of that oleophosphoric acid which forms so distinctive an element of the brain. That a large quantity of binoxide of protein is also present, may be ascribed to the fact that a great part of the sperm-cell consists merely of the protective membrane and its locomotive appendage; the really efficient portion being but the central contents[h].

Evidence of a more conclusive nature – evidence, too, which will show in what direction our argument tends – is seen in the marked antagonism of the nervous and generative systems. Thus, the fact that intense mental application, involving great waste of the nervous tissues, and a corresponding consumption of nervous matter for their repair, is accompanied by a cessation in the production of sperm-cells, gives strong support to the hypothesis that the sperm-cells consist essentially of neurine. And this becomes yet clearer on finding that the converse fact is true – that undue production of sperm-cells involves cerebral inactivity. The first result of a morbid excess in this direction is headache, which may be taken to indicate that the brain is out of repair; this is followed by stupidity; should the disorder continue, imbecility supervenes, ending occasionally in insanity.

That the sperm-cell is co-ordinating matter, and the germ-cell matter to be co-ordinated, is, therefore, an hypothesis not only having much *à priori* probability, but one supported by numerous facts.

§ 13. This hypothesis alike explains, and is confirmed by, the truth, that throughout the vertebrate tribes the degree of fertility varies inversely as the development of the nervous system.

[...]

§ 14. The law which we have thus traced throughout the animal kingdom, and which must alike determine the different fertilities of different species, and the variations of fertility in the same species, we have now to consider in its application to mankind.

From the fact that the human race is in a state of transition, we may suspect that the existing ratio between its ability to multiply, and its ability to maintain life, is not a constant ratio. From the fact that its fertility is at present in excess of what is needful, we may infer that any change in the ratio will probably be towards a diminution of fertility. And from the fact that, on the whole, civilization increases the ability to maintain life, we may perceive that there is at work some influence by which such diminution is necessitated. Before inquiring for this influence, let us consider what directions an increase of ability to maintain life may take – what scope there is for an increase. In some further development of the co-ordinating system, that is, in some greater co-ordination of actions, the increase must of course consist. But there are several kinds of co-ordination; and it will be well to ask of what kind or kinds increase is most requisite, and therefore most likely. For, doubtless, in conformity with the general law of adaptation, increase will take place only where it is demanded.

Will it be in strength? Probably not. Though from prehistoric remains, we may gather that the race has become more bulky, yet the cause of this change seems now diminishing. Mechanical appliances are fast supplanting muscular force, and will most likely continue to do so until they leave to be done by manual labour only as much as is needful for the healthy maintenance of the body at its then attained size.

Will it be in swiftness or agility? Probably not. In the savage these form important elements of the ability to maintain life; but in the civilized man they subserve that end in quite a minor degree, and there seems no circumstance likely to necessitate an increase of them.

Will it be in mechanical skill, that is, in the better co-ordination of complex movements? Most likely in some degree. Awkwardness is continually entailing injuries and loss of life. Moreover, the complicated tools developed by civilization are constantly requiring greater delicacy of manipulation. Already the cerebellum, which is the nervous centre directing compound motions is larger in man than in any other creature except the elephant; and the daily-increasing variety and complexity of the processes he has to perform, and the appliances he has to use, may be expected to cause a further growth of it.

Will it be in intelligence? Largely, no doubt. There is ample room for progress in this direction, and ample demand for it. Our lives are universally shortened by our ignorance. In attaining complete knowledge of our own nature, and of the nature of surrounding things – in ascertaining the conditions of existence to which we must conform, and in discovering means of conforming to them under all variations of seasons and circumstances – we have abundant scope for intellectual culture and urgent need for intellectual development.

Will it be in morality, that is, in greater power of self-regulation? Largely also; perhaps most largely. Normal conduct, or in other words, conduct conducive to the maintenance of perfect and long-continued life, is usually come short of more from defect of will than of knowledge. To the due co-ordination of those complex actions which constitute human life in its civilized form, there goes not only the prerequisite – recognition of the proper course; but the further prerequisite – a due impulse to pursue that course. And on calling to mind our daily failures to fulfil often-repeated resolutions, we shall perceive that lack of the needful desire, rather than lack of the needful insight, is the chief cause of faulty action. A further endowment of those feelings which civilization is developing in us – sentiments responding to the requirements

In the Australian	*75 cubic inches.*
" African	82 "
" Malayan	86 "
" Englishman	96 "

of the social state – emotive faculties that find their gratifications in the duties devolving on us – must be acquired before the crimes, excesses, diseases, improvidences, dishonesties, and cruelties, that now so greatly diminish the duration of life, can cease.

But whether greater co-ordination of actions take place in any or in all of these directions, and in whatever degree or proportions, it is clear that, if it take place at all, it must be at the expense of fertility. Regarded from the abstract point of view, increased ability to maintain life in this case, as in all others, necessarily involves decreased ability to multiply. Or, regarded in the concrete, that further development of the co-ordinating system, which any advance presupposes, implies further decrease in the production of co-ordinating cells.

§ 15. That an enlargement of the nervous centres is going on in mankind, is an ascertained fact. Not alone from a general survey of human progress – not alone from the greater power of self-preservation shown by civilized races, are we left to infer such enlargement; it is proved by actual measurement. The mean capacities of the crania in the leading divisions of the species have been found to be –

showing an increase in the course of the advance from the savage state to our present phase of civilization, amounting to nearly 30 per cent. on the original size. That this increase will be continuous, might be reasonably assumed; and to infer a future decrease of fertility would be tolerably safe, were no further evidence forthcoming. But it may be shown why a greater development of the nervous system *must* take place, and why, consequently, there *must* be a diminution of the present excess of fertility; and further, it may be shown that the sole agency needed to work out this change is – *the excess of fertility itself.*

For, as we all know, this excess of fertility entails a constant pressure of population upon the means of subsistence; and, as long as it exists, must continue to do this. Looking only at the present and the immediate future, it is unquestionably true, that, if unchecked, the rate of increase of people would exceed the rate of increase of food. It is clear that the wants of their redundant numbers constitute the only stimulus mankind have to a greater production of the necessaries of life; for, were not the demand beyond the supply, there would be no motive to increase the supply. Moreover, this excess of demand over supply, and this pressure of population, of which it is the index, cannot be eluded. Though by the emigration that takes place when the pressure arrives at a certain intensity, a partial and temporary relief may be obtained, yet, as by this process all habitable countries must gradually become peopled, it follows, that in the end the pressure, whatever it may then be, must be borne in full.

But this inevitable redundancy of numbers – this constant increase of people beyond the means of subsistence – involving as it does an increasing stimulus to better the modes of producing food and other necessaries – involves also an increasing demand for skill, intelligence, and self-control – involves, therefore, a constant exercise of these, that is – involves a gradual growth of them. Every improvement is at once the product of a higher form of humanity, and demands that higher form of humanity to carry it into practice. The application of science to the arts is simply the bringing to bear greater intelligence for satisfying our wants; and implies continued increase of that intelligence. To get more produce from the acre, the farmer must study chemistry – must adopt new mechanical appliances – and must, by the multiplication of tools and processes, cultivate both his own powers and the powers of his labourers. To meet the requirements of the market, the manufacturer is perpetually improving his old machines, and inventing new ones; and by the premium of high wages incites artizans to acquire greater skill. The daily-widening ramifications of commerce entail upon the merchant a need for more knowledge and more complex calculations; whilst the lessening profits of the ship-owner force him to employ greater science in building, to get captains of higher intelligence, and better crews. In all cases, increase of numbers is the efficient cause. Were it not for the competition this entails, more thought would not daily be brought to bear upon the business of life; greater activity of mind would not be called for; and development of mental power would not take place. Difficulty in getting a living is alike the incentive to a higher education of children, and to a more intense and long-continued application in adults. In the mother it induces foresight, economy, and skilful housekeeping; in the father, laborious days and constant self-denial. Nothing but necessity could make men submit to this discipline, and nothing but this discipline could produce a continued progression. The contrast between a Pacific Islander, all whose wants are supplied by Nature, and an Englishman, who, generation after generation, has had to bring to the satisfaction of his wants ever-increasing knowledge and skill, illustrates at once the need for, and the effects of, such discipline. And this being admitted, it cannot be denied that a further continuance of such discipline, possibly under a yet more intense form, must produce a further progress in the same direction – a further enlargement of the nervous centres, and a further decline of fertility.

And here it must be remarked, that the effect of pressure of population, in increasing the ability to maintain life, and decreasing the ability to multiply, is not a uniform effect, but an average one. In this case, as in many others, Nature secures each step in advance by a succession of trials, which are perpetually repeated, and cannot fail to be repeated, until success is achieved. All mankind in turn subject themselves more or less to the discipline described; they either may or may not advance under it; but, in the nature of things, only those who *do* advance under it eventually survive. For, necessarily, families and races whom this increasing difficulty of getting a living which excess of fertility entails, does not stimulate to improvements in production – that is, to greater mental activity – are on the high road to extinction; and must ultimately be supplanted by those whom the pressure does so stimulate. This truth we have recently seen exemplified in Ireland. And here, indeed, without further illustration, it will be seen that premature death, under all its forms, and from all its causes, cannot fail to work in the same direction. For as those prematurely carried off must, in the average of cases, be those in whom the power of self-preservation is the least, it unavoidably follows, that those left behind to continue the race must be those in whom the power of self-preservation is the greatest – must be the select of their generation. So that, whether the dangers to existence be of the kind produced by excess of fertility, or of any other kind, it is clear, that by the ceaseless exercise of the faculties needed to contend with them, and by the death of all men who fail to contend with them successfully, there is ensured a constant progress towards a higher degree of skill, intelligence, and self-regulation – a better co-ordination of actions – a more complete life.

§ 16. There now remains but to inquire towards what limit this progress tends. Evidently, so long as the fertility of the race is more than sufficient to balance the diminution by deaths, population must continue to increase: so long as population continues to increase, there must be pressure on the means of subsistence: and so long as there is pressure on the means of subsistence, further mental development must go on, and further diminution of fertility must result. Hence, the change can never cease until the rate of multiplication is just equal to the rate of mortality; that is – can never cease until, on the average, each pair brings to maturity but two children. Probably this involves that each pair will rarely produce more than two offspring; seeing that with the greatly-increased ability to preserve life, which the hypothesis presupposes, the amount of infant and juvenile mortality must become very small. Be this as it may, however, it is manifest that, in the end, pressure of population and its accompanying evils will entirely disappear; and will leave a state of things which will require from each individual no more than a normal and pleasurable activity. That this last inference is a legitimate corollary will become obvious on a little consideration. For, a cessation in the decrease of fertility implies a cessation in the development of the nervous system; and this implies that the nervous system has become fully equal to all that is demanded of it – has not to do more than is natural to it. But that exercise of faculties which does not exceed what is natural constitutes gratification. Consequently, in the end, the obtainment of subsistence will require just that kind and that amount of action needful to perfect health and happiness.

Thus do we see how simple are the means by which the greatest and most complex results are worked out. From the point of view now reached, it becomes plain that the necessary antagonism of individuation and reproduction not only fulfils with precision the *à priori* law of maintenance of race, from the monad up to man, but ensures the final attainment of the highest form of this maintenance – a form in which the amount of life shall be the greatest possible, and the births and deaths the fewest possible. In the nature of things, the antagonism could not fail to work out the results we see it working out. The gradual diminution and ultimate disappearance of the original excess of fertility could take place only through the process of civilization; and, at the same time, the excess of fertility has itself rendered the process of civilization inevitable. From the beginning, pressure of population has been the proximate cause of progress. It produced the original diffusion of the race. It compelled men to abandon predatory habits and take to agriculture. It led to the clearing of the earth's surface. It forced men into the social state; made social organization inevitable; and has developed the social sentiments. It has stimulated to progressive improvements in production, and to increased skill and intelligence. It is daily pressing us into closer contact and more mutually-dependent relationships. And after having caused, as it ultimately must, the due peopling of the globe, and the bringing of all its habitable parts into the highest state of culture – after having brought all processes for the satisfaction of human wants to the greatest perfection – after having, at the same time, developed the intellect into complete competency for its work, and the feelings into complete fitness for social life – after having done all this, we see that the pressure of population, as it gradually finishes its work, must gradually bring itself to an end.

G. Woodfall and Son, Printers, Angel Court, Skinner Street, London.

Notes

a Prin. of Phys., p. 907.

b Should it be objected that in the higher plants the sperm-cell and germ-cell differ, though no distinct co-ordinating system exists, it is replied that there *is* co-ordination of actions, though of a feeble kind, and that there must be some agency by which this is carried on.

c It is a significant fact that amongst the diœcious invertebrata, where the nutritive system greatly exceeds the other systems in development, the female is commonly the largest, and often greatly so. In some of the Rotifera the male has no nutritive system at all. See *Prin. of Phys.*, p. 954.

d Prin. of Phys., p. 908.

e Parthenogenesis, pp. 66, 67.

f Lectures on Animal Chemistry. By Dr. Bence Jones. *Medical Times,* Sept. 13th, 1851. See also *Prin. of Phys.*, p. 171.

g Cyclopædia of Anatomy and Physiology, vol. iv. p. 506.

h From a remark of Drs. Wagner and Leuckart this chemical evidence seems to have already suggested the idea that the sperm-cell becomes "metamorphosed into the central parts of the nervous system." But though they reject this assumption, and though the experiments of Mr. Newport clearly render it untenable, yet none of the facts latterly brought to light conflict with the hypothesis that the sperm-cell contains unorganized co-ordinating matter.

26

"PROGRESS: ITS LAW AND CAUSE"

Herbert Spencer

Source: Herbert Spencer, "Progress: Its Law and Cause", *The Westminster Review*, 67 (April 1857), pp. 445–85

ART. V. – PROGRESS: ITS LAW AND CAUSE.

1. *Cosmos: a Sketch of a Physical Description of the Universe*. By Alexander von Humboldt. Translated from the German, by E. C. Otté. 4 vols. London: H. G. Bohn.
2. *Principles of Geology:* or the Modern Changes of the Earth and its Inhabitants considered as illustrative of Geology. By Sir Charles Lyell, M.A., F.R.S. Ninth Edition. London: John Murray.
3. *Principles of Comparative Physiology*. By William B. Carpenter, M.D., F.R.S., F.G.S. Fourth Edition. London: John Churchill.

THE current conception of Progress is somewhat shifting and indefinite. Sometimes it comprehends little more than simple growth – as of a nation in the number of its members and the extent of territory over which it has spread. Sometimes it has reference to quantity of material products – as when the advance of agriculture and manufactures is the topic. Sometimes the superior quality of these products is contemplated; and sometimes the new or improved appliances by which they are produced. When, again, we speak of moral or intellectual progress, we refer to the state of the individual or people exhibiting it; whilst, when the progress of Knowledge, of Science, of Art, is commented upon, we have in view certain abstract results of human thought and action. Not only, however, is the current conception of Progress more or less vague, but it is in great measure erroneous. It takes in not so much the reality of Progress as its accompaniments – not so much the substance as the shadow. That progress in intelligence which takes place during the evolution of the child into the man, or the savage into the philosopher, is commonly regarded as consisting in the greater number of facts known and laws understood: whereas the actual progress consists in those internal modifications of which this increased knowledge is the expression. Social progress is supposed to consist in the produce of a greater quantity and variety of the articles required for the satisfaction of men's wants; in the increasing security of person and property;

 DOI: 10.4324/9781003490548-29

in the widening freedom of action enjoyed: whereas, rightly understood, social progress consists in those changes of structure in the social organism which have entailed these consequences. The current conception is a teleological one. The phenomena are contemplated solely as bearing on human happiness. Only those changes are held to constitute progress which directly or indirectly tend to heighten human happiness. And they are thought to constitute progress simply *because* they tend to heighten human happiness. But rightly to understand Progress, we must inquire what is the nature of these changes, considered apart from our interests. Ceasing, for example, to regard the successive geological modifications that have taken place in the Earth, as modifications that have gradually fitted it for the habitation of Man, and as *therefore* a geological progress, we must seek to determine the character common to these modifications – the law to which they all conform. And similarly in every other case. Leaving out of sight concomitants and beneficial consequences, let us ask what Progress is in itself.

In respect to that progress which individual organisms display in the course of their evolution, this question has been answered by the Germans. The investigations of Wolff, Goethe, and Von Baer, have established the truth that the series of changes gone through during the development of a seed into a tree, or an ovum into an animal, constitute an advance from homogeneity of structure to heterogeneity of structure. In its primary stage, every germ consists of a substance that is uniform throughout, both in texture and chemical composition. The first step in its development is the appearance of a difference between two parts of this substance; or, as the phenomenon is described in physiological language – a differentiation. Each of these differentiated divisions presently begins itself to exhibit some contrast of parts; and by and by these secondary differentiations become as definite as the original one. This process is continuously repeated – is simultaneously going on in all parts of the growing embryo; and by endless multiplication of these differentiations there is ultimately produced that complex combination of tissues and organs constituting the adult animal or plant. This is the course of evolution followed by all organisms whatever. It is settled beyond dispute that organic progress consists in a change from the homogeneous to the heterogeneous.

Now, we propose in the first place to show, that this law of organic progress is the law of all progress. Whether it be in the development of the Earth, in the development of Life upon its surface, in the development of Society, of Government, of Manufactures, of Commerce, of Language, Literature, Science, Art, this same evolution of the simple into the complex, through a process of continuous differentiation, holds throughout. From the earliest traceable cosmical changes down to the latest results of civilization, we shall find that the transformation of the homogeneous into the heterogeneous, is that in which Progress essentially consists.

With the view of showing that *if* the Nebular Hypothesis be true, the genesis of the solar system supplies one illustration of this law; let us assume that the matter of which the sun and planets consist once existed in a diffused form; and that from the gravitation of its atoms there resulted a gradual concentration. By the hypothesis, the solar system, in its nascent state, existed as an indefinitely extended and nearly homogeneous medium – a medium almost homogeneous in density, in temperature, and in other physical attributes. The first advance towards consolidation resulted in a differentiation between the occupied space which the nebulous mass still filled, and the unoccupied space which it previously filled. There simultaneously resulted a contrast in density and a contrast in temperature, between the interior and the exterior of this mass. And at the same time there arose throughout it rotatory movements, whose velocities varied according to their distances from its centre. These differentiations increased in number and degree until there was evolved the organized group of sun, planets, and satellites, which we now know – a group which presents numerous contrasts of structure and action among its members. There are the immense contrasts between the sun and the planets, in bulk and in weight; as well as the subordinate contrasts between one planet and another, and between the planets and their satellites. There is the similarly marked contrast between the sun as almost stationary, and the planets as moving round him with great velocity; whilst there are the secondary contrasts between the velocities and periods of the several planets, and between their simple revolutions and the double ones of their satellites, which have to move round their primaries whilst moving round the sun. There is the yet further strong contrast between the sun and the planets in respect of temperature; and there is reason to suppose that the planets and satellites differ from each other in their proper heat, as well as in the heat they receive from the sun. When we bear in mind that, in addition to these various contrasts, the planets and satellites also differ in respect to their distances from each other and their primary; in respect to the inclinations of their orbits, the inclinations of their axes, their times of rotation on their axes, their specific gravities, and their physical constitutions; we see what a high degree of heterogeneity the solar system exhibits, when compared with the almost complete homogeneity of the nebulous mass out of which it is supposed to have originated.

Passing from this hypothetical illustration, which must be taken for what it is worth, without prejudice to the general argument, let us descend to a more certain order of evidence. It is now generally agreed among geologists that the Earth was at first a mass of molten matter; and that it is still fluid and incandescent at the distance of a few miles beneath its surface. Originally, then, it was homogeneous in consistence, and, in virtue of the circulation that takes place in heated fluids, must have been comparatively homogeneous in temperature; and it must have been surrounded by an atmosphere consisting partly of the elements of air and water, and partly of those various other elements which assume a gaseous form at high temperatures. That slow cooling by radiation, which is still going on at an inappreciable rate, and which, though originally far more rapid than now, necessarily required an immense time to produce any decided change, must ultimately have resulted in the solidification of the portion most able to part with its heat – namely, the surface. In the thin crust thus formed we have the first marked differentiation. A still further cooling, a consequent thickening of this crust, and an accompanying deposition of all solidifiable elements contained in the atmosphere, must ultimately have been followed by the condensation of the water previously existing in a gaseous state. A second marked differentiation must thus have arisen: and as the condensation must have taken place on the coolest parts of the surface – namely, about the poles – there must simultaneously have resulted the first geographical distinction of parts. To these illustrations of increasing heterogeneity, which, though deduced from the known laws of matter will perhaps be regarded as more or less hypothetical, Geology adds an extensive series that have been inductively established. Its investigations show that the Earth has been continually becoming more heterogeneous, in virtue of the multiplication of the strata constituting its crust; further, that it has been becoming more heterogeneous in respect of the composition of these strata, the latter of which, as being formed from the detritus of the older ones, are many of them rendered highly complex by the mixture of materials they contain; that the heterogeneity has been further increased by the action of the Earth's still molten nucleus upon its envelope, whence have resulted not only a great variety of igneous rocks, but the tilting up of sedimentary strata at all angles, the formation of faults and metallic veins, the production of endless dislocations and irregularities. Yet again, geologists teach us that the Earth's surface has been becoming more varied in elevation – that the most ancient mountain systems are the smallest, and the Andes and Himalayas the most modern; whilst in all probability there have been corresponding changes in the bed of the ocean. As a consequence of these ceaseless differentiations, we now find that no considerable portion of the Earth's exposed surface is like any other portion, either in contour, in geologic structure, or in chemical composition; and that in most parts it changes, from mile to mile, in all these characteristics. Moreover, it must not be forgotten that there has been simultaneously going on a gradual differentiation of climates. As fast as the Earth cooled and its crust solidified, there arose appreciable differences in temperature between those parts of its surface most exposed to the sun and those less exposed. Gradually, as the cooling progressed, these differences became more pronounced; until there finally resulted those marked contrasts between regions of perpetual ice and snow, regions where winter and summer alternately reign for periods varying according to the latitude, and regions where summer follows summer with scarcely an appreciable variation. At the same time, the successive elevations and subsidences of different portions of the Earth's crust, tending as they have done to the present irregular distribution of land and sea, have entailed various modifications of climate beyond those dependent on latitude; while a yet further series of such modifications have been produced by increasing differences of elevation in the land, which have in sundry places brought arctic, temperate, and tropical climates to within a few miles of each other. And the general result of these changes is, that not only has every extensive region its own meteorologic conditions, but that every locality in each region differs more or less from others in those conditions, as in its structure, its contour, its soil. Thus, between our existing Earth, the phenomena of whose varied crust neither geographers, geologists, mineralogists, nor meteorologists have yet enumerated, and the molten globe out of which it was evolved, the contrast in heterogeneity is sufficiently striking.

When from the Earth itself we turn to the plants and animals that have lived, or still live, upon its surface, we find ourselves in some difficulty from lack of facts. That every existing organism has been developed out of the simple into the complex, is indeed the first established truth of all; and that every organism that has existed was similarly developed, is an inference which no physiologist will for a moment hesitate to draw. But when we pass from individual forms of life to Life in general, and inquire whether the same law is seen in the *ensemble* of its manifestations, – whether modern plants and animals are of more heterogeneous structure than ancient ones, and whether the Earth's present Flora and Fauna are more heterogeneous than the Flora and Fauna of the past, – we find the evidence so fragmentary, that every conclusion is open to dispute. Two-thirds of the Earth's surface being covered by water; a great part of the exposed land being inaccessible to, or untravelled by, the geologist; the greater part of the remainder having been scarcely more than glanced at; and even the most familiar portions, as England, having been so imperfectly explored that a new series of strata has been added within these four years, – it is manifestly impossible for us to say with any certainty what creatures have, and what have not, existed at any particular period. Considering the perishable nature of many of the lower organic forms, the metamorphosis of many sedimentary strata, and the gaps that occur among the rest, we shall see further reason for distrusting our deductions. On the one hand, the repeated discovery of vertebrate remains in strata previously supposed to contain none, – of reptiles where only fish were thought to exist, – of mammals where it was believed there were no creatures higher than reptiles, – renders it daily more manifest how small is the value of negative evidence. On the other hand, the worthlessness of the assumption that we have discovered the earliest, or anything like the earliest, organic remains, is becoming equally clear. That the oldest known sedimentary rocks have been greatly changed by igneous action, and that still older ones have been totally transformed by it, is becoming undeniable. And the fact that sedimentary strata earlier than any we know, have been melted up, being admitted, it must also be admitted that we cannot say how far back in time this destruction of sedimentary strata has been going on. Thus it becomes manifest that the title *Palæozoic*, as applied to the earliest known fossiliferous strata, involves a *petitio principii;* and that, for aught we know to the contrary, only the last few chapters of the Earth's biological history may have come down to us. On neither side, therefore, is the evidence conclusive. Nevertheless we cannot but think that, scanty as they are, the facts, viewed in their *ensemble*, tend to show both that the more heterogeneous organisms have been evolved in the later geologic periods, and that Life in general has been more heterogeneously manifested as time has advanced. Let us take, in illustration, the one case of the *vertebrata.* The earliest known vertebrate remains are those of Fishes; and Fishes are the most homogeneous of the vertebrata. Later and more heterogeneous are Reptiles. Later still, and more heterogeneous still, are Mammals and Birds. If it be said, as it may fairly be said, that the Palæozoic deposits, not being estuary deposits, are not likely to contain the remains of terrestrial vertebrata, which may nevertheless have existed at that era, we reply that we are merely pointing to the leading facts, *such as they are*. But to avoid any such criticism, let us take the mammalian subdivision only. The earliest known remains of mammals are those of small marsupials, which are the lowest of the mammalian type; whilst, conversely, the highest of the mammalian type – Man – is the most recent. The evidence that the vertebrate fauna, as a whole, has become more heterogeneous, is considerably stronger. To the argument that the vertebrate fauna of the Palæozoic period, consisting, so far as we know, entirely of Fishes, was less heterogeneous than the modern vertebrate fauna, which includes Reptiles, Birds, and Mammals of multitudinous genera, it may be replied, as before, that estuary deposits of the Palæozoic period, could we find them, might contain other orders of vertebrata. But no such reply can be made to the argument that, whereas the marine vertebrata of the Palæozoic period consisted entirely of cartilaginous fishes, the marine vertebrata of later periods include numerous genera of osseous fishes; and that, therefore, the later marine vertebrate faunas are more heterogeneous than the oldest known one. Nor, again, can any such reply be made to the fact that there are far more numerous orders and genera of mammalian remains in the tertiary formations than in the secondary formations. Did we wish merely to make out the best case, we might dwell upon the opinion of Dr. Carpenter, who says that "the general facts of Palæontology appear to sanction the belief, that *the same plan* may be traced out in what may be called *the general life of the globe*, as in *the individual life* of every one of the forms of organized being which now people it." Or we might cite, as decisive, the judgment of Professor Owen, who holds that the earlier examples of each group of creatures severally departed less widely from archetypal generality than the later ones – were severally less unlike the fundamental form common to the group as a whole; that is to say – constituted a less heterogeneous group of creatures; and who further upholds the doctrine of a biological progression. But in deference to an authority for whom we have the highest respect, who considers that the evidence at present obtained does not justify a verdict either way, we are content to leave the question open.

Whether an advance from the homogeneous to the heterogeneous is or is not displayed in the biological history of the globe, it is clearly enough displayed in the progress of the latest and most heterogeneous creature – Man. It is alike true that, during the period in which the Earth has been peopled, the human organism has become more heterogeneous among the civilized divisions of the species; and that the species, as a whole, has been growing more heterogeneous in virtue of the multiplication of races and the differentiation of these races from each other. In proof of the first of these positions, we may cite the fact that, in the relative development of the limbs, the civilized man departs more widely from the general type of the placental mammalia than do the lower human races. Whilst often possessing well-developed body and arms, the Papuan has extremely small legs: reminding us in this respect of the quadrumana, in which there is no great contrast in size between the hind and fore limbs. But in the European, the greater length and massiveness of the legs has become very marked – the fore and hind limbs are relatively more heterogeneous. Again, in the greater ratio which the cranial bones bear to the facial bones, we may see the same truth. Among the vertebrata in general, progress is marked by an increasing heterogeneity in the vertebral column, and more especially in the vertebræ constituting the skull: the higher forms being distinguished by the relatively larger size of the bones which cover the brain, and the relatively smaller size of those which form the jaws, &c. Now, this characteristic, which is more marked in Man than in any other creature, is more marked in the European than in the savage. Judging from the greater extent and variety of faculty he exhibits, we may infer that the civilized man has also a more complex or heterogeneous nervous system than the uncivilized man: and indeed the fact is in part visible in the increased ratio which his cerebrum bears to the subjacent ganglia. If further elucidation be needed, we may find it in every nursery. The infant European has sundry marked points of resemblance to the lower human races; as in the flatness of the alæ of the nose, the depression of its bridge, the divergence and forward opening of the nostrils, the form of the lips, the absence of a frontal sinus, the width between the eyes, the smallness of the legs. Now, as the developmental process by which these characteristics are changed into those of the adult European, is a continuation of that change from the homogeneous to the heterogeneous exhibited during the previous evolution of the embryo, which every physiologist will admit; it follows that the parallel developmental process by which the like characteristics of the barbarous races have been changed into those of the civilized races, has also been a continuation of the change from the homogeneous to the heterogeneous. The truth of the second position – that Mankind, as a whole, have become more heterogeneous – is so obvious as scarcely to need illustration. Every work on Ethnology, by its divisions and subdivisions of races, bears testimony to it. Even were we to admit the hypothesis that Mankind originated from several separate stocks, it would still remain true, that as, from each of these stocks, there have sprung many now widely different tribes, which are proved by philological evidence to have had a common origin, the race as a whole is far less homogeneous than it was at first. Add to which, that we have, in the Anglo-Americans, an example of a new variety arising within these few generations; and that, if we may trust to the descriptions of observers, we are likely soon to have another such example in Australia.

On passing from Humanity under its individual form, to Humanity as socially embodied, we find the general law still more variously exemplified. The change from the homogeneous to the heterogeneous is displayed equally in the evolution of civilization as a whole, and in the progress of every tribe or nation; and is still going on with increasing rapidity. As we see in still existing barbarous tribes, society in its first and lowest form is a homogeneous aggregation of individuals having like powers and performing like functions: the only marked differentiation of function being that which accompanies difference of sex. Every man is warrior, hunter, fisherman, tool-maker, builder; every woman performs the same drudgeries; every family is self-sufficing, and, save for purposes of aggression and defence, might as well live apart from the rest. Very early, however, in the process of social evolution, we find an incipient differentiation between the governing and the governed. Some kind of chieftainship seems almost co-ordinate with the first advance from the state of separate wandering families to that of a nomadic tribe. The authority of the strongest makes itself felt among a body of savages as in a herd of animals, or a posse of schoolboys. At first, however, it is indefinite, uncertain, – is shared by others of scarcely inferior power, and is unaccompanied by any difference in occupation or style of living: the first ruler kills his own game, makes his own weapons, builds his own hut, and, economically considered, does not differ from others of his tribe. Gradually, as the tribe progresses, the contrast between the governing and the governed grows more marked. Supreme power becomes hereditary in one family; the head of that family ceasing to provide for his own wants, is served by others; and he begins to assume the sole office of ruling. At the same time there has been arising a co-ordinate species of government – that of Religion. As all ancient records and traditions prove, the earliest rulers are regarded as divine personages. The maxims and commands they uttered during their lives are held sacred after their deaths, and are enforced by their divinely-descended successors; who in their turns are promoted to the pantheon of the race, there to be worshipped and propitiated along with their predecessors: the most ancient of whom is the supreme god, and the rest subordinate gods. For a long time these connate forms of government – civil and religious – continue closely associated. For many generations the king continues to be the chief priest, and the priesthood to be members of the royal race. For many ages religious law continues to contain more or less of civil regulation, and civil law to possess more or less of religious sanction; and even among the most advanced civilized nations these two controlling agencies are by no means completely differentiated from each other. Having a common root with these, and becoming gradually separate from them, we find yet another controlling agency – that of Manners or ceremonial usages. All titles of honour are originally the names of the god-king; afterwards of God and the king; still later of persons of high rank; and finally come, some of them, to be used between man and man. All forms of complimentary address were primarily the expressions of submission from prisoners to their conqueror, or from subjects to their ruler, either human or divine – expressions that were afterwards used to propitiate subordinate authorities, and gradually descended into ordinary intercourse. All modes of salutation were originally obeisances made before the monarch and used in worship of him after his death. Presently others of the god-descended race were similarly saluted; and by degrees some of the salutations have become the due of all.[a] Thus, no sooner does the originally homogeneous social mass become definitely differentiated into the governed and the governing parts, than this last exhibits an incipient differentiation into religious and secular – Church and State; while at the same time there begins to be differentiated from both, that less concrete species of government which rules the daily intercourse of individuals – a species of government which, as we may see in heralds' colleges, in books of the peerage, in masters of ceremonies, is not without a certain embodiment of its own. Each of these is itself subject to successive differentiations. In the course of ages, there arises, as among ourselves, a highly complex political organization of monarch, ministers, lords and commons, with their subordinate administrative departments, courts of justice, revenue offices, &c., supplemented in the provinces by municipal governments, county governments, parish or union governments – all of them more or less elaborated. By its side there grows up a highly complex religious organization, with its various grades of officials, from archbishops down to sextons, its colleges, convocations, ecclesiastical courts, &c.; to all which must be added the ever-multiplying independent sects, each with its general and local authorities. And at the same time there is developed a highly complex aggregation of customs, manners, and temporary fashions, enforced by society at large, and serving to control those minor transactions between man and man which are not regulated by civil and religious law. Moreover it is to be observed that this ever-increasing heterogeneity in the governmental appliances of each nation, has been accompanied by an increasing heterogeneity in the governmental appliances of different nations: all of which are more or less unlike in their political systems and legislation, in their creeds and religious institutions, in their customs and ceremonial usages.

Simultaneously there has been going on a second differentiation of a still more familiar kind; that, namely, by which the mass of the community has become segregated into distinct classes and orders of workers. While the governing part has been undergoing the complex development above described, the governed part has been undergoing an equally complex development, which has resulted in that minute division of labour characterizing advanced nations. It is needless to trace out this progress from its first stages, up through the caste divisions of the East and the incorporated guilds of Europe, to the elaborate producing and distributing organization existing among ourselves. Political economists have made familiar to all, the evolution which, beginning with a tribe whose members severally perform the same actions each for himself, ends with a civilized community whose members severally perform different actions for each other; and they have further explained the evolution through which the solitary producer of any one commodity, is transformed into a combination of producers who, united under a master, take separate parts in the manufacture of such commodity. But there are yet other and higher phases of this advance from the homogeneous to the heterogeneous in the industrial structure of the social organism. Long after considerable progress has been made in the division of labour among different classes of workers, there is still little or no division of labour among the widely separated parts of the community: the nation continues comparatively homogeneous in the respect that in each district the same occupations are pursued. But when roads and other means of transit become numerous and good, the different districts begin to assume different functions, and to become mutually dependent. The calico manufacture locates itself in this county, the woollen-cloth manufacture in that; silks are produced here, lace there; stockings in one place, shoes in another; pottery, hardware, cutlery, come to have their special towns; and ultimately every locality becomes more or less distinguished from the rest by the leading occupation carried on in it. Nay, more, this subdivision of functions shows itself not only among the different parts of the same nation, but among different nations. That exchange of commodities which free-trade promises so greatly to increase, will ultimately have the effect of specializing, in a greater or less degree, the industry of each people. So that beginning with a barbarous tribe, almost if not quite homogeneous in the functions of its members, the progress has been, and still is, towards an economic aggregation of the whole human race, growing ever more heterogeneous in respect of the separate functions assumed by separate nations, the separate functions assumed by the local sections of each nation, the separate functions assumed by the many kinds of makers and traders in each town, and the separate functions assumed by the workers united in producing each commodity.

Not only is the law thus clearly exemplified in the evolution of the social organism, but it is exemplified with equal clearness in the evolution of all products of human thought and action; whether concrete or abstract, real or ideal. Let us take Language as our first illustration.

The lowest form of language is the exclamation, by which an entire idea is vaguely conveyed through a single sound; as among the lower animals. That human language ever consisted solely of exclamations, and so was strictly homogeneous in respect of its parts of speech, we have no evidence. But that language can be traced down to a form in which nouns and verbs are its only elements, is an established fact. In the gradual multiplication of parts of speech out of these primary ones – in the differentiation of verbs into active and passive, of nouns into abstract and concrete – in the rise of distinctions of mood, tense, person, of number and case – in the formation of auxiliary verbs, of adjectives, adverbs, pronouns, prepositions, articles – in the evolution of those orders, genera, species, and varieties of parts of speech by which civilized races express minute modifications of meaning – we see a change from the homogeneous to the heterogeneous. And it may be remarked, in passing, that it is more especially in virtue of having carried this subdivision of function to a greater extent and completeness, that the English language is superior to all others. Another phase under which we may contemplate the development of language is the differentiation of words of allied meanings. Philology early disclosed the truth that in all languages words may be grouped into families having a common ancestry. An aboriginal name applied indiscriminately to each of an extensive and ill-defined class of things or actions, presently undergoes modifications by which the chief divisions of the class are expressed. These several names springing from the primitive root, themselves become the parents of other names still further modified. And by the aid of those systematic modes which presently arise, of making derivatives and forming compound terms expressing still smaller distinctions and qualifications, there is ultimately developed a tribe of words so heterogeneous in character and meaning, that to the uninitiated it seems incredible they should have had a common origin. Meanwhile from other roots there are being evolved other such tribes, until there results a language of some sixty thousand or more unlike words, signifying as many unlike objects, qualities, acts. Yet another way in which language in general exhibits advance from the homogeneous to the heterogeneous, is in the multiplication of languages. Whether, as Max Müller and Bunsen think, all languages have grown from one stock, or whether, as some philologists say, they have grown from two or more stocks, it is clear that since large families of languages, as the Indo-European, are of one parentage, they have become distinct through a process of continuous divergence. The same diffusion over the Earth's surface which has led to the differentiation of the race, has simultaneously led to a differentiation of their speech: a truth which we see further illustrated in each nation by the peculiarities of dialect found in separate districts. Thus the progress of Language conforms to the general law alike in the evolution of languages, in the evolution of families of words, and in the evolution of parts of speech.

On passing from spoken to written language, we come upon several classes of facts, all having similar implications. Written language is connate with Painting and Sculpture; and at first all three are appendages of Architecture, and have a direct connexion with the primary form of all government – the Theocratic. Merely noting by the way the fact that sundry wild races, as for example the Australians and the tribes of South Africa, are given to depicting personages and events upon the walls of caves, which are probably regarded as sacred places, let us pass to the case of the Egyptians. Among them, as also among the Assyrians, we find mural paintings used to decorate the temple of the god and the palace of the king (which were indeed originally identical); and as such they were governmental appliances in the same sense that state-pageants and religious feasts were. Further, they were governmental appliances in virtue of representing the worship of the god, the triumphs of the god-king, the submission of his subjects, and the punishment of the rebellious. And yet again they were governmental, as being the products of an art reverenced by the people as a sacred mystery. From the habitual use of this pictorial representation there naturally grew up the but slightly modified practice of picture-writing – a practice which was found still extant among the Mexicans at the time they were discovered. By a process of abbreviation analogous to that which has been abundantly exemplified in our own written and spoken language, the most familiar of these pictured figures were successively simplified; and ultimately there grew up a system of symbols, most of which had but a distant resemblance to the things for which they stood. The inference that the hieroglyphics of the Egyptians were thus evolved is confirmed by the fact that the picture-writing of the Mexicans was found to have given birth to a like family of ideographic forms; and among them, as among the Egyptians, these had been partially differentiated into the *kuriological* or imitative, and the *tropical* or symbolic: which were, however, used together in the same record. In Egypt, written language underwent a further differentiation; whence resulted the *hieratic* and the *epistolographic* or *enchorial:* both of which can be clearly affiliated upon the original hieroglyphic. At the same time we find that for the expression of proper names which could not be otherwise conveyed, phonetic symbols were employed; and though it is alleged that the Egyptians never actually achieved complete alphabetic writing, yet it can scarcely be doubted that these phonetic symbols occasionally used in aid of their ideographic ones, were the germs out of which alphabetic writing grew. Once having become separate from hieroglyphics, alphabetic writing itself underwent numerous differentiations – multiplied alphabets were produced; between most of which, however, more or less connexion can still be traced. And in each civilized nation there has now grown up, for the representation of one set of sounds, several sets of written signs used for distinct purposes. Finally, through a yet more important differentiation came printing; which, uniform in kind as it was at first, has since become multiform.

While written language was passing through its earlier stages of development, the mural decoration which formed its root was being differentiated into Painting and Sculpture. The gods, kings, men, and animals represented, were originally marked by indented outlines and coloured. In most cases these outlines were of such depth, and the object they circumscribed so far rounded and marked out in its leading parts, as to form a species of work intermediate between intaglio and bas-relief. In other cases we see an advance upon this: the raised spaces between the figures being chiselled off, and the figures themselves appropriately tinted, a painted bas-relief was produced. The restored Assyrian architecture at Sydenham exhibits this style of art carried to greater perfection – the persons and things represented, though still barbarously coloured, are carved out with more truth and in greater detail: and in the winged lions and bulls used for the angles of gateways, we may see a considerable advance towards a completely sculptured figure; which, nevertheless, is still coloured, and still forms part of the building. But while in Assyria the production of a statue proper seems to have been little, if at all, attempted, we may trace in Egyptian art the gradual separation of the sculptured figure from the wall. A walk through the collection in the British Museum will clearly show this; while it will at the same time afford an opportunity of observing the evident traces which the independent statues bear of their derivation from bas-relief: seeing that nearly all of them not only display that union of the limbs with the body which is the characteristic of bas-relief, but have the back of the statue united from head to foot with a block which stands in place of the original wall. Greece repeated the leading stages of this progress. As in Egypt and Assyria, these twin arts were at first united with each other and with their parent, Architecture, and were the aids of Religion and Government. On the friezes of Greek temples we see coloured bas-reliefs representing sacrifices, battles, processions, games – all in some sort religious. On the pediments we see painted sculptures more or less united with the tympanum, and having for subjects the triumphs of gods or heroes. Even when we come to statues that are definitely separated from the buildings to which they pertain, we still find them coloured; and only in the later periods of Greek civilization does the differentiation of sculpture from painting appear to have become complete. In Christian art we may clearly trace a parallel re-genesis. All early paintings and sculptures throughout Europe were religious in subject – represented Christs, crucifixions, virgins, holy families, apostles, saints. They formed integral parts of church architecture, and were among the means of exciting worship; as in Roman Catholic countries they still are. Moreover, the early sculptures of Christ on the cross, of virgins, of saints, were coloured: and it needs but to call to mind the painted madonnas and crucifixes still abundant in continental churches and highways, to perceive the significant fact that painting and sculpture continue in closest connexion with each other where they continue in closest connexion with their parent. Even when Christian sculpture was pretty clearly differentiated from painting, it was still religious and governmental in its subjects – was used for tombs in churches and statues of kings: while at the same time painting, where not purely ecclesiastical, was applied to the decoration of palaces, and besides representing royal personages, occupied itself almost wholly with sacred legends. Only in quite recent times have painting and sculpture become entirely secular arts. Only within these few centuries has painting been divided into historical, landscape, marine, architectural, genre, animal, still-life, &c., and sculpture grown heterogeneous in respect of the variety of real and ideal subjects with which it occupies itself.

Strange as it seems then, we find it no less true, that all forms of written language, of painting, and of sculpture, have a common root in the politico-religious decorations of ancient temples and palaces. Little resemblance as they now have, the bust that stands on the console, the landscape that hangs against the wall, and the copy of the *Times* lying upon the table, are remotely akin; not only in nature, but by extraction. The brazen face of the knocker which the postman has just lifted, is related not only to the woodcuts of the *Illustrated London News* which he is delivering, but to the characters of the *billet-doux* which accompanies it. Between the painted window, the prayer-book on which its light falls, and the adjacent monument, there is consanguinity. The effigies on our coins, the signs over shops, the figures that fill every ledger, the coat of arms outside the carriage panel, and the placards inside the omnibus, are, in common with dolls, blue-books, paper-hangings, lineally descended from the rude sculpture-paintings in which the Egyptians represented the triumphs and worship of their god-kings. Perhaps no example can be given which more vividly illustrates the multiplicity and heterogeneity of the products that in course of time may arise by successive differentiations from a common stock.

Before passing to other classes of facts, it should be observed that the evolution of the homogeneous into the heterogeneous is displayed not only in the differentiation of Painting and Sculpture from Architecture and from each other, and in the increased variety and specialty of the subjects they embody, but it is further shown in the structure of each separate work. A modern picture or statue is far more heterogeneous in its constitution than an ancient one. An Egyptian sculpture-fresco represents all its figures as on one plane – that is, at the same distance from the eye; and so is less heterogeneous than a painting that represents them as at various distances from the eye. It exhibits all objects as exposed to the same degree of light; and so is less heterogeneous than a painting which exhibits its different objects and different parts of each object as in different degrees of light. It uses scarcely any but the primary colours, and these in their full intensity; and so is less heterogeneous than a painting which, introducing the primary colours but sparingly, employs an endless variety of intermediate tints, each of heterogeneous composition, and differing from the others not only in quality but in intensity. Moreover, we see in these aboriginal works a great uniformity of conception. The same arrangement of figures is continually represented – the same actions, attitudes, faces, dresses. In Egypt the modes of representation were so fixed that it was sacrilege to introduce a novelty; and indeed it could have been only in virtue of a fixed mode of representation that a system of hieroglyphics became possible. The Assyrian bas-reliefs display parallel characters. Deities, kings, attendants, winged-figures, and animals, are severally depicted in like positions, holding like implements, doing like things, and with like expression or non-expression of face. If a palm grove is introduced, all the trees are of the same height, have the same number of leaves, and are equidistant. When water is represented, each wave is a counterpart of the rest; and the fish, almost always of one kind, are evenly distributed over the surface. The beards of the kings, the gods, and the winged-figures, are everywhere similar; as are the manes of the lions, and equally so those of the horses. Hair is represented throughout by one form of curl. The king's beard is quite architecturally built up of compound tiers of uniform curls, alternating with twisted tiers placed in a transverse direction, and arranged with perfect regularity; and the terminal tufts of the bulls' tails are represented in exactly the same manner. Without tracing out the like traits in early Christian art, in which, though less striking, they are still visible, the advance in heterogeneity will be sufficiently manifest on remembering that in the pictures of our own day the composition is endlessly varied; the attitudes, faces, expressions unlike; the subordinate objects different in size, form, position, texture; and more or less of contrast even in the smallest details. Or, if we compare an Egyptian statue, seated bolt upright on a block, with hands on knees, fingers outspread and parallel, eyes looking straight forward, and the two sides perfectly symmetrical in every particular, with a statue of the advanced Greek or the modern school, which is asymmetrical in respect of the position of the head, the body, the limbs, the arrangement of the hair, dress, appendages, and in its relations to neighbouring objects, we shall see the change from the homogeneous to the heterogeneous clearly manifested.

The co-ordinate origin and gradual differentiation of Poetry, Music, and Dancing, supply a series of facts of analogous character. Rhythm in speech, rhythm in sound, and rhythm in motion, were in the beginning but different elements of the same thing, and have only in process of time become separate things. Among various existing barbarous tribes we find them still united. The dances of savages are accompanied by some kind of monotonous chant, the clapping of hands, the striking of rude instruments: there are measured movements, measured words, and measured tones; and the whole ceremony, usually having reference to war or sacrifice, is of governmental character. In the early records of the historic races we similarly find these three forms of metrical action united in religious festivals. In the Hebrew writings we read that the triumphal ode composed by Moses on the defeat of the Egyptians, was sung to an accompaniment of dancing and timbrels. The Israelites danced and sung "at the inauguration of the golden calf.

And as it is generally agreed that this representation of the Deity was borrowed from the mysteries of Apis, it is probable that the dancing was copied from that of the Egyptians on those occasions." There was an annual dance in Shiloh on the sacred festival; and David danced before the ark. Again, in Greece the like relation is everywhere seen: the original type being there, as probably in other cases, a simultaneous chanting and mimetic representation of the life and adventures of the god. The Spartan dances were accompanied by hymns and songs; and in general the Greeks had "no festivals or religious assemblies but what were accompanied with songs and dances" – both of them being forms of worship used before altars. Among the Romans, too, there were sacred dances: the Salian and Lupercalian being named as of that kind. And even in Christian countries, as at Limoges, in comparatively recent times, the people have danced in the choir in honour of a saint. The incipient separation of these once united arts from each other and from religion, was early visible in Greece. Probably diverging from dances partly religious partly warlike, as the Corybantian, came the war dances proper, of which there were various kinds; and from these resulted secular dances. Meanwhile Music and Poetry, though still united, came to have an existence separate from dancing. The aboriginal Greek poems, religious in their subject matter, were not recited but chanted; and though at first the chant of the poet was accompanied by the dance of the chorus, it ultimately grew into independence. Later still, when the poem had been differentiated into epic and lyric – when it became the custom to sing the lyric and recite the epic – poetry proper was born. As during the same period musical instruments were being multiplied, we may presume that music came to have an existence apart from words. And both of them were simultaneously assuming other forms besides the religious. Facts having like implications might be cited from the histories of later times and peoples: as the practices of our own early minstrels, who sang to the harp heroic narratives versified by themselves to music of their own composition: thus uniting the now separate offices of poet, composer, vocalist, and instrumentalist. But, without further illustration, the common origin and gradual differentiation of Dancing, Poetry, and Music will be sufficiently manifest.

The advance from the homogeneous to the heterogeneous is displayed not only in the separation of these arts from each other and from religion, but also in the multiplied differentiations which each of them afterwards undergoes. Not to dwell upon the numberless kinds of dancing that have, in course of time, come into use, – and not to occupy space in detailing the progress of poetry, as seen in the development of the various forms of metre, of rhyme, and of general organization, – let us confine our attention to music as a type of the group. As argued by Dr. Burney, and as implied by the customs of still extant barbarous races, the first musical instruments were, without doubt, percussive – sticks, calabashes, tom-toms, – and were used simply to mark the time of the dance; and in this constant repetition of the same sound, we see music in its most homogeneous form. The Egyptians had a lyre with three strings; the early lyre of the Greeks had four, constituting their tetrachord; in course of some centuries lyres of seven and eight strings were employed; and, by the expiration of a thousand years, they had advanced to their "great system" of the double octave: through all which changes there of course arose a greater heterogeneity of melody. Simultaneously there came into use the different modes – Dorian, Ionian, Phrygian, Æolian, and Lydian – answering to our keys: and of these there were ultimately fifteen. As yet, however, there was but little heterogeneity in the time of their music. Instrumental music during this period being merely the accompaniment of vocal music, and vocal music being completely subordinated to words, – the singer being also the poet, chanting his own compositions and making the lengths of his notes agree with the feet of his verses, – there necessarily resulted a tiresome uniformity of measure, which, as Dr. Burney says, "no resources of melody could disguise." Lacking the complex rhythm obtained by our equal bars and unequal notes, the only rhythm was that produced by the quantity of the syllables, and was of necessity comparatively monotonous. And further, it may be observed that the chant thus resulting, being like recitative, was much less clearly differentiated from ordinary speech than is our modern song. Nevertheless, in virtue of the extended range of notes in use, the variety of modes, the occasional variations of time consequent on changes of metre, and the multiplication of instruments, music had, towards the close of Greek civilization, attained to considerable heterogeneity – not indeed as compared with our music, but as compared with that which preceded it. As yet, however, there existed nothing but melody: harmony was unknown. It was not until Christian church-music had reached some development, that music in parts was evolved; and then it came into existence through a very unobtrusive differentiation. Difficult as it may be to conceive *à priori* how the advance from melody to harmony could take place without a sudden leap, it is none the less true that it did so. The circumstance which prepared the way for it, was the employment of two choirs singing alternately the same air. Afterwards it became the practice – very possibly first suggested by a mistake – for the second choir to commence before the first had ceased; thus producing a fugue: and with the simple airs then in use, a harmonious fugue might not improbably thus result. The idea having once been given, the composing of airs productive of fugal harmony would naturally grow up; as in some way it *did* grow up out of this alternate choir-singing. And from the fugue to concerted music of two, three, four, and more parts, the transition was easy. Without tracing in detail the increasing complexity that resulted from introducing notes of various lengths, from the multiplication of keys, from the use of accidentals, from varieties of time, and so forth, it needs but to contrast music as it is, with music as it was, to see how immense is the increase of heterogeneity. We see this if, looking at music in its *ensemble*, we enumerate its many different genera and species – if we consider the divisions into vocal, instrumental, and mixed; and their subdivisions into music for different voices and different instruments: if we observe the many forms of sacred music, from the simple hymn, the chant, the canon, motet, anthem, &c., up to the oratorio; and the still more numerous forms of secular music, from the ballad up to the serenata, from the instrumental solo up to the symphony. Again, the same thing is seen on comparing any one sample of aboriginal music with a sample of modern music – even an ordinary song for the piano; which we find to be relatively highly heterogeneous, not only in respect of the varieties of pitch and length of the notes, the number of different notes sounding at the same instant in company with the voice, and the variations of strength with which they are sounded and sung, but in respect of the changes of key, the changes of time, the changes of *timbre* of the voice, and the many other modifications of expression. While between the old monotonous dance-chant and a grand opera of our own day, with its endless orchestral complexities and vocal combinations, the contrast in heterogeneity is so extreme that it seems scarcely credible that the one should have been the ancestor of the other.

Were they needed, many further illustrations might be cited. Going back to the early time when the deeds of the god-king, chanted and mimetically represented in dances round his altar, were further narrated in picture-writings on the walls of temples and palaces, and so constituted a rude literature, we might trace the development of Literature through phases in which, as in the Hebrew Scriptures, it presented in one work theology, cosmogony, history, biography, civil law, ethics, poetry; through other phases in which, as in the Iliad, the religious, martial, historical, the epic, dramatic, and lyric elements are similarly commingled; down to its present heterogeneous development, in which its divisions and subdivisions are so numerous and varied as to defy complete classification. Or we might trace out the evolution of Science; beginning with the era in which it was not yet differentiated from Art, and was, in union with Art, the handmaid of Religion; passing through the era in which the sciences were so few and rudimentary, as to be simultaneously cultivated by the same philosophers; and ending with the era in which the genera and species are so numerous that few can enumerate them, and no one can adequately grasp even one genus. Or we might do the like with Architecture, with the Drama, with Dress. But doubtless the reader is already weary of illustrations; and our promise has been amply fulfilled. We believe we have shown beyond question, that that which the German physiologists have found to be the law of organic development, is the law of all development. The advance from the simple to the complex, through a process of successive differentiations, is seen alike in the earliest changes of the Universe to which we can reason our way back, and in the earliest changes which we can inductively establish; it is seen in the geologic and climatic evolution of the Earth, and of every single organism on its surface; it is seen in the evolution of Humanity, whether contemplated in the civilized individual, or in the aggregation of races; it is seen in the evolution of Society in respect both of its political and economical organization; and it is seen in the evolution of all those endless concrete and abstract products of human activity which constitute the environment of our daily life. From the remotest past which Science can fathom, down to the novelties of yesterday, that in which Progress essentially consists, is the transformation of the homogeneous into the heterogeneous.

And now, from this uniformity of procedure, may we not infer some fundamental necessity whence it results? May we not rationally seek for some all-pervading principle which determines this all-pervading process of things? Does not the universality of the *law* imply a universal *cause?*

That we can fathom such cause, noumenally considered, is not to be supposed. To do this would be to solve that ultimate mystery which must ever transcend human intelligence. But it still may be possible for us to reduce the law of all Progress, above established, from the condition of an empirical generalization, to the condition of a rational generalization. Just as it was possible to interpret Kepler's laws as necessary consequences of the law of gravitation; so it may be possible to interpret this law of Progress, in its multiform manifestations, as the necessary consequence of some similarly universal principle. As gravitation was assignable as the *cause* of each of the groups of phenomena which Kepler formulated; so may some equally simple attribute of things be assignable as the cause of each of the groups of phenomena formulated in the foregoing pages. We may be able to affiiliate all these varied and complex evolutions of the homogeneous into the heterogeneous, upon certain simple facts of immediate experience, which, in virtue of endless repetition, we regard as necessary.

The probability of a common cause, and the possibility of formulating it, being granted, it will be well, before going further, to consider what must be the general characteristics of such cause, and in what direction we ought to look for it. We can with certainty predict that it has a high degree of generality; seeing that it is common to such infinitely varied phenomena: just in proportion to the universality of its application must be the abstractness of its character. We need not expect to see in it an obvious solution of this or that form of Progress; because it equally refers to forms of Progress bearing little apparent resemblance to them: its association with multiform orders of facts, involves its dissociation from any particular order of facts. Being that which determines Progress of every kind – astronomic, geologic, organic, ethnologic, social, economic, artistic, &c. – it must be concerned with some fundamental attribute possessed in common by all these; and must be expressible in terms of this fundamental attribute. The only obvious respect in which all kinds of Progress are alike, is, that they are modes of *change;* and hence, in some characteristic of changes in general, the desired solution will probably be found. We may suspect *à priori* that in some law of change lies the explanation of this universal transformation of the homogeneous into the heterogeneous.

Thus much premised, we pass at once to the statement of the law, which is this: – *Every active force produces more than one change – every cause produces more than one effect.*

Before this law can be duly comprehended, a few examples must be contemplated. When we strike one body against another, that which we usually regard as the effect, is a change of position or motion in one or both bodies. But a moment's thought shows us that this is a careless and very incomplete view of the matter. Besides the visible mechanical result, sound is produced; or, to speak accurately – a vibration in one or both bodies, and in the surrounding air: and under some circumstances we call this the effect. Moreover, the air has not only been made to vibrate, but has had sundry currents caused in it by the transit of the bodies. Further, there is a disarrangement of the particles of the two bodies in the neighbourhood of their point of collision; amounting in some cases to a visible condensation. Yet more, this condensation is accompanied by the evolution of a certain amount of heat. In some cases a spark – that is, light – results, from the incandescence of a portion struck off: and sometimes this incandescence is associated with chemical combination. Thus, by the original mechanical force expended in the collision, at least five, and often more, different kinds of changes have been produced. Take, again, the lighting of a candle. Primarily this is a chemical change consequent on a rise of temperature. The process of combination having once been set going by extraneous heat, there is a continued evolution of carbonic acid, water, &c. – in itself a result more complex than the extraneous heat that first caused it. But accompanying this process of combination there is a production of heat; there is a production of light; there is an ascending column of hot gases generated; there are currents established in the surrounding air. Moreover, the decomposition of one force into many forces does not end here: each of the several changes produced becomes the parent of further changes. The carbonic acid given off will by and by enter into combination with some base; or under the influence of sunshine give up its carbon to the leaf of a plant. The water will modify the hygrometric condition of the surrounding air; or, if the current of hot gases containing it come against a cold body, will be condensed: altering the temperature, and perhaps the chemical state, of the surface it covers. The heat evolved melts the subjacent tallow, and expands the neighbouring air. The light, falling on various substances, calls forth from them reactions by which it is modified; and so various colours are produced. Similarly even with these secondary actions, which may be traced out into ever-multiplying ramifications, until they become too minute to be appreciated. And thus it is with all changes whatever. No case can be named in which an active force does not evolve forces of several kinds, and each of these, other groups of forces. Universally the effect is more complex than the cause.

Doubtless the reader already foresees the course of our argument. This multiplication of results, which is displayed in every event of to-day, has been going on from the beginning, and is true of the grandest phenomena of the universe as of the most insignificant. From the law that every active force produces more than one change, it is an inevitable corollary that throughout all time there has been an ever-growing complication of things. Starting with the ultimate fact that every cause produces more than one effect, we may readily see that throughout creation there must necessarily have gone on, and must still go on, a never-ceasing transformation of the homogeneous into the heterogeneous. But let us trace out this truth in detail.

Without committing ourselves to it as more than a speculation, though a highly probable one, let us again commence with the evolution of the solar system out of a nebulous medium.[b] From the mutual attraction of the atoms of a diffused mass whose form is unsymmetrical, there results not only condensation but rotation – gravitation simultaneously generates both the centripetal and the centrifugal forces. While the condensation and the rate of rotation are progressively increasing, the approximation of the atoms necessarily generates a progressively increasing temperature. As this temperature rises to incandescence, light begins to be evolved; and ultimately there results a revolving sphere of fluid matter radiating intense heat and light – a sun. There are good reasons for believing that, in consequence of the high tangential velocity, and consequent centrifugal force, acquired by the peripheral portions of the condensing nebulous mass, the periodical detachment of rotating rings is a necessary result; and that, from the breaking up of these nebulous rings, there must arise masses which in the course of their condensation repeat the actions of the parent mass, and so produce planets and their satellites – an inference strongly supported by the still extant rings of Saturn. Should it hereafter be satisfactorily demonstrated that planets and satellites were thus generated, a striking illustration will be afforded of the highly heterogeneous effects resulting from the primary homogeneous cause; but it will sufficiently serve our present purpose to point to the fact that from the mutual attraction of the particles of an irregular nebulous mass there result condensation, rotation, heat, and light.

It follows as a corollary from the Nebular Hypothesis, that the Earth must originally have been in a state of incandescence; and whether the Nebular Hypothesis be true or not, this original incandescence of the Earth is now inductively established – or if not established, at least rendered so highly probable that it is a generally admitted geological doctrine. Let us look first at the astronomical attributes of this once molten globe. From its rotation there result the oblateness of its form, the alternations of day and night, and (under the influence of the moon) the tides, aqueous and atmospheric. From the inclination of its axis of rotation, there result the precession of the equinoxes and the many differences of the seasons, both simultaneous and successive, that pervade its surface. Thus the multiplication of effects is obvious. Several of the differentiations consequent on the gradual cooling of the Earth have been already noticed – as the formation of a crust, the solidification of sublimed elements, the precipitation of water, &c., – and we here again refer to them merely to point out that they are simultaneous results of the one cause, diminishing temperature. Let us now, however, observe the multiplied changes subsequently arising from the continuance of this one cause. The cooling of the Earth necessarily involves its contraction. Hence the solid crust first formed is presently too large for the shrinking nucleus; and as it cannot support itself, inevitably follows the nucleus. But a spheroidal envelope cannot sink down into contact with a smaller internal spheroid, without undergoing disruption: it must run into wrinkles as the rind of an apple does when the bulk of its interior decreases from evaporation. As the cooling progresses and the envelope thickens, the ridges consequent on these contractions necessarily become greater, rising ultimately into hills and mountains; and the later systems of mountains thus produced must not only be higher, as we find them to be, but they must be longer, as we also find them to be. Thus, leaving out of view other modifying forces, we see what immense heterogeneity of surface has resulted from the one cause, loss of heat – a heterogeneity which the telescope shows us to be paralleled on the face of the moon, where aqueous and atmospheric agencies have been absent. But we have yet to notice another species of heterogeneity of surface similarly and simultaneously caused. While the Earth's crust was still thin, the ridges consequent on its contraction must not only have been small, but the spaces between these ridges must have rested with great evenness upon the subjacent liquid spheroid; and the water in those arctic and antarctic regions in which it first condensed, must have been evenly distributed. But as fast as the crust grew thicker and gained corresponding strength, the lines of fracture from time to time caused in it, must have occurred at greater distances apart; the intermediate surfaces must have followed the contracting nucleus with less uniformity; and there must have resulted larger areas of land and water. If any one, after wrapping up an orange in wet tissue paper, and observing not only how comparatively small are the wrinkles, but how evenly the intervening spaces lie upon the surface of the orange, will then wrap it up in thick cartridge-paper, and note both the greater height of the ridges and the much larger spaces throughout which the paper does not touch the orange, he will be able to realize the fact, that as the Earth's solid envelope grew thicker, the areas of elevation and depression must have become greater. In place of islands more or less homogeneously scattered over an all-embracing sea, there must have gradually arisen heterogeneous arrangements of continent and ocean, such as we now know. Yet further, this double change in the extent and in the elevation of the exposed lands, involved yet another species of heterogeneity – that of coast-line. A tolerably even surface raised out of the ocean must have a simple, regular sea-margin; but a surface varied by table-lands and intersected by mountain-chains of different heights, must, when raised out of the ocean, have an outline extremely irregular both in its leading features and in its details. Thus we see how enormous is the accumulation of geological and geographical results slowly brought about by the one cause – contraction of the Earth.

When we pass from the agency which geologists term igneous, to aqueous and atmospheric agencies, we see the like evergrowing complications of effects. The denuding actions of air and water have, from the beginning, been modifying every exposed surface; everywhere simultaneously producing many different changes. Oxidation, heat, wind, frost, rain, glaciers, rivers, tides, waves, have been unceasingly producing disintegration; varying in kind and amount according to local circumstances. Acting upon a simple tract of granite, they here work scarcely an appreciable effect; there cause exfoliations of the surface, and a consequent accumulation of *débris* and boulders; and elsewhere, after decomposing the feldspar into a white clay, carry away this and the accompanying quartz and mica, and deposit them in separate beds, fluviatile and marine. When the exposed land consists of several different formations, sedimentary and igneous, the denudation produces changes proportionably more heterogeneous. The formations being disintegrable in different degrees, an increased irregularity of surface is produced. The areas drained by different rivers being differently constituted, these rivers carry down to the sea different combinations of ingredients; and so sundry new strata of distinct composition are formed. And here indeed we may see very simply illustrated the truth, which we shall presently have to trace out in its more complex manifestations, that in proportion to the heterogeneity of the object or objects on which any force expends itself, is the heterogeneity of the results. A continent of complex structure, exposing numerous strata irregularly distributed, raised to various levels, tilted up at all angles, must, under the same denuding influences, give origin to immensely multiplied results: each district must be differently modified; each river must carry down a different kind of detritus; each deposit must be differently distributed by the involved currents, tidal and other, which wash the contorted shores; and this multiplication of results must manifestly be greatest where the complexity of the surface is greatest.

It is out of the question here to trace in detail the genesis of those endless complications described by Geology and Physical Geography: else we might show how the general truth, that every active force produces more than one change, is exemplified in the highly involved flow of the tides, in the ocean currents, in the winds, in the distribution of rain, in the distribution of heat, and so forth. But not to dwell upon these, let us, for the fuller elucidation of this truth in relation to the inorganic world, consider what would be the consequences of some extensive cosmical revolution – say the subsidence of Central America. The immediate results of the disturbance would themselves be sufficiently complex. Besides the numberless dislocations of strata, the ejections of igneous matter, the propagation of earthquake vibrations thousands of miles around, the loud explosions, and the escape of gases; there would be the rush of the Atlantic and Pacific Oceans to supply the vacant space, the subsequent recoil of enormous waves, which would traverse both these oceans and produce myriads of changes along their shores, the corresponding atmospheric waves complicated by the currents surrounding each volcanic vent, and the extensive electrical discharges with which such disturbances are accompanied. But these temporary effects would be insignificant compared with the permanent ones. The complex currents of the Atlantic and Pacific would be altered in direction and amount. The distribution of heat achieved by these ocean currents would be different from what it is. The arrangement of the isothermal lines, not only on the neighbouring continents, but even throughout Europe, would be changed. The tides would flow differently from what they do now. There would be more or less modification of the winds in their periods, strengths, directions, qualities. Rain would fall scarcely anywhere at the same times and in the same quantities as at present. In short, the meteorological conditions thousands of miles off, on all sides, would be more or less revolutionized. Thus, without taking into account the infinitude of modifications which these changes of climate would produce upon the flora and fauna, both of land and sea, the reader will sufficiently realize the immense heterogeneity of the results wrought out by one force, when that force expends itself upon a previously complicated area; and he will readily draw the unavoidable corollary that from the beginning the complication has advanced at an increasing rate.

Before going on to show how organic progress also depends upon the universal law that every force produces more than one change, we have to notice the manifestation of this law in yet another species of inorganic progress – namely, chemical. The same general causes that have wrought out the heterogeneity of the Earth, physically considered, have simultaneously wrought out its chemical heterogeneity. Without dwelling upon the general fact that the forces which have been increasing the variety and complexity of geological formations, have, at the same time, been bringing into contact elements not previously exposed to each other under conditions favourable to union, and so have been adding to the number of chemical compounds, let us pass to the more important complications that have resulted from the cooling of the Earth. There is every reason to believe that at an extreme heat the elements cannot exist in combination at all. Even under such heat as can be artificially produced, some extremely powerful affinities yield, as for instance, that of oxygen for hydrogen; and the great majority of chemical compounds are decomposed at much lower temperatures. But without insisting upon the highly probable inference, that when the Earth was in its first state of incandescence there were no chemical combinations at all, it will suffice our purpose to point to the unquestionable fact that the compounds that can exist at the highest temperatures, and which must, therefore, have been the first to make their appearance as the Earth cooled, are those of the simplest constitutions. The protoxides – including under that head the alkalies, earths, &c. – are, as a class, the most stable compounds we know: most of them resisting decomposition by any heat we can generate. These, consisting severally of one atom of each component element, are combinations of the simplest order – are but one degree less homogeneous than the elements themselves. More heterogeneous than these, less stable, and therefore later in the Earth's history, are the deutoxides, tritoxides, peroxides, &c.; in which two, three, four, or more atoms of oxygen are united with one atom of metal or other element. Higher than these in heterogeneity are the hydrates; in which an oxide of hydrogen, united with an oxide of some other element, forms a combination whose atoms severally contain at least four ultimate atoms of three different kinds. Yet more heterogeneous and less stable still, are the salts; which present us with compound atoms each made up of five, six, seven, eight, ten, twelve, or more atoms, of three, if not more, kinds. Then there are the hydrated salts, of a yet greater heterogeneity, which undergo partial decomposition at much lower temperatures. After them come the further-complicated supersalts and double salts, having a stability again decreased; and so throughout. Without entering into qualifications for which we lack space, we believe no chemist will deny it to be a general law of these inorganic combinations that, *other things equal*, the stability decreases as the complexity increases. And then when we pass to the compounds of organic chemistry, we find this general law still further exemplified: we find much greater complexity and much less stability. An atom of albumen, for instance, consists of 482 ultimate atoms of five different kinds. Fibrine, still more intricate in constitution, contains in each atom, 298 atoms of carbon, 49 of nitrogen, 2 of sulphur, 228 of hydrogen, and 92 of oxygen – in all, 660 atoms; or, more strictly speaking – equivalents. And these two substances are so unstable as to decompose at quite ordinary temperatures; as that to which the outside of a joint of roast meat is exposed. Thus it becomes manifest that the present chemical heterogeneity of the Earth's surface has arisen by degrees, as the progressive refrigeration has permitted; and that it has exhibited itself in three forms – first, in the multiplication of chemical compounds; second, in the greater number of different elements contained in the more modern of these compounds; and third, in the higher and more varied multiples in which these more numerous elements combine.

To say that this advance in chemical heterogeneity is due to the one cause, diminution of the Earth's temperature, would be to say too much; for it is clear that aqueous and atmospheric agencies have been concerned; and, further, that the affinities of the elements themselves are implied. The cause has all along been a composite one: the cooling of the Earth having been simply the most general of the concurrent causes, or assemblage of conditions. And here, indeed, it may be remarked that in the several classes of facts already dealt with (excepting perhaps the first), and still more in those with which we shall presently deal, the causes are more or less compound; as indeed are nearly all causes with which we are acquainted. Scarcely any change can with logical accuracy be wholly ascribed to one agency, to the neglect of the permanent or temporary conditions under which only this agency produces the change. But as it does not materially affect our argument, we prefer, for simplicity's sake, to use throughout the popular mode of expression. Perhaps it will be further objected, that to assign loss of heat as the cause of any changes, is to attribute these changes not to a force, but to the absence of a force. And this is true. Strictly speaking, the changes should be attributed to that force, or those forces, which come into action when the antagonist force is withdrawn. But though there is an inaccuracy in saying that the freezing of water is due to the loss of its heat, no practical error arises from it; nor will a parallel laxity of expression vitiate our statements respecting the multiplication of effects. Indeed, the objection serves but to draw attention to the general fact, that not only does the exertion of a force produce more than one change, but the withdrawal of a force produces more than one change. And this suggests that perhaps the most correct statement of the general principle would be its most abstract statement – every change is followed by more than one other change.

Returning to the thread of our exposition, we have next to trace out, in organic progress, this same all-pervading principle. And here, where the evolution of the homogeneous into the heterogeneous was first observed, the production of many changes by one cause is most difficult to demonstrate. The development of a seed into a plant, or an ovum into an animal, is so gradual, while the forces which determine it are so involved, and at the same time so unobtrusive, that it is scarcely possible to detect that multiplication of effects which is elsewhere so obvious. Nevertheless, guided by indirect evidence, we may pretty safely reach the conclusion that here too the law holds. Observe, first, how numerous are the effects which any marked change will produce upon an adult organism – a human being, for instance. An alarming sound or sight, besides the immediate impressions on the organs of sense and the nerves, may produce a muscular start, a scream, a distortion of the face, a trembling consequent upon a general muscular relaxation, a burst of perspiration, an excited action of the heart, a rush of blood to the brain, followed possibly by arrest of the heart's action and by syncope; and if the system be feeble, an indisposition with its long train of complicated symptoms may set in. Similarly in cases of disease. A minute portion of the small-pox virus introduced into the system, will, in a severe case, cause, during the first stage, rigors, heat of skin, accelerated pulse, furred tongue, loss of appetite, thirst, epigastric uneasiness, vomiting, headache, pains in the back and limbs, muscular weakness, convulsions, delirium, &c.; in the second stage, cutaneous eruption, itching, tingling, sore throat, swelled fauces, salivation, cough, hoarseness, dyspnœa, &c.; and in the third stage, œdematous inflammations, pneumonia, pleurisy, diarrhœa, inflammation of the brain, ophthalmia, erysipelas, &c.: each of which enumerated symptoms is itself more or less complex. Medicines, special foods, better air, might in like manner be instanced as producing multiplied results. Now it needs only to consider that the heterogeneity of change thus wrought by a simple force upon an adult organism, will be in part parallel in an embryo organism, to understand how here also, the evolution of the homogeneous into the heterogeneous may be consequent upon the production of many effects by one cause. The external heat and other agencies which determine the first complications of the germ, may, by acting upon these, superinduce further complications; upon these still higher and more numerous ones; and so on continually – each organ as it is developed, serving by its actions and reactions upon the rest, to initiate further complexities. The first pulsations of the fœtal heart must simultaneously affect the nutrition of all the tissues. The growth of each tissue, involving as it does the abstraction from the blood of special proportions of elements, must modify the constitution of the blood; and so must modify the nutrition of all the other tissues. The action of the heart, implying as it does a certain waste, necessitates the addition to the blood of effete matters, which must influence the rest of the system, and perhaps, as some think, themselves determine the formation of excretory organs. The nervous connexions established among the viscera must further multiply their mutual influences: and so continually. Indeed we may find *à priori* reason to think that the evolution proceeds after this manner. For since it is now known that no germ, animal or vegetable, contains the slightest rudiment, trace, or indication of the future organism – now that the microscope has shown us that the process first set up in every fertilized germ, is a process of repeated spontaneous fissions ending in the production of a mass of seemingly homogeneous cells, not one of which exhibits any special character; there seems no alternative but to suppose that the incipient or partial organization at any moment subsisting in a growing embryo, is itself transformed by environing influences into the succeeding phase of organization, and this into the next, until, through everincreasing complexities, the ultimate form is reached. Thus, though the subtilty of the forces and the slowness of the results, render it impossible for us directly to show that the successive stages of increasing heterogeneity through which every growing embryo passes, are severally consequent upon the production of many changes by one force, yet, *indirectly*, we have strong evidence that they are so. We have observed how multitudinous are the effects which one cause may generate in an adult organism; the abstract necessity that an analogous multiplication of effects must happen in the unfolding organism, we have contemplated in sundry illustrative cases; and we have seen that structureless as every germ originally is, the development of an organism out of it is otherwise incomprehensible. Not indeed that we are thus enabled really to explain the production of any plant or animal. We are still in the dark respecting those mysterious properties in virtue of which the germ, when subjected to fit influences, undergoes the special changes initiating the series of developmental transformations. All which we aim to show, is, that given a germ possessing these mysterious properties, the evolution of a heterogeneous organism from it, probably depends upon that multiplication of effects which we have seen to be the cause of progress in general, so far as we have yet traced it.

When, leaving the development of single plants and animals, we pass to that of the Earth's flora and fauna, the course of our argument again becomes clear and simple. Though, as was admitted in the first part of this article, the fragmentary facts of Palæontology at present accumulated, do not clearly warrant us in saying that, in the lapse of geologic time, there have been evolved more heterogeneous individual organisms, and more heterogeneous assemblages of organisms, yet we shall now see that there *must* ever have been a tendency towards these results. We shall find that the production of many effects by one cause, which, as already shown, has been all along increasing the physical heterogeneity of the Earth, has further involved an increasing heterogeneity in its flora and fauna, individually and collectively. An illustration will make this clear. Suppose that by a succession of upheavals, occurring, as they are now known to do, at long intervals, the East Indian Archipelago were to be step by step raised into a continent, and a chain of mountains formed along the axis of elevation. By the first of these changes the plants and animals inhabiting Borneo, Sumatra, New Guinea, and the rest, would be subjected to slightly modified sets of conditions. The climate in general would be altered in temperature, in humidity, and in its periodical variations; whilst the local differences would be multiplied. These modifications would affect, perhaps inappreciably, the entire flora and fauna of the region. The change of level would produce additional modifications; varying in different species, and also in different members of the same species, according to their proximity to the axis of elevation. Certain kinds of plants, growing only on the sea-shore in special localities, might disappear entirely. Others, living only in swamps of a certain humidity, would, if they survived at all, probably undergo visible changes of appearance. While still greater alterations would occur in the plants gradually spreading over the lands newly raised above the sea. The animals and insects living upon these modified plants, would themselves be in some degree modified by change of food, as well as by change of climate; and the modification would be more marked where, from the dwindling or disappearance of one species of plant, an allied species was eaten. In the lapse of the many generations arising before the next upheaval, the sensible or insensible alteration thus produced in each species would become organized – there would be a more or less complete adaptation to the new conditions. The next upheaval would superinduce additional organic changes, involving more distinct divergences from the primary structures: and so repeatedly. But now let it be observed that the revolution thus resulting would not be a substitution of a thousand more or less modified species for the thousand original species; but in place of the thousand original species there would arise several thousand species, or varieties, or changed forms. Each species being distributed over an area of some extent, and tending continually to colonize the new area exposed, its different members would be subject to different sets of changes. Plants and animals spreading towards the equator would not be affected in the same way with others spreading from it. Those spreading towards the new shores would undergo changes unlike the changes undergone by those spreading into the mountains. Thus, each original race of organisms that survived, would become the root from which diverged several races differing more or less from it and from each other; and while some of these might subsequently disappear, probably more than one would survive into the next geologic period: the very dispersion itself increasing the chances of survival. Not only would there be thus caused certain modifications consequent on change of physical conditions and kind of nutriment, but also in some cases other modifications consequent upon change of habits. The fauna of each island, peopling, step by step, the newly elevated tracts, would eventually come in contact with the faunas of other islands; and some members of these other faunas would differ more or less from any creatures before seen. Herbivores meeting with new beasts of prey, would, in some cases, be led into modes of defence or escape somewhat differing from those previously used; and simultaneously the beasts of prey would modify their modes of pursuit and attack. We know that when circumstances demand it, such changes of habit *do* take place in animals; and we know that if the new habits become the dominant ones, they must eventually in some degree alter the organization. Observe now, however, a further consequence. The process involves not simply a tendency towards the differentiation of each race of organisms into several races; but it involves a tendency to the occasional production of a somewhat higher organism. These divergent varieties of any species which have been caused by fresh physical conditions and habits of life, will exhibit changes quite indefinite in kind and degree; and changes that do not necessarily constitute an advance. Probably in most cases the modified type will neither be more nor less heterogeneous than the original one. In some cases the habits of life adopted being simpler than before, a less heterogeneous structure will result: there will be a retrogradation. But it *must* now and then occur, that some divergent branch of a species, falling into circumstances which give it somewhat more complex experiences, and demand actions somewhat more involved, will have certain of its organs further differentiated in proportionately small degrees, – will become slightly more heterogeneous. Thus, in the natural course of things, there will from time to time arise an increased heterogeneity both of the Earth's flora and fauna, and of individual races included in them. Omitting detailed explanations, and allowing for the qualifications which cannot here be specified, we think it is clear that geological mutations have all along tended to complicate the forms of life, whether regarded separately or collectively. The same general causes which have determined the evolution of the Earth's crust from the simple into the complex, have simultaneously determined a parallel evolution of the Life upon its surface. In this case, as in previous ones, we see that the transformation of the homogeneous into the heterogeneous is consequent upon the universal principle, that every active force produces more than one change.

The deduction which we have here drawn from the established truths of geology and the general laws of life, gains immensely in weight on finding it to be in perfect harmony with an induction drawn from direct experience. Just that divergence of many races from one race, which we inferred must have been continually occurring during geologic time, we know to have taken place, during the pre-historic and historic periods, in man and domestic animals. And just that multiplication of effects which we concluded must have produced the first, we see has produced the last. Single causes, as famine, pressure of population, war, have periodically led to further dispersions of mankind and of dependent creatures: each such dispersion initiating new modifications, new varieties of type. Whether all the human races be or be not derived from one stock, philological evidence makes it clear that whole groups of races now easily distinguishable from each other, were originally one race, – that the diffusion of one race in sundry directions into different climates and conditions of existence, has simultaneously produced many modified forms of it. Similarly with domestic animals. Though in some cases – as that of dogs – community of origin will perhaps be disputed, yet in other cases – as that of the sheep or the cattle of our own country – it will not be questioned that local differences of climate, food, and treatment, have transformed one original breed into numerous breeds now become so far distinct as to produce unstable hybrids. Moreover, through the complication of effects flowing from single causes, we here find, what we before inferred, not only an increase of general heterogeneity but also of special heterogeneity. While of the divergent divisions and subdivisions of the human race, many have undergone modifications of detail not constituting an advance; while in some the type may have degraded; in others it has become decidedly more heterogeneous. The civilized European departs more widely from the vertebrate archetype than does the savage. Thus, both the law and the cause of progress, which, from lack of evidence, can be but hypothetically substantiated in respect of the earlier forms of life on our globe, can be actually substantiated in respect of the latest forms.

If the advance of Man towards greater heterogeneity is traceable to the production of many effects by one cause, still more clearly may the advance of Society towards greater heterogeneity be similarly explained. Consider the growth of an industrial organization. When, as must occasionally happen, some individual of a tribe displays unusual aptitude for making a particular article of general use – a weapon, for instance – which was before made by each man for himself, there arises an inevitable tendency towards the differentiation of that individual into a maker of such article. His companions – warriors and hunters all of them, – severally feel the importance of having the best weapons that can be made; and are therefore certain to offer strong inducements to the skilled individual to make weapons for them. He, on the other hand, having not only an unusual faculty, but an unusual liking for making such weapons (the talent and the desire for any special occupation being usually associated), is predisposed to fulfil these commissions on the offer of an adequate reward; especially as his love of distinction is also gratified. This first specialization of function, once commenced, tends ever to become more decided. On the side of the weapon-maker continued practice gives increased skill – increased superiority to his products: on the side of his clients, cessation of practice entails decreased skill. Thus the influences that determine this division of labour grow stronger in both ways; and the incipient heterogeneity is, on the average of cases, likely to become permanent for that generation, if no longer. Observe now, however, that this process not only differentiates the social mass into two parts, the one monopolizing, or almost monopolizing, the performance of a certain function, and the other having lost the habit, and in some measure the power, of performing that function; but it tends to initiate other differentiations. The advance we have described implies the introduction of barter, – the maker of weapons has, on each occasion, to be paid in such other articles as he agrees to take in exchange. But he will not habitually take in exchange one kind of article, but many kinds. He does not want mats only, or skins, or fishing gear, but he wants all these; and on each occasion will bargain for the particular articles he most needs. What follows? If among the several members of the tribe there exist any slight differences of skill in the manufacture of these various things, as there are almost sure to do, the weapon-maker will take from each one the thing which that one excels in making: he will exchange for mats with him whose mats are superior, and will bargain for the fishing gear of whoever has the best. But he who has bartered away his mats or his fishing gear, must make other mats or fishing gear for himself; and in so doing must, in some degree, further develop his aptitude. Thus it necessarily results that the small specialities of faculty possessed by various members of the tribe, will tend to become more decided. If such transactions are from time to time repeated, the increased specializations may become appreciable. And whether or not there ensues a distinct differentiation of one or more other individuals into makers of particular articles, it is clear that incipient differentiations take place throughout the tribe: the one original cause produces not only the first dual effect, but a number of secondary dual effects analogous in kind, but minor in degree. This process, of which traces may be seen among groups of schoolboys, cannot well produce any permanent effects in an unsettled tribe; but where there grows up a fixed and multiplying community, these differentiations become established, and increase with each generation. A larger population, involving a greater demand for every commodity, intensifies the functional activity of each specialized person or class; and this renders the specialization more definite where it already exists, and establishes it where it is but incipient. By increasing the pressure on the means of subsistence, a larger population again augments these results; seeing that each person is forced more and more to confine himself to that which he can do best, and by which he can gain most. This industrial progress, facilitating future production, opens the way for a further growth of population, which reacts as before: in all which the multiplication of effects is manifest. Presently, under these same stimuli, new occupations arise. Competing workers, ever aiming to produce improved articles, occasionally discover better processes or raw materials. In weapons and cutting tools, the substitution of bronze for stone entails upon him who first makes it a great increase of demand – so great an increase that he presently finds all his time occupied in making the bronze for the articles he sells, and is obliged to depute the fashioning of these articles to others: and, eventually, the making of bronze, thus gradually differentiated from a pre-existing occupation, becomes an occupation by itself. But now mark the ramified changes which follow this change. Bronze soon replaces stone, not only in the particular articles in which it was first used, but in many others – in arms, tools, and utensils of various kinds; and so affects the manufacture of these things. Further, it affects the processes which these utensils subserve, and the resulting products – modifies buildings, carvings, dress, personal decorations. Yet, again, it initiates sundry manufactures which were before impossible, from lack of a material fit for the requisite tools. And all these alterations react on the people – increase their manipulative skill, their intelligence, their comfort, – refine their habits and tastes. Thus the evolution of a homogeneous society into a heterogeneous one, is very clearly consequent on the general principle, that many effects are produced by one cause.

Our limits will not allow us to trace out this process in its higher complications; else might we show how the localization of special industries in special parts of a kingdom, as well as the minute subdivision of labour in the production of each commodity, are similarly determined. Or, turning to a somewhat different order of illustrations, we might dwell on the multitudinous changes – material, intellectual, moral, – caused by printing; or the further extensive series of changes wrought by gunpowder. But leaving the intermediate phases of social development, let us take a few illustrations from its most recent and its passing phases. To trace out the effects of steam-power in its manifold applications to mining, navigation, and manufactures of all kinds, would carry us into unmanageable detail. Let us confine ourselves to the latest embodiment of steam-power – the locomotive engine. This, as the proximate cause of our railway system, has changed the face of the country, the course of trade, and the habits of the people. Consider, first, the complicated sets of changes that precede the making of every railway – the provisional arrangements, the meetings, the registration, the trial section, the parliamentary survey, the lithographed plans, the books of reference, the local deposits and notices, the application to Parliament, the passing Standing-Orders Committee, the first, second, and third readings: each of which brief heads indicates a multiplicity of transactions, and the development of sundry occupations – as those of engineers, surveyors, lithographers, parliamentary agents, share-brokers; and the creation of sundry others – as those of traffic-takers, reference-takers. Consider, next, the yet more marked changes implied in railway construction – the cuttings, embankings, tunnellings, diversions of roads; the building of bridges and stations; the laying down of ballast, sleepers, and rails; the making of engines, tenders, carriages, and waggons: which processes, acting upon numerous trades, increase the importation of timber, the quarrying of stone, the manufacture of iron, the mining of coal, the burning of bricks; institute a variety of special manufactures weekly advertised in the *Railway Times;* and, finally, open the way to sundry new occupations, as those of drivers, stokers, cleaners, plate-layers, &c., &c. And then consider the changes, more numerous and involved still, which railways in action produce upon the community at large. The organization of every business is more or less modified: facility of communication makes it better to do directly what was before done by proxy; agencies are established where previously they would not have paid; goods are obtained from distant wholesale houses instead of near retail ones; and commodities are used which distance previously rendered inaccessible. Again, the rapidity and small cost of carriage tend to specialize more than ever the industries of different districts – to confine each manufacture to the parts in which, from local advantages, it can be best carried on. Further, the diminished cost of carriage facilitating distribution, equalizes prices, and also, on the average, lowers prices: thus bringing sundry articles within the means of those before unable to buy them, and so increasing their comforts and improving their habits. At the same time the practice of travelling is immensely extended. Classes who never before thought of it, take annual trips to the sea; visit their distant relations; make tours; and so we are benefited in body, feelings, and intellect. Moreover, the more prompt transmission of letters and of news produces further changes – makes the pulse of the nation faster. Yet more, there arises an extensive dissemination of cheap literature through railway book-stalls, and of advertisements in railway carriages: both of them aiding ulterior progress. And all the innumerable changes here briefly indicated are consequent on the invention of the locomotive engine. The social organism has been rendered more heterogeneous in virtue of the various new occupations introduced, and the many old ones further specialized; prices in every place have been altered; each trader has, more or less, modified his way of doing business; and almost every member of the community has been affected in his actions, thoughts, emotions.

Illustrations to the same effect might be indefinitely accumulated. That every influence brought to bear upon society works multiplied effects; and that increase of heterogeneity is a consequence of this multiplication of effects; may be seen in the history of every trade, every custom, every belief. But it is needless to give additional evidence of this. The only further fact demanding notice, is, that we here see still more clearly than ever, the truth before pointed out, that in proportion as the area on which any force expends itself becomes heterogeneous, the results are in a still higher degree multiplied in number and kind. While among the primitive tribes to whom it was first known, caoutchouc initiated but few changes, among ourselves the changes have been so many and varied that the history of them occupies a volume.[c] Upon the small, homogeneous community inhabiting one of the Hebrides, the electric telegraph would produce, were it used, scarcely any results; but in England the results it produces are multitudinous. The comparatively simple organization under which our ancestors lived five centuries ago, could have undergone but few modifications from an event like the recent one at Canton; but now, the legislative decision respecting it sets up many hundreds of complex modifications, each of which will be the parent of numerous future ones.

Space permitting, we could willingly have pursued the argument in relation to all the subtler results of civilization. As before, we showed that the law of Progress to which the organic and inorganic worlds conform, is also conformed to by Language, Sculpture, Music, &c.; so might we here show that the cause which we have hitherto found to determine Progress holds in these cases also. We could demonstrate in detail how, in Science, an advance in one division presently advances other divisions – how Astronomy has been immensely developed through discoveries in Optics, whilst other optical discoveries have initiated Microscopic Anatomy, and greatly aided the growth of Physiology – how Chemistry has simultaneously opened the way to advance in our knowledge of Electricity, Magnetism, Biology, Geology – how Electricity has reacted on Chemistry and Magnetism, developed our views of Light and Heat, and disclosed sundry laws of nervous action. In Literature the same truth might be exhibited in the manifold effects resulting from the primitive mystery-play, not only as originating the modern drama, but as affecting through it other kinds of poetry and fiction; or in the continually multiplied forms of periodical literature that have descended from the first newspaper, and which have severally acted and reacted on other forms of literature and on each other. The influence which a new school of Painting – as that of the pre-Raffaelites – exercises upon other schools; the hints which all kinds of pictorial art are deriving from Photography; the complex results of new critical doctrines, as those of Mr. Ruskin, might severally be dwelt upon as displaying the like multiplication of effects. But it would needlessly tax the reader's patience to pursue, in their many ramifications, these various changes; here become so involved and subtle as to be followed out with some difficulty.

Without further accumulation of evidence, we venture to think our case is made out. The many imperfections of statement which brevity has necessitated, do not, we believe, militate against the truth of the propositions enunciated. The qualifications here and there demanded would not, if made, affect the inferences. Though in one instance, in which sufficient evidence is not attainable, we have been unable to show that the law of Progress applies; yet there is high probability that the same generalization holds which holds throughout the rest of creation. Though, in tracing the genesis of Progress, we have frequently spoken of very complex causes as if they were simple ones; it still remains true that such causes are far less complex than their results. Detailed criticisms cannot affect our main position. Endless facts go to show that every kind of progress is from the homogeneous to the heterogeneous; and that it is so because each change is followed by many changes. And it is significant that where the facts are most accessible and abundant, there are these truths most manifest.

However, to avoid committing ourselves to more than is yet proved, we must be content with saying that such are the law and the cause of all progress that is known to us. Should the Nebular Hypothesis ever be established, then it will become manifest that the Universe at large, like every organism, was once homogeneous; that as a whole, and in every detail, it has unceasingly progressed towards greater heterogeneity; and that its heterogeneity is still increasing. It will be seen that as in each phenomenon of to-day, so from the beginning, the decomposition of every expended force into several forces has been perpetually producing a higher complication; that the increase of heterogeneity so brought about is still going on, and must continue to go on; and that thus Progress is not an accident, not a thing within human control, but a beneficent necessity.

A few words must be added on the ontological bearings of our argument. Probably not a few will conclude that here is an attempted solution of the great questions with which Philosophy in all ages has perplexed itself. Let none thus deceive themselves. Only such as know not the scope and the limits of Science can fall into so grave an error. After all that has been said, the ultimate mystery of things remains just as it was. The explanation of that which is explicable, does but bring out into greater clearness the inexplicableness of that which remains behind. However we may succeed in reducing the equation to its lowest terms, we are not thereby enabled to determine the unknown quantity: on the contrary, it only becomes more manifest that the unknown quantity can never be found. We feel ever more and more certain that fearless inquiry tends continually to give a firmer basis to all true Religion. The timid sectarian, alarmed at the progress of knowledge, obliged to abandon one by one the superstitions of his ancestors, and daily finding sundry of his cherished beliefs more and more shaken, secretly fears that all things may some day be explained; and has a corresponding dread of Science: thus evincing the profoundest of all infidelity – the fear lest the truth be bad. On the other hand, the sincere man of science, content fearlessly to follow wherever the evidence leads him, becomes by each new inquiry more profoundly convinced that the Universe is an insoluble problem. Alike in the external and the internal worlds, he sees himself in the midst of perpetual changes, of which he can discover neither the beginning nor the end. If, tracing back the genesis of things, he allows himself to entertain the still unproved hypothesis that all matter once existed in a diffused form, he finds it utterly impossible to conceive how this came to be so; and equally, if he speculates on the future, he can assign no limit to the grand succession of phenomena ever evolving themselves before him. On the other hand, if he looks inward, he perceives that both terminations of the thread of consciousness are beyond his grasp: he cannot remember when or how consciousness commenced, and he cannot examine the consciousness that at any moment exists; for only a state of consciousness that is already past can become the object of thought, and never one which is passing. When, again, he turns from the succession of phenomena, external or internal, to their essential nature, he is equally at fault. Though he may succeed in resolving all properties of objects into manifestations of force, he is not thereby enabled to realize what force is; but finds, on the contrary, that the more he thinks about it, the more he is baffled. Similarly, though analysis of mental actions may finally bring him down to sensations as the original materials out of which all thought is woven, he is none the forwarder; for he cannot in the least comprehend sensation – cannot even conceive how sensation is possible. Inward and outward things he thus discovers to be alike inscrutable in their ultimate genesis and nature. He sees that the Materialist and Spiritualist controversy is a mere war of words; the disputants being equally absurd – each believing he understands that which it is impossible for any human being to understand. In all directions his investigations eventually bring him face to face with the unknowable; and he ever more clearly perceives it to be the unknowable. He learns at once the greatness and the littleness of human intellect – its power in dealing with all that comes within the range of experience; its impotence in dealing with all that transcends experience. He feels, with a vividness which no others can, the utter incomprehensibleness of the simplest fact, considered in itself. He alone truly *sees* that absolute knowledge is impossible. He alone *knows* that under all things there lies an impenetrable mystery.

Notes

a For detailed proof of these assertions the reader is referred to an article on *Manners and Fashion,* published in No. X. of this Journal, for April, 1854.

b The idea that the Nebular Hypothesis has been disproved because what were thought to be existing nebulæ have been resolved into clusters of stars, is almost beneath notice. *A priori* it was highly improbable, if not impossible, that nebulous masses should still remain uncondensed, while others have been condensed millions of years ago.

c "Personal Narrative of the Origin of the Caoutchouc, or India-Rubber Manufacture in England." By Thomas Hancock.

Editorial Headnote

27. Baden Powell, *Essays in the Spirit of the Inductive Philosophy, The Unity of Worlds, and the Philosophy of Creation* (London: Longman, Green, Brown, & Longmans, 1855), Essay 3, "On the Philosophy of Creation," pp. 329–30, 337, 342–3, 349–51, 369–72, 376–7, 380, 391, 392–5, 398–402, 404–5, 408–18, 424.

Baden Powell (1796–1860) was a liberal Anglican priest and mathematician and was Savilian Chair of Geometry at Oxford from 1827 until his death in 1860. He was among the liberal theologians who wrote the controversial *Essays and Reviews* which was published in 1860 and which arguably caused as much of a theological stir as the publication of Darwin's *Origin of Species*. In his contribution to that work, Baden Powell denied the rationality of belief in miracles in a uniform and lawful world, and heralded *Origin* as a "masterly volume."[1]

However, Powell's sympathies for evolutionary ideas had a longer history. In his *Unity of Worlds* (1855), and clearly influenced by Charles Lyell's *Principles of Geology* (1830–33), Powell outlined his support for the uniformity of nature and rejected divine intervention in the working out of God's creation as heretical.[2] He also voiced clear support for the kind of evolutionary ideas outlined by the recently published *Vestiges of the Natural History of Creation* as being more consistent with this view of a lawful creator God than with the notion of special creation.[3] To have such a prominent and respected theologian state his views so openly likely served to open the minds of other liberal-minded theologians to Darwin's argument.

Following the publication of *Origin* in 1859, Powell wrote to Darwin, drawing his attention to his own ideas as outlined in *Unity of Worlds*, as well as to the work of other naturalists, whom Powell believed had anticipated aspects of Darwin's argument. Darwin responded, saying that he had intended to include notice of Powell's work as well as that of other earlier naturalists in a historical preface that he had begun but not finished, an omission he corrected in the third and subsequent editions of *Origin* by adding a "historical sketch" in which he pointed out brief notes to earlier evolutionists.[4] There, he noted:

> The "Philosophy of Creation" has been treated in a masterly manner by the Rev. Baden Powell, in his "Essays on the Unity of Worlds," 1855. Nothing can be more striking than the manner in which he shows that the introduction of new species is 'a regular, not a casual phenomenon,' or, as Sir John Herschel expresses it, 'a natural in contra-distinction to a miraculous process.'[5]

Notes

1 Baden Powell, "On the Study of the Evidences of Christianity," in *Essays and Reviews* (London: John W. Parker, 1860), pp. 94–144, on p. 139.

2 For more on Lyell, see extracts and contextual notes in Caden C. Testa and Piers J. Hale, (eds.), *Evolution Before Darwin* (London: Routledge, 2025); also, in greater depth, Martin J. S. Rudwick, *Worlds Before Adam. The Reconstruction of Geohistory in the Age of Reform* (Chicago: University of Chicago Press, 2008).

3 For more on Robert Chambers and his anonymously published work, *Vestiges of the Natural History of Creation*, see James Secord, *Victorian Sensation. The Extraordinary Publication, Reception, and Secret Authorship of The Vestiges of the Natural History of Creation* (Chicago: University of Chicago Press, 2000).

4 Curtis N. Johnson, *Darwin's Historical Sketch. An Examination of the 'Preface' to the Origin of Species* (Oxford: Oxford University Press, 2020), p. 6. Johnson's work on the 'historical sketch' that Darwin inserted into the *Origin of Species* explores the importance and influence of Baden Powell's work at greater length.

5 Charles Darwin, *On the Origin of Species by Means of Natural Selection, Or, the Preservation of the Favoured Races in the Struggle for Life* (London: John Murray, 1861).

27

ESSAY 3, "ON THE PHILOSOPHY OF CREATION"

Baden Powell

Source: Baden Powell, Essay 3, "On the Philosophy of Creation", *Essays in the Spirit of the Inductive Philosophy, The Unity of Worlds, and the Philosophy of Creation* (London: Longman, Green, Brown, & Longmans, 1855), pp. 329–30, 337, 342–3, 349–51, 369–72, 376–7, 380, 391, 392–5, 398–402, 404–5, 408–18, 424

THE PHILOSOPHY OF CREATION.

027 "The Revered Baden Powell"

Image in the public domain.

DOI: 10.4324/9781003490548-30

INTRODUCTION.

THE question of "creation," – whether in the higher sense, of the first origination of the material universe, and of all physical causes, – or in the secondary and more accessible meaning, of the earliest history of the cosmical arrangements of stellar or planetary systems, – and more particularly of our own globe, of its physical revolutions, and of the successive introduction of new forms of organised life on its surface, – has at the present day excited an unprecedented degree of general interest.

[...]

§ I. – THE EVIDENCE DERIVED FROM GEOLOGY.

[...]

When we trace backward, by the light of fossil remains, the succession of varied forms of organised existence which have tenanted the surface of our globe during the incalculably vast periods of past time, the fact of their presenting apparently very different characters in different epochs naturally led geologists and naturalists to speculate on the question, whether those variations could be reduced to anything like *a determinate order or law of succession;* and probably the most prevalent opinion has *some* elevations to be washed down, and form the deposited beds.

[...]

There is one material consideration the force of which, *on any view of the question of progression*, it is impossible to overlook – the fact of the *central heat* of the earth, with its undeniable consequences.[a] A hot body in free space must cool; and if now cool at the surface and hot within, the earth must have cooled from a hotter state, and must once have been intensely hot; and, by the same rule, once in fusion, or in vapour. Here, then, there *must* have been a series of *progressive changes;* and if the cooling process *had not reached* its present apparent state of equilibrium *before* organised life began, it would be a natural consequence that some marked changes in animated nature could not but have accompanied those changes in temperature, and have followed a like determinate order. Whether the present equilibrium had been attained before the commencement of life may still be a question; yet, considering the enormous length of time through which organisation has certainly existed, it is difficult not to suppose that some part of the series at least must have reached back into the period of perceptible cooling, so that influences of terrestrial temperature may have been not without their effect on the changes of species.

[...]

Among existing animals, it is now generally allowed that the arrangement of species in the scale of organisation is not that of simple ascent in any line, or even in several branching in any one direction; it is more properly compared by Professor Owen[b] to a "net-work;" – every species being connected with others by a variety of ramifications, and not simply by ascent or descent in a scale.[c]

And as to the law of changes of species in *past epochs*, in the very imperfect knowledge we at present possess of it, it is at least clear that, so far from a regular *advance* from lower to higher forms, in many instances there appears rather a deterioration and degradation of character in the progress of time towards the existing state of things. But what seems most material towards the probable ultimate enunciation of a more true and general expression of the case, is the law of *combination and separation* of characters; that is, a combination of the characteristics of several species, or even genera or orders, in the same individual in one period, to be developed separately in different species in a succeeding era; and this in such distribution as to present appearances of advance in some respects, along with degradation in others; as if, in the functions of vitality, the principle of "division of labour" had been gradually introduced. Of this combination of characters in an individual species, examples are familiar to every student of geology.

[...]

Speaking of the *tertiary* formations, Sir C. Lyell observes, "There are usually so many species in common to the groups which stand next in succession, as to show that there is no great chasm, no signs of a crisis, when one class of organic beings was annihilated to give place suddenly to another. This analogy, therefore, derived from a period of the earth's history which can best be compared with the present state of things, and more thoroughly investigated than any other, leads us to the conclusion that the *extinction* and CREATION of species has been, and is, the result of a slow and gradual change in the organic world."[d]

Again, he argues at length, from the actual causes which determine the conditions of successive deposits, that, assuming, as we thus must do, the "fluctuations in the animate world to be brought about by the slow and successive *removal and creation of species*," yet, from the local nature of the formations, we *cannot expect* to find conditions such as shall enable us to trace the "*gradual passage* from one state of organic life to another."[e]

Lamarck, indeed, held that there may be some gaps in the series greater than we can attribute to mere want of evidence, or hope to see filled up by future discoveries; yet he conceived that the difficulty might be obviated from the consideration of the counteracting influence of a variety of external causes, which are perpetually interfering with the regular order of succession. If these interfering causes did not exist, we might expect an exact continuity of forms; but by these immensely varied agencies of external and local conditions, the progress of some races may be retarded, and that of others accelerated; so that at length wide breaks of continuity may necessarily appear after a long lapse of time.

In many cases too, it must be recollected, that the apparent interruption is confined to certain classes of animals only, and does not extend to others; chiefly among the higher forms; while in the lower, during the same periods, no such interruption occurs, some of them being persistent through many epochs.

But supposing the existence of such apparent gaps or breaches of continuity granted, and that we failed to explain them by any such theoretical suggestions, although we may not yet have hit upon the true explanation or traced *the particular law* in this case, we are sure that *some law* is really involved even in a seeming infraction of a regular series, and cannot doubt that future discovery will ultimately disprove or explain the apparent anomaly.[f] So strong is the inductive assurance of this, that we may safely allow any such apparent exceptions to await their solution without in the least influencing our opinion of the soundness of the broad principle of the continuity of physical causes: a principle of that truly philosophical character which no apparent exception in detail can subvert, or make really inapplicable or unfruitful.

No inductive inquirer can bring himself to believe in the existence of any *real hiatus* in the continuity of physical laws in past eras more than in the existing order of things; or to imagine that changes, however seemingly abrupt, can have been brought about except by the gradual agency of some regular causes. On such principles the whole superstructure of rational geology entirely reposes; to deny them in any instance would be to endanger all science.

[...]

But however little we know of the laws or causes of these changes, one thing is perfectly clear, *the introduction of new species was a regular, not a casual phenomenon;* it was not one preceding or transcending the order of nature; it was a case occurring in the midst of ordinary operations going on in accordance with ordinary causes.

[...]

§ II. – THE EVIDENCE DERIVED FROM PHYSIOLOGY.

GEOLOGY is essentially dependent on physiology; hence any argument derived from the former science, as bearing on the evidence of "creation" in the organic world, must be in some degree an application of the latter; as, indeed, is manifest throughout the foregoing remarks. But some questions are involved in the present inquiry, which depend on a more particular reference to points of *pure physiology;* and to these the present section relates.

[...]

indicated the archetypal form, belonging to them in common with the former, up to a certain stage, beyond which their peculiarity of character is superinduced.[g]

In advancing from these researches on *what is*, to their application to *what has been*, and with a view to the discussions which have arisen on such questions, it becomes necessary, in the first instance, to advert briefly to some considerations as to the nature and distinctions of *species* in general.

[...]

another. There are, also, cases in which certain deviations from the original type are more marked, and continue to affect several or even many generations constituting "*a variety*."

These varieties, after a longer or shorter period, *in many cases* cease to be continued; and it is probable that the same external conditions, which are favourable to the original type, are less so to the variety, which is thus more easily checked in its increase, or at length extinguished. But it is a subject on which nothing is known as to the real causes which may give rise to such changes, and on which, therefore, it is clearly *unwarrantable to dogmatise*, or to reason upon such failure as if it were a *necessary* law.

There are also cases in which varieties have been found to continue so long, and to maintain so completely distinct a character, that it has become difficult, if not impossible, to determine whether they do not constitute a *sub-species*. And so far as any speculation can be carried, on a subject so little understood, it would seem most probable that wherever such permanency has been attained, there has existed some peculiarity in external conditions which has been the determining cause for the perpetuation of such variety, just as some other external conditions of an opposite kind are in other cases unfavourable, and cause its declension or extinction. It is therefore a fair inference, that if the favourable conditions were continued, and the variety were locally isolated from the rest of the species, it would become a permanent type or species.

Within certain limits we observe species *fixed* at the present day: we have, in some instances, proofs from historical monuments and preserved remains that they have not altered within very high limits of antiquity. Some writers refer, for example, to the mummies of Egypt, reaching back to an interval of 3000 years (or, indeed, as much longer as their chronology may dictate); but, in fact, we can go much higher, since we have undoubted evidence of *some* existing species having remained permanent during the countless ages since the tertiary deposits up to the present time.

This, however, proves little, since the point to be explained is, that associated with these are found other older and extinct species closely allied to, but different from those which now exist; and the question is as to the relation between the existing species and its ancient allied form, or generally between any species existing in one formation or period, and the one analogous, or nearest allied to it, in the next earlier formation in which it had no place.

Again, it is the fact that there are at present numerous instances in which naturalists are at issue as to what constitutes a *variety* or a *species;* and the very difficulty of determining whether any given character is or is not permanent, is alone sufficient to show how little we ought to press any such principle as the basis of a universal conclusion – still less as partaking of the character of a great and necessary law of nature, on which we can satisfactorily reason from the present to the past, or infer the perpetuity of distinctions so little settled.

The distinctions admitted as those of *species* are indeed in some cases much more minute than those which, in other instances, are considered to mark only *varieties*. An American butterfly (the Vanesso Atalanta) is held to be a distinct species, yet its characteristic is only a single spot on the wing; nevertheless the entire black skin of the Negro and the white of the European mark only varieties.

In a report formally put forth by the British Association, it is expressly set down as a DESIDERATUM to discover some sure mode of *distinguishing real species from local varieties*. "One school," the reporter observes, "attribute all specific distinctions to the influence of external agents; another regard the most trivial circumstances as fixed;" but he recommends a just mean as the preferable view.[h]

Professor Pictet also enlarges on the difficulty of distinguishing species, and gives instances of such ambiguity of character.[i]

On this subject I will again cite the authority of Dr. Carpenter: – "The uncertainty of the limits of species is daily becoming more and more evident; and every naturalist is aware that a very large number of races are usually considered as having a distinct origin, when they are nothing more than permanent varieties of a common stock."[j]

[...]

Much discussion (as is well known) has arisen on the question whether the *different races of men* are *varieties* of one species, or *distinct species:* and it seems to be at present the prevailing opinion that they are *varieties* merely.[k] But the question *how*, by what steps or processes, did such large and fundamental differences arise? entails more important consequences than many, in their zeal to maintain a single origin, seem to perceive. It is clear that these differences are fully as great as those which in many other cases are allowed to constitute distinct species.

If in the case of man they have occurred as transitional varieties, how comes it that they have become so inveterately permanent? And if those changes have all occurred within the lapse of a few thousand years of the received chronology, it cannot with any reason be denied that similar changes might occur among inferior animals, and become just as permanent. And if so, changes to an indefinitely greater extent might occur in indefinite lapse of time. If these changes take place by the gradual operation of natural causes, it would be preposterous to deny the possibility of equal or greater changes by equally natural causes in other species in equal or greater periods of time. The advocates of the fixity of species would argue that the single spot on a butterfly's wing, which constitutes a species, never has changed, and never can change, without a miracle; and yet the vast differences between a European and a Negro or Australian are mere modifications of one parent stock by natural causes in the lapse of a few thousand years!

The peculiar characters of the Negro race are recorded, as prominently marked as at present, in the ancient Egyptian paintings, which may go back 3000 years or more.[l] Here, then, is a *variety* which has been *permanent* for at least that long period; a period, too, which has been expressly relied on by many to prove the permanence of *species* by appeal to these very monuments. And then we have to ask, How long must it have taken, at this rate of imperceptible progress, to have been developed out of the original stock?

Another instance has been much dwelt upon, the so called "varieties" of the dog, presumed to be derived from a common stock; but how long since, is undetermined. Yet in these varieties (in which even the form of the cranium greatly differs) it would be difficult to deny that the distinctive characters are permanent, at least in some of the more marked instances, and under the continuance of the same external conditions; and that each race, when preserved isolated under such conditions, would remain permanently distinct.

Much stress has also been laid by some on the asserted sterility of hybrids; though, in truth, it affects very little the general question; while its very limited evidence dependent only on a few isolated facts, occurring in a state of domestication, is utterly insufficient for the foundation of any general law. The cases commonly referred to should be regarded by an unprejudiced mind as probably exceptional, under peculiar conditions, and not to be dogmatised upon, as involving any real and necessary law of organised existence. As there are limits beyond which union will not take place, so within these there may very probably be *certain limits* of still nearer affinity, *beyond which* sterility in the offspring prevails, but which have not yet been determined. The recurrence to the original type often observed, only proves that conditions are not favourable to the continuance of the variety.[m] And of the very positive assertions so liberally made in these and the like cases, it is to be observed that they are, at best, merely *empirical* conclusions, wholly unsupported by any wide analogies, or explained by any known causes which can confer on them the character of real natural principles.

Yet the immutability of species, as something essential to their nature and inherent in it, has been upheld by a large section of naturalists – and still more strenuously by some who are not naturalists – in this country, with a degree of positiveness and even vehemence, which the mere *negative* character of the evidence could never justify, and which it would be difficult to account for, so far as any arguments of a philosophical nature may be supposed to influence the opinion.

[...]

On these questions I have much satisfaction in referring to the opinions of Dr. Carpenter[n]. put forth with equal candour and freedom from prejudice, and with a union of caution and enlargement of view which eminently commends them to the convictions of the reader.

He observes[o]. "Our belief that the new beings formed by the process of reproduction always closely resemble the parent stock, is certainly founded upon a limited induction from observations made upon the higher classes of plants and animals. Reasons have already been given for the opinion that the same germ may assume very dissimilar forms according to the circumstances under which it is developed: and knowing, as we do, how readily the simpler classes of organised beings are affected by changes in their external conditions, it is not difficult to admit the possibility of their forms being thus greatly modified, as well as of the continued propagation of the varieties thus produced."

As to the origin of varieties, it has been admitted by most physiologists as a general principle, that those peculiar "vital forces" (of whatever nature they may intrinsically be) are always the main acting cause, but are subject to modifications from external causes: and hence it is the preservation or interruption of the balance between these antagonist causes which determines the formation or the modification of the type, so as in the one case to keep it up, in the other to produce *varieties*.

[...]

Cuvier, while he professed a rejection of all hypothetical speculation, appears to have been strongly prepossessed with one hypothesis – that of the essential and eternal *immutability of species*. At any rate, under the sanction of his name it has been since maintained by many of his followers with a degree of positiveness not easy to account for on any merely philosophical grounds.[p]

But an opposite opinion began to be taken up by those of a different school, intimately connected with the more speculative and transcendental views already alluded to.

When Geoffroy and his coadjutors were engaged in upholding the unity of composition of all animated structures, it was regarded by them as a natural consequence that, as all the details which mark different species were the mere modifications of greater or less development of the different parts of the primary plan common to them all, so those developed parts had no essential or permanent place in the nature of the species. Hence, naturally arose the further idea of the possible migration or transition from one species to another, or rather from one species to a *new* modification.

Thus the theory of "unity of composition" was, in the minds of many, closely allied to that of "transmutation," which seemed to be a sort of natural sequel to it, and was, among a large school of the continental naturalists, associated with the advancing prospects thus held out to their view of the system of nature, and the attainment of a more transcendental theory of her operations.

From the principle of unity of composition as applied to the existing animal world, it was by a natural extension that this school of physiologists were led to infer that there had been an equally close analogy preserved in *time*, and that there had been a continuous succession of the several species of the animal world, of which only a few detached fragments are preserved to us, as disclosed by geological research: a succession, as they contended, which took place in the natural course of reproduction by continual small deviations from a primitive type, according to the influence of external conditions which varied from one epoch to another, yet were determined alike by fixed and regulated laws, from the present era backwards into the abysses of past time. They conceived that in none of the varied forms of organisation which we trace has there been any new fundamental principle introduced; all are but modifications of forms now existing. Thus throughout all nature, present and past, external forms are mere accidents; development of parts in excess or in defect as changing causes led to such necessity, and modifications of parts according to the functions required to be exercised under the particular conditions. The fundamental unity of principle is that which alone is permanent and invariable, and which admits of endless adaptations according to the varying conditions of existence.

The speculations of Lamarck were founded, in the first instance, on observation of the fact of the near approximation of different species. He contends that the farther our researches extend the greater we find the difficulty of distinguishing species. Apparent interruptions in the continuity of the series are continually being filled up by new discoveries, and what were supposed broad lines of separation effaced. "Everything," he observes, "passes by indivisible shades into something else."

De Maillet (under the anagram of Telliamed) and Lamarck were perhaps those who followed out speculations on the transmutation of species to the greatest extent. Not contenting themselves with asserting cautiously the philosophical grounds on which the close unity of all organised structure is supported, they pursued their hypothesis into details of the most minute, and often most extraordinary, kind. Lamarck's theory, in particular, in truth presents many salient points easily open to attack, and some which are readily susceptible of being held up to ridicule. His principle, that organisation is the result of function, not function of organisation, seems, at first sight, somewhat like a mystified version of that very doctrine of final causes to which many of his school so strongly object. At any rate his theory was carried out to such unwarrantable lengths in many instances, as to discredit even the more sober primary assumptions on which it was founded, in the eyes of his opponents; while, as might be expected of so imaginative a speculation, increasing knowledge of facts has led to its abandonment, at least in its full extent, on the part of many who yet strenuously uphold the same broader principles. Thus Dr. Knox[q]. one of the most zealous supporters of the principle of transmutation in this country, speaks very slightingly of Lamarck, and regards his theory (in its extent and detail, at least) as of little weight or authority at the present day: and the necessity of great modifications in the theory is admitted by the author of the "Vestiges."[r]

But in order to be applicable to the facts of the ancient earth, any process of transmutation must be imagined to extend, not only to minor features, but to a total change even of the characters which mark whole species and genera, so that entire classes and orders may in the course of indefinite progression disappear, and be replaced by others of a different kind. Lamarck endeavoured to support his theory by adducing instances actually occurring within our experience, of slight variations in species; but of more extensive changes confessedly none could be adduced from *existing* facts. Yet it was argued in such cases we must bear in mind the necessity for the introduction of another important element – *the influence of time.* Thus the allegation of actual instances of slight changes in finite times is unanswerable as far as it goes; but the absence of such evidence for greater changes in *finite* times is no argument for their non-occurrence in *indefinitely extended* duration.

As in general, from connecting the conceptions of physical causes with that of immense duration of time, we obtain a very different view of the magnitude of the effects they may produce; so the continental school of transmutation regard all differences in the succession of species of past existence as rather of a chronological than of an essentially physiological kind, due to the lapse of time rather than the introduction of new physiological elements. Thus, De Blainville says that "species mark an epoch in time, not a distinction in animal nature."[s]

But a large section of geologists and physiologists, including many names of the highest distinction, have been strongly opposed to all speculative theories as to the probable origin of those changes in species which geology discloses. This, in many instances, has been the result simply of a rigid – perhaps over rigid – adherence to the rule of appealing to *facts* only, and allowing of no *hypothesis;* a rule which, however strictly inductive, if carried to an extreme would defeat the grand purpose of induction; and partly from imagining that, because the precise theory of development from lower to higher forms is untenable, therefore all hypotheses of the same kind are inadmissible.

Professor Pictet[t] denies the transmutation theory, on the ground that to produce such changes the "powers of nature" must be supposed to have been much greater in the earlier periods than we now find them to be; not allowing, apparently, for the indefinitely long time they had to work in; and, in opposition, he is disposed to maintain a series of sudden introductions of new species, which would imply powers greater still: though he afterwards seems not to consider this as quite satisfactory.

Professor Sedgwick, while he upholds the *progressive* scale of organisation in past epochs, yet earnestly repudiates the development theory, maintaining that the changes of form "mark a gradual evolution of creative power, manifested by a gradual ascent towards a higher type of being . . . . . But the elevation of the fauna of successive periods was not made by transmutation, but by creative additions."[u]

But the question is as to the *nature* and *law* of those additions, and in what particular *sense* or *manner* they were "creative;" whether they were not made according to some determinate law, and in some fixed relation to those which preceded them or were most nearly allied to them. The question is one demanding calm and patient philosophical analysis, and will be the more fitly and worthily discussed the more expanded and generalised the views adopted of geological and cosmogonical order and progress, the more free the inquirer may be from bias and prepossession of other kinds.

The term "creation" indeed, especially as respects new species, seems now, by common consent, to be adopted among geologists as a mere *term of convenience*, to signify simply the fact of origination of a particular form of animal or vegetable life, without implying anything as to the *precise mode* of such origination – as simply involving the assertion that a period can be assigned at which that species appears, and before which we have no evidence of its appearance. In this sense there can be no objection to its use, but it should be carefully guarded against possible misapplication.

Dr. Knox observes, "It was the opinion of Geoffroy that there never had been but one creation: this (he adds) is also my own opinion. I believe all animals to be descended from primitive forms of life forming an integral part of the globe itself, and that the successive varieties of animals and plants which the dissection of the strata of the earth clearly sets forth are due to the occurrence of geological epochs, of the power of which we cannot form any true conception."[v] This might seem like a leaning to the convulsionist theory, were it not more clearly explained in the following passage: –

"We know not, then, the cause of the specific and generic differences in animals, nor why such differences continue fixed for a period – the historic period, for example; they depend, no doubt, on secondary laws, which some future Newton may discover."[w]

The subject of Professor Owen's investigation, before referred to, acquires a higher interest when viewed in reference to the question of the progression of animal forms in past epochs of our globe; and to this view he has pointed[x] as appearing to assign a certain conformity in the order of ancient existence with that of development of the archetype, as indicated by these anatomical researches.

Professor Owen, however, is specially desirous to be understood as applying his conclusions solely to the *order* and *law* of succession, without any attempt to assign a *cause or to trace its origin.* But, though anxiously disclaiming the charge of being supposed in the slightest degree to support the development theory, he nevertheless expresses himself in the most guarded and strictly philosophical language in reference to the possible modes of explaining past changes of species, and proposes no opposite hypothesis of "creations": – "To what actual or secondary causes the orderly succession and progression of such organic phenomena may have been committed we are as yet ignorant."[y] I would only venture to add, that it must have been committed to *some* regularly-ordained causes is surely the verdict of all inductive philosophy.

Sir C. Lyell, though strongly opposing Lamarck's theory of development, yet freely admits, in a geological point of view, that "*If* the doctrine of changes operating in an indefinite lapse of time *be tenable*, we are at once presented with a principle of incessant change in the organic world, and no degree of dissimilarity in the plants and animals which may formerly have existed, and are found fossil, would entitle us to conclude that they may not have been the prototypes and progenitors of the species now living."[z]

[...]

On the whole, then, comparing the limited extent and purely empirical nature of our knowledge of *species* in the existing state of things, with the positive evidence of past changes, it would seem that the more correct statement of the general fact would be simply that *species* (*within certain limits of deviation*) *are permanent during very long periods, but beyond those periods a change, in some sense, occurs*, and this bears some relation to changes of external conditions. But under the same change of conditions one species may be highly susceptible of, and sensitive to, the influence of that change, while another may be insensible to it. Thus one may remain permanent, while another may undergo change, or be exterminated.

Notes

a See especially Mr. Hopkins's Address to the Geological Society, Anniversary, 1852.

b Lecture, British Association, Liverpool, 1854.

c The same difficulty in the definition of *higher* or *lower* organisation has been felt also by other naturalists. That species may be higher in certain respects and lower in others, is also dwelt upon and illustrated by Professor Pictet, "Traité Elémentaire de Paléontologie," &c. (See Geol. Quarterly Journal, No. V. p. 50.)

And Oken, in proposing his new scheme of classification founded on the modifications of the organs of sense (Physio-Philosophy, § 3065.), confesses the difficulty of arranging animals on any satisfactory principle in the relation of *higher* and *lower* organisation. (§ 3561.)

In his own system, he expressly notices that, while each class stands above another, yet *in each* the lower animals are inferior to the higher animals in the next below. (§ 3582.)

d Principles of Geology, p. 179. 8th Edition. To the same purport, see also Sir H. De La Beche, "Researches on Theoretical Geology," p. 365. To these testimonies I would add one, of even a more decided character, from the very able anniversary address of

Mr. W. J. Hamilton, 1855; but being unwilling to spoil so admirable a passage by abridgment, as it is too long for insertion here, I have given it in the Appendix, No. XII.

e Ib. p. 184.

f As an instance I may observe that, while this work was in the press, there was announced the discovery of *mamnalian* remains in the Purbeck beds, thus *filling up the hiatus* between the hitherto enigmatical solitary marsupials of the oolite, with the tertiary epochs, – having insectivorous teeth, and associated with masses of fossil insects! – Geol. Quart. Journal., vol. x. 420. 475., and xi. 51.

g To Mr. Huxley I am indebted for a valuable original sketch of these investigations, from which, by his permission, I am enabled to present to the reader a very copious extract in the Appendix No. VII.

h British Association Report, 1844, p. 219., On Ornithology, by the late H. E. Strickland, Esq.

i Geological Quarterly Journal, No. v. p. 48.

j Principles of Physiology, Art. 542.

k For example, Dr. Pickering, after an extensive investigation, concludes that there are either *eleven* species, or only *one,* of the human race. But upon further examination he decides in favour of *one,* and thinks the original seat of man was in Africa. (The Races of Men, &c., by C. Pickering, M.D. Lond. 1850.)

Others have assigned *six,* or other numbers of species: the author of the "Vestiges" supposes *two* local centres necessary. (P. 220. 6th Ed.)

l See Kenrick on Primeval History, p. 20.

m On the subject of hybrids, some important remarks will be found in Dr. Carpenter's Physiology, Art. 545.

n A number of important examples are collected by Dr. Carpenter, "Physiology," Art. 547, 548.

I do not here profess to go into details of particular instances, but many such, of transmigrations of existing species, which have been collected, are unassailable evidence as far as they go. (See for example, "Vestiges," p. 136. *et seq.* 6th Ed.) See also a paper in the British Association Report, 1852, sectional proceedings, p. 68., by Major Munro, F. L. S., who limits transmutation to plants of the same genus.

o Physiology, Art. 517.

p Dr. Knox, however, the personal friend of Cuvier, states that, at least latterly, he was much inclined to modify his opinions on this point, and quotes one passage in which he says, "Nous ne croyons pas même à la possibilité d'une apparition successive des formes diverses." – Great Anat. &c. p. 44., also p. 27. The expression, however, is perhaps ambiguous.

q Knox, p. 72.

r Vestiges, p. 148. 6th Ed.

s Knox, p. 207.

t Geological Quarterly Journal, No. V. pp. 53. 56.

u See "Studies of Cambridge," 5th Ed. Introduction, pp. 44. 154. 216.

v Knox, p. 109.

w Knox, p. 109.

x On Limbs, p. 86.

y On Limbs, p. 86.

z Principles of Geology, p. 545.

Editorial Headnote

28. Wallace, Alfred Russel, (1855), "On the Law Which has Regulated the Introduction of New Species" *Annals and Magazine of Natural History*, (September 1855), 16, pp. 184–96.

Alfred Russel Wallace (1823–1913) is now recognized as the co-discoverer of the theory of evolution by natural selection. The story of how Wallace came very close to pre-empting Darwin's publication of his own ideas on the topic has been told elsewhere.[1] Suffice to say here that extracts from both Darwin's and Wallace's works were presented to the Linnean Society of London in 1858 in such a way as to show that Darwin had arrived at his ideas first, even though they were published together and their joint contribution recognized. Here, though, we turn to Wallace's 1855 essay "On the Law Which has Regulated the Introduction of New Species." Although Darwin read the essay, he thought that Wallace fell quite far short of his own ideas on the formation of new species – noting in the margin of his own copy "nothing very new." He also noted Wallace's frequent reference to the "great laws of creation" and thus seems to have been uncertain as to the extent to which Wallace was really offering a natural rather than a supernatural account of species. Indeed, as the historian John van Wyhe has argued, Darwin most likely read Wallace's essay as a critique of Lyell's ideas about the geographical distribution of species and species succession rather than as an argument about the origin of new species *per se*.[2] It is thus uncertain whether, as many historians have claimed, it was in response to reading this paper that Charles Lyell urged Darwin to publish in May 1856.[3]

What is certain, though, is that Wallace was firm in his belief that species evolved when he wrote this; indeed, he had become converted to evolutionary or transmutationist thinking at least as early as 1845, the year in which he read the then-anonymous transmutationist work *The Vestiges of the Natural History of Creation* (1844). Wallace had long made the species question the goal of his own research. In this essay, written while Wallace was in Sarawak, he argued that the introduction of new species had occurred gradually over a vast period of time and that "the present condition of the organic world is clearly derived by a natural process of gradual extinction and creation of species from that of the latest geological periods."[4] Wallace's cautious articulation of this, though, seemed to suggest lines for future research "and giving a direction to future researches" rather than offering a coherent argument about how species came into being.[5] Since Darwin thought that Wallace did little more than underline the importance of his own ongoing work, he was thus very much surprised by Wallace's subsequent paper "On the Tendency of Varieties to Depart Indefinitely from the Original Type," which arrived by mail in 1858.[6] Wallace's main thesis in this 1855 paper, which has subsequently become known as Wallace's 'Sarawak Law,' was that "Every species has come into existence coincident both in time and space with a pre-existing closely allied species." He clearly left much of his then-current thinking on species unsaid.

Notes

1 Adrian Desmond and James Moore, *Darwin. The Life of a Tormented Evolutionist* London & New York: W. W. Norton & Co., 1991); *Janet Browne, Charles Darwin. The Power of Place: A Biography* (New York: Alfred A. Knopf, 2002), pp. 14–17.
2 John van Wyhe, "The impact of A. R. Wallace's Sarawak Law paper reassessed," *Studies in the History and Philosophy of the Biological and Biomedical Sciences*, 60, (2016): 55–66.
3 Charles Lyell to Charles Darwin, 1 May 1856, in Burkhardt et. al., *The Correspondence of Charles Darwin*, vol. 6, (Cambridge: Cambridge University Press, 1990), pp. 89–92.
4 See text that follows.
5 See text that follows.
6 Wallace, Alfred Russel, "On the Tendency of Varieties to Depart Indefinitely from the Original Type", *Proceedings of the Linnean Society*, August 1858; this text is reprinted in Caden C. Testa and Piers J. Hale, *Evolution Before Darwin* (London: Routledge, 2024).

28

"ON THE LAW WHICH HAS REGULATED THE INTRODUCTION OF NEW SPECIES"

Alfred Russel Wallace

Source: Alfred Russel Wallace, "On the Law Which has Regulated the Introduction of New Species", *Annals and Magazine of Natural History*, 16 (September 1855), 16, pp. 184–96

I.
ON THE LAW WHICH HAS REGULATED THE INTRODUCTION OF NEW SPECIES.

Geographical Distribution dependent on Geologic Changes.

EVERY naturalist who has directed his attention to the subject of the geographical distribution of animals and plants, must have been interested in the singular facts which it presents. Many of these facts are quite different from what would have been anticipated, and have hitherto been considered as highly curious, but quite inexplicable. None of the explanations attempted from the time of Linnæus are now considered at all satisfactory; none of them have given a cause sufficient to account for the facts known at the time, or comprehensive enough to include all the new facts which have since been, and are daily being added. Of late years, however, a great light has been thrown upon the subject by geological investigations, which have shown that the present state of the earth and of the organisms now inhabiting it, is but the last stage of a long and uninterrupted series of changes which it has undergone, and consequently, that to endeavour to explain and account for its present condition without any reference to those changes (as has frequently been done) must lead to very imperfect and erroneous conclusions.

The facts proved by geology are briefly these: – That during an immense, but unknown period, the surface of the earth has undergone successive changes; land has sunk beneath the ocean, while fresh land has risen up from it; mountain chains have been elevated; islands have been formed into continents, and continents submerged till they have become islands; and these changes have taken place, not once merely, but perhaps hundreds, perhaps thousands of times: – That all these operations have been

DOI: 10.4324/9781003490548-31

more or less continuous, but unequal in their progress, and during the whole series the organic life of the earth has undergone a corresponding alteration. This alteration also has been gradual, but complete; after a certain interval not a single species existing which had lived at the commencement of the period. This complete renewal of the forms of life also appears to have occurred several times: – That from the last of the geological epochs to the present or historical epoch, the change of organic life has been gradual: the first appearance of animals now existing can in many cases be traced, their numbers gradually increasing in the more recent formations, while other species continually die out and disappear, so that the present condition of the organic world is clearly derived by a natural process of gradual extinction and creation of species from that of the latest geological periods. We may therefore safely infer a like gradation and natural sequence from one geological epoch to another.

Now, taking this as a fair statement of the results of geological inquiry, we see that the present geographical distribution of life upon the earth must be the result of all the previous changes, both of the surface of the earth itself and of its inhabitants. Many causes, no doubt, have operated of which we must ever remain in ignorance, and we may, therefore, expect to find many details very difficult of explanation, and in attempting to give one, must allow ourselves to call into our service geological changes which it is highly probable may have occurred, though we have no direct evidence of their individual operation.

The great increase of our knowledge within the last twenty years, both of the present and past history of the organic world, has accumulated a body of facts which should afford a sufficient foundation for a comprehensive law embracing and explaining them all, and giving a direction to new researches. It is about ten years since the idea of such a law suggested itself to the writer of this essay, and he has since taken every opportunity of testing it by all the newly-ascertained facts with which he has become acquainted, or has been able to observe himself. These have all served to convince him of the correctness of his hypothesis. Fully to enter into such a subject would occupy much space, and it is only in consequence of some views having been lately promulgated, he believes, in a wrong direction, that he now ventures to present his ideas to the public, with only such obvious illustrations of the arguments and results as occur to him in a place far removed from all means of reference and exact information.

A Law deduced from well-known Geographical and Geological Facts.

The following propositions in Organic Geography and Geology give the main facts on which the hypothesis is founded.

Geography.

1 Large groups, such as classes and orders, are generally spread over the whole earth, while smaller ones, such as families and genera, are frequently confined to one portion, often to a very limited district.

2 In widely distributed families the genera are often limited in range; in widely distributed genera, well marked groups of species are peculiar to each geographical district.
3 When a group is confined to one district, and is rich in species, it is almost invariably the case that the most closely allied species are found in the same locality or in closely adjoining localities, and that therefore the natural sequence of the species by affinity is also geographical.
4 In countries of a similar climate, but separated by a wide sea or lofty mountains, the families, genera and species of the one are often represented by closely allied families, genera and species peculiar to the other.

Geology.

5 The distribution of the organic world in time is very similar to its present distribution in space.
6 Most of the larger and some small groups extend through several geological periods.
7 In each period, however, there are peculiar groups, found nowhere else, and extending through one or several formations.
8 Species of one genus, or genera of one family occurring in the same geological time are more closely allied than those separated in time.
9 As generally in geography no species or genus occurs in two very distant localities without being also found in intermediate places, so in geology the life of a species or genus has not been interrupted. In other words, no group or species has come into existence twice.
10 The following law may be deduced from these facts: – *Every species has come into existence coincident both in space and time with a pre-existing closely allied species.*

This law agrees with, explains and illustrates all the facts connected with the following branches of the subject: – 1st. The system of natural affinities. 2nd. The distribution of animals and plants in space. 3rd. The same in time, including all the phænomena of representative groups, and those which Professor Forbes supposed to manifest polarity. 4th. The phænomena of rudimentary organs. We will briefly endeavour to show its bearing upon each of these.

The Form of a true system of Classification determined by this Law.

If the law above enunciated be true, it follows that the natural series of affinities will also represent the order in which the several species came into existence, each one having had for its immediate antitype a closely allied species existing at the time of its origin. It is evidently possible that two or three distinct species may have had a common antitype, and that each of these may again have become the antitypes from which other closely allied species were created. The effect of this would be, that so long as each species has had but one new species formed on its model, the line of affinities will be simple, and may be represented by placing the several species in direct succession in a straight line. But if two or more species have been independently formed on the plan of a common antitype, then the series of affinities will be compound, and can only be represented by a forked or many branched line. Now, all attempts at a Natural classification and arrangement of organic beings show, that both these plans have obtained in creation. Sometimes the series of affinities can be well represented for a space by a direct progression from species to species or from group to group, but it is generally found impossible so to continue. There constantly occur two or more modifications of an organ or modifications of two distinct organs, leading us on to two distinct series of species, which at length differ so much from each other as to form distinct genera or families. These are the parallel series or representative groups of naturalists, and they often occur in different countries, or are found fossil in different formations. They are said to have an analogy to each other when they are so far removed from their common antitype as to differ in many important points of structure, while they still preserve a family resemblance. We thus see how difficult it is to determine in every case whether a given relation is an analogy or an affinity, for it is evident that as we go back along the parallel or divergent series, towards the common antitype, the analogy which existed between the two groups becomes an affinity. We are also made aware of the difficulty of arriving at a true classification, even in a small and perfect group; – in the actual state of nature it is almost impossible, the species being so numerous and the modifications of form and structure so varied, arising probably from the immense number of species which have served as antitypes for the existing species, and thus produced a complicated branching of the lines of affinity, as intricate as the twigs of a gnarled oak or the vascular system of the human body. Again, if we consider that we have only fragments of this vast system, the stem and main branches being represented by extinct species of which we have no knowledge, while a vast mass of limbs and boughs and minute twigs and scattered leaves is what we have to place in order, and determine the true position each originally occupied with regard to the others, the whole difficulty of the true Natural System of classification becomes apparent to us.

We shall thus find ourselves obliged to reject all those systems of classification which arrange species or groups in circles, as well as those which fix a definite number for the divisions of each group. The latter class have been very generally rejected by naturalists, as contrary to nature, notwithstanding the ability with which they have been advocated; but the circular system of affinities seems to have obtained a deeper hold, many eminent naturalists having to some extent adopted it. We have, however, never been able to find a case in which the circle has been closed by a direct and close affinity. In most cases a palpable analogy has been substituted, in others the affinity is very obscure or altogether doubtful. The complicated branching of the lines of affinities in extensive groups must also afford great facilities for giving a show of probability to any such purely artificial arrangements. Their death – blow was given by the admirable paper of the lamented Mr. Strickland, published in the "Annals of Natural History," in which he so clearly showed the true synthetical method of discovering the Natural System.

Geographical Distribution of Organisms.

If we now consider the geographical distribution of animals and plants upon the earth, we shall find all the facts beautifully in accordance with, and readily explained by, the present hypothesis. A country having species, genera, and whole families peculiar to it, will be the necessary result of its having been isolated for a long period, sufficient for many series of species to have been created on the type of pre-existing ones, which, as well as many of the earlier-formed species, have become extinct, and thus made the groups appear isolated. If in any case the antitype had an extensive range, two or more groups of species might have been formed, each varying from it in a different manner, and thus producing several representative or analogous groups. The Sylviadæ of Europe and the Sylvicolidæ of North America, the Heliconidæ of South America and the Euplœas of the East, the group of Trogons inhabiting Asia, and that peculiar to South America, are examples that may be accounted for in this manner.

Such phænomena as are exhibited by the Galapagos Islands, which contain little groups of plants and animals peculiar to themselves, but most nearly allied to those of South America, have not hitherto received any, even a conjectural explanation. The Galapagos are a volcanic group of high antiquity, and have probably never been more closely connected with the continent than they are at present. They must have been first peopled, like other newly-formed islands, by the action of winds and currents, and at a period sufficiently remote to have had the original species die out, and the modified prototypes only remain. In the same way we can account for the separate islands having each their peculiar species, either on the supposition that the same original emigration peopled the whole of the islands with the same species from which differently modified prototypes were created, or that the islands were successively peopled from each other, but that new species have been created in each on the plan of the pre-existing ones. St. Helena is a similar case of a very ancient island having obtained an entirely peculiar, though limited, flora. On the other hand, no example is known of an island which can be proved geologically to be of very recent crigin (late in the Tertiary, for instance), and yet possesses generic or family groups, or even many species peculiar to itself.

When a range of mountains has attained a great elevation, and has so remained during a long geological period, the species of the two sides at and near their bases will be often very different, representative species of some genera occurring, and even whole genera being peculiar to one side only, as is remarkably seen in the case of the Andes and Rocky Mountains. A similar phænomenon occurs when an island has been separated from a continent at a very early period. The shallow sea between the Peninsula of Malacca, Java, Sumatra and Borneo was probably a continent or large island at an early epoch, and may have become submerged as the volcanic ranges of Java and Sumatra were elevated. The organic results we see in the very considerable number of species of animals common to some or all of these countries, while at the same time a number of closely allied representative species exist peculiar to each, showing that a considerable period has elapsed since their separation. The facts of geographical distribution and of geology may thus mutually explain each other in doubtful cases, should the principles here advocated be clearly established.

In all those cases in which an island has been separated from a continent, or raised by volcanic or coralline action from the sea, or in which a mountain-chain has been elevated in a recent geological epoch, the phænomena of peculiar groups or even of single representative species will not exist. Our own island is an example of this, its separation from the continent being geologically very recent, and we have consequently scarcely a species which is peculiar to it; while the Alpine range, one of the most recent mountain elevations, separates faunas and floras which scarcely differ more than may be due to climate and latitude alone.

The series of facts alluded to in Proposition (3), of closely allied species in rich groups being found geographically near each other, is most striking and important. Mr. Lovell Reeve has well exemplified it in his able and interesting paper on the Distribution of the Bulimi. It is also seen in the Humming-birds and Toucans, little groups of two or three closely allied species being often found in the same or closely adjoining districts, as we have had the good fortune of personally verifying. Fishes give evidence of a similar kind: each great river has its peculiar genera, and in more extensive genera its groups of closely allied species. But it is the same throughout Nature; every class and order of animals will contribute similar facts. Hitherto no attempt has been made to explain these singular phænomena, or to show how they have arisen. Why are the genera of Palms and of Orchids in almost every case confined to one hemisphere? Why are the closely allied species of brown-backed Trogons all found in the East, and the green-backed in the West? Why are the Macaws and the Cockatoos similarly restricted? Insects furnish a countless number of analogous examples; – the Goliathi of Africa, the Ornithopteræ of the Indian Islands, the Heliconidæ of South America, the Danaidæ of the East, and in all, the most closely allied species found in geographical proximity. The question forces itself upon every thinking mind, – why are these things so? They could not be as they are had no law regulated their creation and dispersion. The law here enunciated not merely explains, but necessitates the facts we see to exist, while the vast and long-continued geological changes of the earth readily account for the exceptions and apparent discrepancies that here and there occur. The writer's object in putting forward his views in the present imperfect manner is to submit them to the test of other minds, and to be made aware of all the facts supposed to be inconsistent with them. As his hypothesis is one which claims acceptance solely as explaining and connecting facts which exist in nature, he expects facts alone to be brought to disprove it, not *à priori* arguments against its probability.

Geological Distribution of the Forms of Life.

The phænomena of geological distribution are exactly analogous to those of geography. Closely allied species are found associated in the same beds, and the change from species to species appears to have been as gradual in time as in space. Geology, however, furnishes us with positive proof of the extinction and production of species, though it does not inform us how either has taken place. The extinction of species, however, offers but little difficulty, and the *modus operandi* has been well illustrated by Sir C. Lyell in his admirable "Principles." Geological changes, however gradual, must occasionally have modified external conditions to such an extent as to have rendered the existence of certain species impossible. The extinction would in most cases be effected by a gradual dying-out, but in some instances there might have been a sudden destruction of a species of limited range. To discover how the extinct species have from time to time been replaced by new ones down to the very latest geological period, is the most difficult, and at the same time the most interesting problem in the natural history of the earth. The present inquiry, which seeks to eliminate from known facts a law which has determined, to a certain degree, what species could and did appear at a given epoch, may, it is hoped, be considered as one step in the right direction towards a complete solution of it.

High Organization of very ancient Animals consistent with this Law.

Much discussion has of late years taken place on the question, whether the succession of life upon the globe has been from a lower to a higher degree of organization. The admitted facts seem to show that there has been a general, but not a detailed progression. Mollusca and Radiata existed before Vertebrata, and the progression from Fishes to Reptiles and Mammalia, and also from the lower mammals to the higher, is indisputable. On the other hand, it is said that the Mollusca and Radiata of the very earliest periods were more highly organized than the great mass of those now existing, and that the very first fishes that have been discovered are by no means the lowest organised of the class. Now it is believed the present hypothesis will harmonize with all these facts, and in a great measure serve to explain them; for though it may appear to some readers essentially a theory of progression, it is in reality only one of gradual change. It is, however, by no means difficult to show that a real progression in the scale of organization is perfectly consistent with all the appearances, and even with apparent retrogression, should such occur.

Returning to the analogy of a branching tree, as the best mode of representing the natural arrangement of species and their successive creation, let us suppose that at an early geological epoch any group (say a class of the Mollusca) has attained to a great richness of species and a high organization. Now let this great branch of allied species, by geological mutations, be completely or partially destroyed. Subsequently a new branch springs from the same trunk, that is to say, new species are successively created, having for their antitypes the same lower organized species which had served as the antitypes for the former group, but which have survived the modified conditions which destroyed it. This new group being subject to these altered conditions, has modifications of structure and organization given to it, and becomes the representative group of the former one in another geological formation. It may, however, happen, that though later in time, the new series of species may never attain to so high a degree of organization as those preceding it, but in its turn become extinct, and give place to yet another modification from the same root, which may be of higher or lower organization, more or less numerous in species, and more or less varied in form and structure than either of those which preceded it. Again, each of these groups may not have become totally extinct, but may have left a few species, the modified prototypes of which have existed in each succeeding period, a faint memorial of their former grandeur and luxuriance. Thus every case of apparent retrogression may be in reality a progress, though an interrupted one: when some monarch of the forest loses a limb, it may be replaced by a feeble and sickly substitute. The foregoing remarks appear to apply to the case of the Mollusca, which, at a very early period, had reached a high organization and a great development of forms and species in the testaceous Cephalopoda. In each succeeding age modified species and genera replaced the former ones which had become extinct, and as we approach the present æra, but few and small representatives of the group remain, while the Gasteropods and Bivalves have acquired an immense preponderance. In the long series of changes the earth has undergone, the process of peopling it with organic beings has been continually going on, and whenever any of the higher groups have become nearly or quite extinct, the lower forms which have better resisted the modified physical conditions have served as the antitypes on which to found the new races. In this manner alone, it is believed, can the representative groups at successive periods, and the risings and fallings in the scale of organization, be in every case explained.

Objections to Forbes' Theory of Polarity.

The hypothesis of polarity, recently put forward by Professor Edward Forbes to account for the abundance of generic forms at a very early period and at present, while in the intermediate epochs there is a gradual diminution and impoverishment, till the minimum occurred at the confines of the Palæozoic and Secondary epochs, appears to us quite unnecessary, as the facts may be readily accounted for on the principles already laid down. Between the Palæozoic and Neozoic periods of Professor Forbes, there is scarcely a species in common, and the greater part of the genera and families also disappear to be replaced by new ones. It is almost universally admitted that such a change in the organic world must have occupied a vast period of time. Of this interval we have no record; probably because the whole area of the early formations now exposed to our researches was elevated at the end of the Palæozoic period, and remained so through the interval required for the organic changes which resulted in the fauna and flora of the Secondary period. The records of this interval are buried beneath the ocean which covers three-fourths of the globe. Now it appears highly probable that a long period of quiescence or stability in the physical conditions of a district would be most favourable to the existence of organic life in the greatest abundance, both as regards individuals and also as to variety of species and generic group, just as we now find that the places best adapted to the rapid growth and increase of individuals also contain the greatest profusion of species and the greatest variety of forms, – the tropics in comparison with the temperate and arctic regions. On the other hand, it seems no less probable that a change in the physical conditions of a district, even small in amount if rapid, or even gradual if to a great amount, would be highly unfavourable to the existence of individuals, might cause the extinction of many species, and would probably be equally unfavourable to the creation of new ones. In this too we may find an analogy with the present state of our earth, for it has been shown to be the violent extremes and rapid changes of physical conditions, rather than the actual mean state in the temperate and frigid zones, which renders them less prolific than the tropical regions, as exemplified by the great distance beyond the tropics to which tropical forms penetrate when the climate is equable, and also by the richness in species and forms of tropical mountain regions which principally differ from the temperate zone in the uniformity of their climate. However this may be, it seems a fair assumption that during a period of geological repose the new species which we know to have been created would have appeared, that the creations would then exceed in number the extinctions, and therefore the number of species would increase. In a period of geological activity, on the other hand, it seems probable that the extinctions might exceed the creations, and the number of species consequently diminish. That such effects did take place in connexion with the causes to which we have imputed them, is shown in the case of the Coal formation, the faults and contortions of which show a period of great activity and violent convulsions, and it is in the formation immediately succeeding this that the poverty of forms of life is most apparent. We have then only to suppose a long period of somewhat similar action during the vast unknown interval at the termination of the Palæozoic period, and then a decreasing violence or rapidity through the Secondary period, to allow for the gradual repopulation of the earth with varied forms, and the whole of the facts are explained.[a] We thus have a clue to the increase of the forms of life during certain periods, and their decrease during others, without recourse to any causes but those we know to have existed, and to effects fairly deducible from them. The precise manner in which the geological changes of the early formations were effected is so extremely obscure, that when we can explain important facts by a retardation at one time and an acceleration at another of a process which we know from its nature and from observation to have been unequal, – a cause so simple may surely be preferred to one so obscure and hypothetical as polarity.

I would also venture to suggest some reasons against the very nature of the theory of Professor Forbes. Our knowledge of the organic world during any geological epoch is necessarily very imperfect. Looking at the vast numbers of species and groups that have been discovered by geologists, this may be doubted; but we should compare their numbers not merely with those that now exist upon the earth, but with a far larger amount. We have no reason for believing that the number of species on the earth at any former period was much less than at present; at all events the aquatic portion, with which geologists have most acquaintance, was probably often as great or greater. Now we know that there have been many complete changes of species; new sets of organisms have many times been introduced in place of old ones which have become extinct, so that the total amount which have existed on the earth from the earliest geological period must have borne about the same proportion to those now living, as the whole human race who have lived and died upon the earth, to the population at the present time. Again, at each epoch, the whole earth was no doubt, as now, more or less the theatre of life, and as the successive generations of each species died, their exuviæ and preservable parts would be deposited over every portion of the then existing seas and oceans, which we have reason for supposing to have been more, rather than less, extensive than at present. In order then to understand our possible knowledge of the early world and its inhabitants, we must compare, not the area of the whole field of our geological researches with the earth's surface, but the area of the examined portion of each formation separately with the whole earth. For example, during the Silurian period all the earth was Silurian, and animals were living and dying, and depositing their remains more or less over the whole area of the globe, and they were probably (the species at least) nearly as varied in different latitudes and longitudes as at present. What proportion do the Silurian districts bear to the whole surface of the globe, land and sea (for far more extensive Silurian districts probably exist beneath the ocean than above it), and what portion of the known Silurian districts has been actually examined for fossils? Would the area of rock actually laid open to the eye be the thousandth or the ten-thousandth part of the earth's surface? Ask the same question with regard to the Oolite or the Chalk, or even to particular beds of these when they differ considerably in their fossils, and you may then get some notion of how small a portion of the whole we know.

But yet more important is the probability, nay almost the certainty, that whole formations containing the records of vast geological periods are entirely buried beneath the ocean, and for ever beyond our reach. Most of the gaps in the geological series may thus be filled up, and vast numbers of unknown and unimaginable animals, which might help to elucidate the affinities of the numerous isolated groups which are a perpetual puzzle to the zoologist, may there be buried, till future revolutions may raise them in their turn above the waters, to afford materials for the study of whatever race of intelligent beings may then have succeeded us. These considerations must lead us to the conclusion, that our knowledge of the whole series of the former inhabitants of the earth is necessarily most imperfect and fragmentary, – as much so as our knowledge of the present organic world would be, were we forced to make our collections and observations only in spots equally limited in area and in number with those actually laid open for the collection of fossils. Now, the hypothesis of Professor Forbes is essentially one that assumes to a great extent the completeness of our knowledge of the whole series of organic beings which have existed on the earth. This appears to be a fatal objection to it, independently of all other considerations. It may be said that the same objections exist against every theory on such a subject, but this is not necessarily the case. The hypothesis put forward in this paper depends in no degree upon the completeness of our knowledge of the former condition of the organic world, but takes what facts we have as fragments of a vast whole, and deduces from them something of the nature and proportions of that whole which we can never know in detail. It is founded upon isolated groups of facts, recognizes their isolation, and endeavours to deduce from them the nature of the intervening portions.

Rudimentary Organs

Another important series of facts, quite in accordance with, and even necessary deductions from, the law now developed, are those of rudimentary organs. That these really do exist, and in most cases have no special function in the animal œconomy, is admitted by the first authorities in comparative anatomy. The minute limbs hidden beneath the skin in many of the snake-like lizards, the anal hooks of the boa constrictor, the complete series of jointed finger-bones in the paddle of the Manatus and whale, are a few of the most familiar instances. In botany a similar class of facts has been long recognised. Abortive stamens, rudimentary floral envelopes and undeveloped carpels, are of the most frequent occurrence. To every thoughtful naturalist the question must arise, What are these for? What have they to do with the great laws of creation? Do they not teach us something of the system of Nature? If each species has been created independently, and without any necessary relations with pre-existing species, what do these rudiments, these apparent imperfections mean? There must be a cause for them; they must be the necessary results of some great natural law. Now, if, as it has been endeavoured to be shown, the great law which has regulated the peopling of the earth with animal and vegetable life is, that every change shall be gradual; that no new creature shall be formed widely differing from anything before existing; that in this, as in everything else in Nature, there shall be gradation and harmony, – then these rudimentary organs are necessary, and are an essential part of the system of Nature. Ere the higher Vertebrata were formed, for instance, many steps were required, and many organs had to undergo modifications from the rudimental condition in which only they had as yet existed. We still see remaining an antitypal sketch of a wing adapted for flight in the scaly flapper of the penguin, and limbs first concealed beneath the skin, and then weakly protruding from it, were the necessary gradations before others should be formed fully adapted for locomotion.[b] Many more of these modifications should we behold, and more complete series of them, had we a view of all the forms which have ceased to live. The great gaps that exist between fishes, reptiles, birds, and mammals would then, no doubt, be softened down by intermediate groups, and the whole organic world would be seen to be an unbroken and harmonious system.

Conclusion.

It has now been shown, though most briefly and imperfectly, how the law that "*Every species has come into existence coincident both in time and space with a pre-existing closely allied species*," connects together and renders intelligible a vast number of independent and hitherto unexplained facts. The natural system of arrangement of organic beings, their geographical distribution, their geological sequence, the phænomena of representative and substituted groups in all their modifications, and the most singular peculiarities of anatomical structure, are all explained and illustrated by it, in perfect accordance with the vast mass of facts which the researches of modern naturalists have brought together, and, it is believed, not materially opposed to any of them. It also claims a superiority over previous hypotheses, on the ground that it not merely explains, but necessitates what exists. Granted the law, and many of the most important facts in Nature could not have been otherwise, but are almost as necessary deductions from it, as are the elliptic orbits of the planets from the law of gravitation.

Notes

a Professor Ramsay has since shown that a glacial epoch probably occurred at the time of the Permian formation, which will more satisfactorily account for the comparative poverty of species.

b The theory of Natural Selection has now taught us that these are not the steps by which limbs have been formed and that most rudimentary organs have been produced by abortion, owing to disuse, as explained by Mr. Darwin.

Editorial Headnote

29. Alfred Russel Wallace, "On the Tendency of Varieties to Depart Indefinitely from the Original Type", *Journal of the Proceedings of the Linnean Society*, August 1858, pp. 53–62.

Alfred Russel Wallace (1823–1913) has long been considered as very much a secondary figure to Charles Darwin in the history of evolutionary thought. However, even though Wallace himself ceded priority to Darwin in the theorizing of natural selection as the mechanism at the heart of speciation, he arrived at his own very similar theory quite independently and wrote up this essay, which he sent to Darwin in 1858. Darwin was already hard at work writing up his own thoughts on species, but it was the arrival of this essay by Wallace that prompted him to quickly write up his notes in what he thought of as a mere abstract of the longer work he had initially intended to write. It is this 'abstract' that we now know as *On the Origin of Species*.

Upon the advice of the geologist Charles Lyell and the botanist Joseph Dalton Hooker, two very respected and established men of science, it was arranged that extracts from Darwin's work be presented alongside Wallace's essay at the next meeting of the Linnean Society of London. Importantly, however, an extract from a letter that Darwin had written to the American botanist Asa Gray in 1857, in which he had outlined the basis of his theory of natural selection, was also presented. Darwin's priority in their joint discovery was thus secured.

29

"ON THE TENDENCY OF VARIETIES TO DEPART INDEFINITELY FROM THE ORIGINAL TYPE"

Alfred Russel Wallace

Source: Alfred Russel Wallace, "On the Tendency of Varieties to Depart Indefinitely from the Original Type", *Journal of the Proceedings of the Linnean Society*, (August 1858), pp. 53–62

II.

ON THE TENDENCY OF VARIETIES TO DEPART INDEFINITELY FROM THE ORIGINAL TYPE.

Instability of Varieties supposed to prove the permanent distinctness of Species.

One of the strongest arguments which have been adduced to prove the original and permanent distinctness of species is, that *varieties* produced in a state of domesticity are more or less unstable, and often have a tendency, if left to themselves, to return to the normal form of the parent species; and this instability is considered to be a distinctive peculiarity of all varieties, even of those occurring among wild animals in a state of nature, and to constitute a provision for preserving unchanged the originally created distinct species.

In the absence or scarcity of facts and observations as to *varieties* occurring among wild animals, this argument has had great weight with naturalists, and has led to a very general and somewhat prejudiced belief in the stability of species. Equally general, however, is the belief in what are called "permanent or true varieties," – races of animals which continually propagate their like, but which differ so slightly (although constantly) from some other race, that the one is considered to be a *variety* of the other. Which is the *variety* and which the original *species*, there is generally no means of determining, except in those rare cases in which the one race has been known to produce an offspring unlike itself and resembling the other. This, however, would seem quite incompatible with the "permanent invariability of species," but the difficulty is overcome by assuming that such varieties have strict limits, and can never again vary further from the original

DOI: 10.4324/9781003490548-32

type, although they may return to it, which, from the analogy of the domesticated animals, is considered to be highly probable, if not certainly proved.

It will be observed that this argument rests entirely on the assumption, that *varieties* occurring in a state of nature are in all respects analogous to or even identical with those of domestic animals, and are governed by the same laws as regards their permanence or further variation. But it is the object of the present paper to show that this assumption is altogether false, that there is a general principle in nature which will cause many *varieties* to survive the parent species, and to give rise to successive variations departing further and further from the original type, and which also produces, in domesticated animals, the tendency of varieties to return to the parent form.

The Struggle for Existence.

The life of wild animals is a struggle for existence. The full exertion of all their faculties and all their energies is required to preserve their own existence and provide for that of their infant offspring. The possibility of procuring food during the least favourable seasons, and of escaping the attacks of their most dangerous enemies, are the primary conditions which determine the existence both of individuals and of entire species. These conditions will also determine the population of a species; and by a careful consideration of all the circumstances we may be enabled to comprehend, and in some degree to explain, what at first sight appears so inexplicable – the excessive abundance of some species, while others closely allied to them are very rare.

The Law of Population of Species.

The general proportion that must obtain between certain groups of animals is readily seen. Large animals cannot be so abundant as small ones; the carnivora must be less numerous than the herbivora; eagles and lions can never be so plentiful as pigeons and antelopes; and the wild asses of the Tartarian deserts cannot equal in numbers the horses of the more luxuriant prairies and pampas of America. The greater or less fecundity of an animal is often considered to be one of the chief causes of its abundance or scarcity; but a consideration of the facts will show us that it really has little or nothing to do with the matter. Even the least prolific of animals would increase rapidly if unchecked, whereas it is evident that the animal population of the globe must be stationary, or perhaps, through the influence of man, decreasing. Fluctuations there may be; but permanent increase, except in restricted localities, is almost impossible. For example, our own observation must convince us that birds do not go on increasing every year in a geometrical ratio, as they would do, were there not some powerful check to their natural increase. Very few birds produce less than two young ones each year, while many have six, eight, or ten; four will certainly be below the average; and if we suppose that each pair produce young only four times in their life, that will also be below the average, supposing them not to die either by violence or want of food. Yet at this rate how tremendous would be the increase in a few years from a single pair! A simple calculation will show that in fifteen years each pair of birds would have increased to nearly ten millions![a] whereas we have no reason to believe that the number of the birds of any country increases at all in fifteen or in one hundred and fifty years. With such powers of increase the population must have reached its limits, and have become stationary, in a very few years after the origin of each species. It is evident, therefore, that each year an immense number of birds must perish – as many in fact as are born; and as on the lowest calculation the progeny are each year twice as numerous as their parents, it follows that, whatever be the average number of individuals existing in any given country, *twice that number must perish annually*, – a striking result, but one which seems at least highly probable, and is perhaps under rather than over the truth. It would therefore appear that, as far as the continuance of the species and the keeping up the average number of individuals are concerned, large broods are superfluous. On the average all above *one* become food for hawks and kites, wild cats or weasels, or perish of cold and hunger as winter comes on. This is strikingly proved by the case of particular species; for we find that their abundance in individuals bears no relation whatever to their fertility in producing offspring.

Perhaps the most remarkable instance of an immense bird population is that of the passenger pigeon of the United States, which lays only one, or at most two eggs, and is said to rear generally but one young one. Why is this bird so extraordinarily abundant, while others producing two or three times as many young are much less plentiful? The explanation is not difficult. The food most congenial to this species, and on which it thrives best, is abundantly distributed over a very extensive region, offering such differences of soil and climate, that in one part or another of the area the supply never fails. The bird is capable of a very rapid and long-continued flight, so that it can pass without fatigue over the whole of the district it inhabits, and as soon as the supply of food begins to fail in one place is able to discover a fresh feeding-ground. This example strikingly shows us that the procuring a constant supply of wholesome food is almost the sole condition requisite for ensuring the rapid increase of a given species, since neither the limited fecundity, nor the unrestrained attacks of birds of prey and of man are here sufficient to check it. In no other birds are these peculiar circumstances so strikingly combined. Either their food is more liable to failure, or they have not sufficient power of wing to search for it over an extensive area, or during some season of the year it becomes very scarce, and less wholesome substitutes have to be found; and thus, though more fertile in offspring, they can never increase beyond the supply of food in the least favourable seasons.

Many birds can only exist by migrating, when their food becomes scarce, to regions possessing a milder, or at least a different climate, though, as these migrating birds are seldom excessively abundant, it is evident that the countries they visit are still deficient in a constant and abundant supply of wholesome food. Those whose organization does not permit them to migrate when their food becomes periodically scarce, can never attain a large population. This is probably the reasons why woodpeckers are scarce with us, while in the tropics they are among the most abundant of solitary birds. Thus the house sparrow is more abundant than the redbreast, because its food is more constant and plentiful, – seeds of grasses being preserved during the winter, and our farm-yards and stubble-fields furnishing an almost inexhaustible supply. Why, as a general rule, are aquatic, and especially sea birds, very numerous in individuals? Not because they are more prolific than others, generally the contrary; but because their food never fails, the seashores and river-banks daily swarming with a fresh supply of small mollusca and crustacea. Exactly the same laws will apply to mammals. Wild cats are prolific and have few enemies; why then are they never as abundant as rabbits? The only intelligible answer is, that their supply of food is more precarious. It appears evident, therefore, that so long as a country remains physically unchanged, the numbers of its animal population cannot materially increase. If one species does so, some others requiring the same kind of food must diminish in proportion. The numbers that die annually must be immense; and as the individual existence of each animal depends upon itself, those that die must be the weakest – the very young, the aged, and the diseased – while those that prolong their existence can only be the most perfect in health and vigour – those who are best able to obtain food regularly, and avoid their numerous enemies. It is, as we commenced by remarking, "a struggle for existence," in which the weakest and least perfectly organized must always succumb.

The Abundance or Rarity of a Species dependent upon its more or less perfect Adaptation to the Conditions of Existence.

It seems evident that what takes place among the individuals of a species must also occur among the several allied species of a group, – viz., that those which are best adapted to obtain a regular supply of food, and to defend themselves against the attacks of their enemies and the vicissitudes of the seasons, must necessarily obtain and preserve a superiority in population; while those species which from some defect of power or organization are the least capable of counteracting the vicissitudes of food-supply, &c., must diminish in numbers, and, in extreme cases, become altogether extinct. Between these extremes the species will present various degrees of capacity for ensuring the means of preserving life; and it is thus we account for the abundance or rarity of species. Our ignorance will generally prevent us from accurately tracing the effects to their causes; but could we become perfectly acquainted with the organization and habits of the various species of animals, and could we measure the capacity of each for performing the different acts necessary to its safety and existence under all the varying circumstances by which it is surrounded, we might be able even to calculate the proportionate abundance of individuals which is the necessary result.

If now we have succeeded in establishing these two points – 1st, *that the animal population of a country is generally stationary, being kept down by a periodical deficiency of food, and other checks;* and, 2nd, *that the comparative abundance or scarcity of the individuals of the several species is entirely due to their organization and resulting habits, which, rendering it more difficult to procure a regular supply of food and to provide for their personal safety in some cases than in others, can only be balanced by a difference in the population which have to exist in a given area* – we shall be in a condition to proceed to the consideration of *varieties*, to which the preceding remarks have a direct and very important application.

Useful Variations will tend to Increase; useless or hurtful Variations to Diminish.

Most or perhaps all the variations from the typical form of a species must have some definite effect, however slight, on the habits or capacities of the individuals. Even a change of colour might, by rendering them more or less distinguishable, affect their safety; a greater or less development of hair might modify their habits. More important changes, such as an increase in the power or dimensions of the limbs or any of the external organs, would more or less affect their mode of procuring food or the range of country which they could inhabit. It is also evident that most changes would affect, either favourably or adversely, the powers of prolonging existence. An antelope with shorter or weaker legs must necessarily suffer more from the attacks of the feline carnivora; the passenger pigeon with less powerful wings would sooner or later be affected in its powers of procuring a regular supply of food; and in both cases the result must necessarily be a diminution of the population of the modified species. If, on the other hand, any species should produce a variety having slightly increased powers of preserving existence, that variety must inevitably in time acquire a superiority in numbers. These results must follow as surely as old age, intemperance, or scarcity of food produce an increased mortality. In both cases there may be many individual exceptions; but on the average the rule will invariably be found to hold good. All varieties will therefore fall into two classes – those which under the same conditions would never reach the population of the parent species, and those which would in time obtain and keep a numerical superiority. Now, let some alteration of physical conditions occur in the district – a long period of drought, a destruction of vegetation by locusts, the irruption of some new carnivorous animal seeking "pastures new" – any change in fact tending to render existence more difficult to the species in question, and tasking its utmost powers to avoid complete extermination; it is evident that, of all the individuals composing the species, those forming the least numerous and most feebly organized variety would suffer first, and, were the pressure severe, must soon become extinct. The same causes continuing in action, the parent species would next suffer, would gradually diminish in numbers, and with a recurrence of similar unfavourable conditions might also become extinct. The superior variety would then alone remain, and on a return to favourable circumstances would rapidly increase in numbers and occupy the place of the extinct species and variety.

Superior Varieties will ultimately Extirpate the original Species.

The *variety* would now have replaced the *species*, of which it would be a more perfectly developed and more highly organized form. It would be in all respects better adapted to secure its safety, and to prolong its individual existence and that of the race. Such a variety *could not* return to the original form; for that form is an inferior one, and could never compete with it for existence. Granted, therefore, a "tendency" to reproduce the original type of the species, still the variety must ever remain preponderant in numbers, and under adverse physical conditions *again alone survive*. But this new, improved, and populous race might itself, in course of time, give rise to new varieties, exhibiting several diverging modifications of form, any of which, tending to increase the facilities for preserving existence, must, by the same general law, in their turn become predominant. Here, then, we have *progression and continued divergence* deduced from the general laws which regulate the existence of animals in a state of nature, and from the undisputed fact that varieties do frequently occur. It is not, however, contended that this result would be invariable; a change of physical conditions in the district might at times materially modify it, rendering the race which had been the most capable of supporting existence under the former conditions now the least so, and even causing the extinction of the newer and, for a time, superior race, while the old or parent species and its first inferior varieties continued to flourish. Variations in unimportant parts might also occur, having no perceptible effect on the life-preserving powers; and the varieties so furnished might run a course parallel with the parent species, either giving rise to further variations or returning to the former type. All we argue for is, that certain varieties have a tendency to maintain their existence longer than the original species, and this tendency must make itself felt; for though the doctrine of chances or averages can never be trusted to on a limited scale, yet, if applied to high numbers, the results come nearer to what theory demands, and, as we approach to an infinity of examples, become strictly accurate. Now the scale on which nature works is so vast – the numbers of individuals and the periods of time with which she deals approach so near to infinity, than any cause, however slight, and however liable to be veiled and counteracted by accidental circumstances, must in the end produce its full legitimate results.

The Partial Reversion of Domesticated Varieties explained.

Let us now turn to domesticated animals, and inquire how varieties produced among them are affected by the principles here enunciated. The essential difference in the condition of wild and domestic animals is this, – that among the former, their well-being and very existence depend upon the full exercise and healthy condition of all their senses and physical powers, whereas, among the latter, these are only partially exercised, and in some cases are absolutely unused. A wild animal has to search, and often to labour, for every mouthful of food – to exercise sight, hearing, and smell in seeking it, and in avoiding dangers, in procuring shelter from the inclemency of the seasons, and in providing for the subsistence and safety of its offspring. There is no muscle of its body that is not called into daily and hourly activity; there is no sense or faculty that is not strengthened by continual exercise. The domestic animal, on the other hand, has food provided for it, is sheltered, and often confined, to guard it against the vicissitudes of the seasons, is carefully secured from the attacks of its natural enemies, and seldom even rears its young without human assistance. Half of its senses and faculties become quite useless, and the other half are but occasionally called into feeble exercise, while even its muscular system is only irregularly brought into action.

Now when a variety of such an animal occurs, having increased power or capacity in any organ or sense, such increase is totally useless, is never called into action, and may even exist without the animal ever becoming aware of it. In the wild animal, on the contrary, all its faculties and powers being brought into full action for the necessities of existence, any increase becomes immediately available, is strengthened by exercise, and must even slightly modify the food, the habits, and the whole economy of the race. It creates as it were a new animal, one of superior powers, and which will necessarily increase in numbers and outlive those which are inferior to it.

Again, in the domesticated animal all variations have an equal chance of continuance; and those which would decidedly render a wild animal unable to compete with its fellows and continue its existence are no disadvantage whatever in a state of domesticity. Our quickly fattening pigs, short-legged sheep pouter pigeons, and poodle dogs could never have come into existence in a state of nature, because the very first step towards such inferior forms would have led to the rapid extinction of the race; still less could they now exist in competition with their wild allies. The great speed but slight endurance of the race horse, the unwieldly strength of the ploughman's team, would both be useless in a state of nature. If turned wild on the pampas, such animals would probably soon become extinct, or under favourable circumstances might each gradually lose those extreme qualities which would never be called into action, and in a few generations revert to a common type, which must be that in which the various powers and faculties are so proportioned to each other as to be best adapted to procure food and secure safety, – that in which by the full exercise of every part of its organisation the animal can alone continue to live. Domestic varieties, when turned wild, *must* return to something near the type of the original wild stock, *or become altogether extinct*.[b]

We see, then, that no inferences as to the permanence of varieties in a state of nature can be deduced from the observations of those occurring among domestic animals. The two are so much opposed to each other in every circumstance of their existence, that what applies to the one is almost sure not to apply to the other. Domestic animals are abnormal, irregular, artificial; they are subject to variations which never occur and never can occur in a state of nature: their very existence depends altogether on human care; so far are many of them removed from that just proportion of faculties, that true balance of organisation, by means of which alone an animal left to its own resources can preserve its existence and continue its race.

Lamarck's Hypothesis very different from that now advanced.

The hypothesis of Lamarck – that progressive changes in species have been produced by the attempts of animals to increase the development of their own organs, and thus modify their structure and habits – has been repeatedly and easily refuted by all writers on the subject of varieties and species, and it seems to have been considered that when this was done the whole question has been finally settled; but the view here developed renders such hypothesis quite unnecessary, by showing that similar results must be produced by the action of principles constantly at work in nature. The powerful retractile talons of the falcon- and the cat-tribes have not been produced or increased by the volition of those animals; but among the different varieties which occurred in the earlier and less highly organized forms of these groups, *those always survived longest which had the greatest facilities for seizing their prey*. Neither did the giraffe acquire its long neck by desiring to reach the foliage of the more lofty shrubs, and constantly stretching it neck for the purpose, but because any varieties which occurred among its antitypes with a longer neck than usual *at once secured a fresh range of pasture over the same ground as their shorter-necked companions, and on the first scarcity of food were thereby enabled to outlive them*. Even the peculiar colours of many animals, more especially of insects, so closely resembling the soil or leaves or bark on which they habitually reside, are explained on the same principle; for though in the course of ages varieties of many tints may have occurred, *yet those races having colours best adapted to concealment from their enemies would inevitably survive the longest*. We have also here an acting cause to account for that balance so often observed in nature, – a deficiency in one set of organs always being compensated by an increased development of some others – powerful wings accompanying weak feet, or great velocity making up for the absence of defensive weapons; for it has been shown that all varieties in which an unbalanced deficiency occurred could not long continue their existence. The action of this principle is exactly like that of the centrifugal governor of the steam engine, which checks and corrects any irregularities almost before they become evident; and in like manner no unbalanced deficiency in the animal kingdom can ever reach any conspicuous magnitude, because it would make itself felt at the very first step, by rendering existence difficult and extinction almost sure soon to follow. An origin such as is here advocated will also agree with the peculiar character of the modifications of form and structure which obtain in organized beings – the many lines of divergence from a central type, the increasing efficiency and power of a particular organ through a succession of allied species, and the remarkable persistence of unimportant parts, such as colour, texture of plumage and hair, form of horns or crests, through a series of species differing considerably in more essential characters. It also furnishes us with a reason for that "more specialized structure" which

Professor Owen states to be a characteristic of recent compared with extinct forms, and which would evidently be the result of the progressive modification of any organ applied to a special purpose in the animal economy.

Conclusion.

We believe we have now shown that there is a tendency in nature to the continued progression of certain classes of *varieties* further and further from the original type – a progression to which there appears no reason to assign any definite limits – and that the same principle which produces this result in a state of nature will also explain why domestic varieties have a tendency, when they become wild, to revert to the original type. This progression, by minute steps, in various directions, but always checked and balanced by the necessary conditions, subject to which alone existence can be preserved, may, it is believed, be followed out so as to agree with all the phænomena presented by organized beings, their extinction and succession in past ages, and all the extraordinary modifications of form, instinct and habits which they exhibit.

Notes

a This is under estimated. The number would really amount to more than two thousand millions!

b That is, they will vary, and the variations which tend to adapt them to the wild state, and therefore approximate them to wild animals, will be preserved. Those individuals which do not vary sufficiently will perish.

Part 4

THE HISTORICAL SKETCH

Editorial Headnote

30. Charles Darwin, "Preface" to On the Origin of Species by means of natural Selection, or, the preservation of the favoured races in the struggle for life (London: John Murray, 1872).

The preface to the *Origin of Species*, which historians have long referred to as Darwin's "Historical Sketch," made its first appearance in the third edition of *Origin*, published in 1861.[1] In it Darwin included a brief overview of earlier evolutionary thinkers – an acknowledgement of predecessors that some critics (both friendly and otherwise) had suggested was a notable absence from the first and second editions of this work. Across the years and editions of *Origin*, Darwin edited and expanded the list of those he acknowledged as forebears, including in the sixth and final edition mention of some thirty-five people. As the historian Curtis N. Johnson has argued, the list of those Darwin acknowledged includes some obscure thinkers whom Darwin made little use of in the course of developing his own ideas, and it is also notable that he packed his commentary on all thirty-five men into a mere eleven pages. The purpose of the sketch was thus to respond to criticism that he had not adequately acknowledged the work of earlier thinkers but also to make clear that Darwin could still claim priority for his main idea, that of the role of natural selection in the formation of new species. Thus, Darwin's account of these thinkers is somewhat brief and defensive of the originality of his own contribution, even while he acknowledged that one or two thinkers had indeed anticipated him.[2]

Notes

1 The history of the writing of the sketch has been given extensive analysis by the historian of science, Curtis N. Johnson. See Curtis N. Johnson, *Darwin's Historical Sketch. An Examination of the 'Preface' to the Origin of Species*, (Oxford: Oxford University Press 2020).

2 In this regard see, for example, the generosity of Darwin's acknowledgment of Patrick Matthew's priority claim regarding natural selection in Matthew, P. 1860. "Nature's law of selection." *Gardeners' Chronicle and Agricultural Gazette* (7 April): 312–13, and his admission that William Charles Wells "distinctly recognizes the principle of natural selection" in his own earlier work, "An Account of a White Female, Whose Skin Resembles that of a Negro." Copies of these texts are provided in this volume, Caden C. Testa and Piers J. Hale, *Evolution Before Darwin* (London: Routledge, 2025).

30

"AN HISTORICAL SKETCH"

Charles Darwin

Source: Charles Darwin, "An Historical Sketch," in *On the Origin of Species by means of natural Selection, or, the preservation of the favoured races in the struggle for life* 6th ed. (London: John Murray, 1872)

AN HISTORICAL SKETCH OF THE PROGRESS OF OPINION ON THE ORIGIN OF SPECIES, PREVIOUSLY TO THE PUBLICATION OF THE FIRST EDITION OF THIS WORK.

I WILL here a give a brief sketch of the progress of opinion on the Origin of Species. Until recently the great majority of naturalists believed that species were immutable productions, and had been separately created. This view has been ably maintained by many authors. Some few naturalists, on the other hand, have believed that species undergo modification, and that the existing forms of life are the descendants by true generation of pre-existing forms. Passing over allusions to the subject in the classical writers,[a] the first author who in modern times has treated it in a scientific spirit was Buffon. But as his opinions fluctuated greatly at different periods, and as he does not enter on the causes or means of the transformation of species, I need not here enter on details.

Lamarck was the first man whose conclusions on the subject excited much attention. This justly-celebrated naturalist first published his views in 1801; he much enlarged them in 1809 in his 'Philosophie Zoologique,' and subsequently, in 1815, in the Introduction to his 'Hist. Nat. des Animaux sans Vertébres.' In these works he upholds the doctrine that all species, including man, are descended from other species. He first did the eminent service of arousing attention to the probability of all change in the organic, as well as in the inorganic world, being the result of law, and not of miraculous interposition. Lamarck seems to have been chiefly led to his conclusion on the gradual change of species, by the difficulty of distinguishing species and varieties, by the almost perfect gradation of forms in certain groups, and by the analogy of domestic productions. With respect to the means of modification, he attributed something to the direct action of the physical conditions of life, something to the crossing of already existing forms, and much

DOI: 10.4324/9781003490548-34

to use and disuse, that is, to the effects of habit. To this latter agency he seems to attribute all the beautiful adaptations in nature; – such as the long neck of the giraffe for browsing on the branches of trees. But he likewise believed in a law of progressive development; and as all the forms of life thus tend to progress, in order to account for the existence at the present day of simple productions, he maintains that such forms are now spontaneously generated.[b]

Geoffroy Saint-Hilaire, as is stated in his 'Life,' written by his son, suspected, as early as 1795, that what we call species are various degenerations of the same type. It was not until 1828 that he published his conviction that the same forms have not been perpetuated since the origin of all things. Geoffroy seems to have relied chiefly on the conditions of life, or the "*monde ambiant*" as the cause of change. He was cautious in drawing conclusions, and did not believe that existing species are now undergoing modification; and, as his son adds, "C'est donc un problème à réserver entièrement à l'avenir, supposé même que l'avenir doive avoir prise sur lui."

In 1813, Dr. W. C. Wells read before the Royal Society 'An Account of a White Female, part of whose skin resembles that of a Negro'; but his paper was not published until his famous 'Two Essays upon Dew and Single Vision' appeared in 1818. In this paper he distinctly recognises the principle of natural selection, and this is the first recognition which has been indicated; but he applies it only to the races of man, and to certain characters alone. After remarking that negroes and mulattoes enjoy an immunity from certain tropical diseases, he observes, firstly, that all animals tend to vary in some degree, and, secondly, that agriculturists improve their domesticated animals by selection; and then, he adds, but what is done in this latter case "by art, seems to be done with equal efficacy, though more slowly, by nature, in the formation of varieties of mankind, fitted for the country which they inhabit. Of the accidental varieties of man, which would occur among the first few and scattered inhabitants of the middle regions of Africa, some one would be better fitted than the others to bear the diseases of the country. This race would consequently multiply, while the others would decrease; not only from their inability to sustain the attacks of disease, but from their incapacity of contending with their more vigorous neighbours. The colour of this vigorous race I take for granted, from what has been already said, would be dark. But the same disposition to form varieties still existing, a darker and a darker race would in the course of time occur: and as the darkest would be the best fitted for the climate, this would at length become the most prevalent, if not the only race, in the particular country in which it had originated." He then extends these same views to the white inhabitants of colder climates. I am indebted to Mr. Rowley, of the United States, for having called my attention, through Mr. Brace, to the above passage in Dr. Well's work.

The Hon. and Rev. W. Herbert, afterwards Dean of Manchester, in the fourth volume of the 'Horticultural Transactions,' 1822, and in his work on the 'Amaryllidaceæ' (1837, p. 19, 339), declares that "horticultural experiments have established, beyond the possibility of refutation, that botanical species are only a higher and more permanent class of varieties." He extends the same view to animals. The Dean believes that single species of each genus were created in an originally highly plastic condition, and that these have produced, chiefly by intercrossing, but likewise by variation, all our existing species.

In 1826 professor Grant, in the concluding paragraph in his well-known paper ('Edinburgh Philosophical Journal,' vol. xiv. p. 283) on the Spongilla, clearly declares his belief that species are descended from other species, and that they become improved in the course of modification. This same view was given in his 55th Lecture, published in the 'Lancet' in 1834.

In 1831 Mr. Patrick Matthew published his work on 'Naval Timber and Arboriculture,' in which he gives precisely the same view on the origin of species as that (presently to be alluded to) propounded by Mr. Wallace and myself in the 'Linnean Journal,' and as that enlarged in the present volume. Unfortunately the view was given by Mr. Matthew very briefly in scattered passages in an Appendix to a work on a different subject, so that it remained unnoticed until Mr. Matthew himself drew attention to it in the 'Gardeners' Chronicle,' on April 7th, 1860. The differences of Mr. Matthew's view from mine are not of much importance: he seems to consider that the world was nearly depopulated at successive periods, and then re-stocked; and he gives as an alternative, that new forms may be generated "without the presence of any mould or germ of former aggregates." I am not sure that I understand some passages; but it seems that he attributes much influence to the direct action of the conditions of life. He clearly saw, however, the full force of the principle of natural selection.

The celebrated geologist and naturalist, Von Buch, in his excellent 'Description Physique des Isles Canaries' (1836, p. 147), clearly expresses his belief that varieties slowly become changed into permanent species, which are no longer capable of intercrossing.

Rafinesque, in his 'New Flora of North America,' published in 1836, wrote (p. 6) as follows: – "All species might have been varieties once, and many varieties are gradually becoming species by assuming constant and peculiar characters"; but farther on (p. 18) he adds, "except the original types or ancestors of the genus."

In 1843–44 Professor Haldeman ('Boston Journal of Nat. Hist. U. States,' vol. iv. p. 468) has ably given the arguments for and against the hypothesis of the development and modification of species: he seems to lean towards the side of change.

The 'Vestiges of Creation' appeared in 1844. In the tenth and much improved edition (1853) the anonymous author says (p. 155): – The proposition determined on after much consideration is, that the several series of animated beings, from the simplest and oldest up to the highest and most recent, are, under the providence of God, the results, *first*, of an impulse which has been imparted to the forms of life, advancing them, in definite times, by generation, through grades of organisation terminating in the highest dicotyledons and vertebrata, these grades being few in number, and generally marked by intervals of organic character, which we find to be a practical difficulty in ascertaining affinities; *second*, of another impulse connected with the vital forces, tending, in the course of generations, to modify organic structures in accordance with external circumstances, as food, the nature of the habitat, and the meteoric agencies, these being the 'adaptations' of the natural theologian." The author apparently believes that organisation progresses by sudden leaps, but that the effects produced by the conditions of life are gradual. He argues with much force on general grounds that species are not immutable productions. But I cannot see how the two supposed "impulses" account in a scientific sense for the numerous and beautiful co-adaptations which we see throughout nature; I cannot see that we thus gain any insight how, for instance, a woodpecker has become adapted to its peculiar habits of Life. The work, from its powerful and brilliant style, though displaying in the earlier editions little accurate knowledge and a great want of scientific caution, immediately had a very wide circulation. In my opinion it has done excellent service in this country in calling attention to the subject, in removing prejudice, and in thus preparing the ground for the reception of analogous views.

In 1846 the veteran geologist M. J. d'Omalius d'Halloy published in an excellent though short paper ('Bulletins de l'Acad. Roy. Bruxelles, tom. xiii. p. 581), his opinion that it is more probable that new species have been produced by descent with modification than that they have been separately created: the author first promulgated this opinion in 1831.

Professor Owen, in 1849 ('Nature of Limbs,' p. 86), wrote as follows: – "The archetypal idea was manifested in the flesh under diverse such modifications, upon this planet, long prior to the existence of those animal species that actually exemplify it. To what natural laws or secondary causes the orderly succession and progression of such organic phenomena may have been committed, we, as yet, are ignorant." In his Address to the British Association, in 1858, he speaks (p. li.) of "the axiom of the continuous operation of creative power, or of the ordained becoming of living things." Farther on (p. xc.), after referring to geographical distribution, he adds, 'These phenomena shake our confidence in the conclusion that the Apteryx of New Zealand and the Red Grouse of England were distinct creations in and for those islands respectively. Always, also, it may be well to bear in mind that by the word 'creation' the zoologist means 'a process he knows not what.'" He amplifies this idea by adding, that when such cases as that of the Red Grouse are "enumerated by the zoologist as evidence of distinct creation of the bird in and for such islands, he chiefly expresses that he knows not how the Red Grouse came to be there, and there exclusively; signifying also, by this mode of expressing such ignorance, his belief that both the bird and the islands owed their origin to a great first Creative Cause." If we interpret these sentences given in the same Address, one by the other, it appears that this eminent philosopher felt in 1858 his confidence shaken that the Apteryx and the Red Grouse first appeared in their respective homes, "he knew not how," or by some process "he knew not what."

This Address was delivered after the papers, by Mr. Wallace and myself on the Origin of Species, presently to be referred to, had been read before the Linnean Society. When the first edition of this work was published, I was so completely deceived, as were many others, by such expressions as "the continuous operation of creative power," that I included Professor Owen with other palæontologists as being firmly convinced of the immutability of species; but it appears ('Anat. of Vertebrates,' vol. iii. p. 796) that this was on my part a preposterous error. In the last edition of this work I inferred, and the inference still seems to me perfectly just, from a passage beginning with the words "no doubt the type-form," &c. (Ibid. vol. i. p. xxxv.), that Professor Owen admitted that natural selection may have done something in the formation of new species; but this it appears (Ibid. vol. iii. p. 798) is inaccurate and without evidence. I also gave some extracts from a correspondence between Professor Owen and the Editor of the 'London Review,' from which it appeared manifest to the Editor as well as to myself, that Professor Owen claimed to have promulgated the theory of natural selection before I had done so; and I expressed my surprise and satisfaction at this announcement; but as far as it is possible to understand certain recently published passages (Ibid. vol. iii. p. 798), I have either partially or wholly again fallen into error. It is consolatory to me that others find Professor Owen's controversial writings as difficult to understand and to reconcile with each other, as I do. As far as the mere enunciation of the principle of natural selection is concerned, it is quite immaterial whether or not Professor Owen preceded me, for both of us, as shown in this historical sketch, were long ago preceded by Dr. Wells and Mr. Matthews.

M. Isidore Geoffroy Saint-Hilaire, in his Lectures delivered in 1850 (of which a Résumé appeared in the 'Revue et Mag. de Zoolog.,' Jan. 1851), briefly gives his reason for believing that specific characters "sont fixés, pour chaque espèce, tant qu'elle se perpétue au milieu des mêmes circonstances: ils se modifient, si les circonstances ambiantes viennent à changer." "En résumé, *l'observation* des animaux sauvages démontre déjà la variabilité *limitée* des espèces. Les *expériences* sur les animaux sauvages devenus domestiques, et sur les animaux domestiques redevenus sauvages, la démontrent plus clairement encore. Ces mêmes expériences prouvent, de plus, que les différences produites peuvent être de *valeur générique*." In his 'Hist. Nat. Générale' (tom. ii. p. 430, 1859) he amplifies analogous conclusions.

From a circular lately issued it appears that Dr. Freke, in 1851 ('Dublin Medical Press,' p. 322), propounded the doctrine that all organic beings have descended from one primordial form. His grounds of belief and treatment of the subject are wholly different from mine; but as Dr. Freke has now (1861) published his Essay on 'the Origin of Species by means of Organic Affinity,' the difficult attempt to give any idea of his views would be superfluous on my part.

Mr. Herbert Spencer, in an Essay (originally published in the 'Leader,' March 1852, and republished in his 'Essays' in 1858), has contrasted the theories of the Creation and the Development of organic beings with remarkable skill and force. He argues from the analogy of domestic productions, from the changes which the embryos of many species undergo, from the difficulty of distinguishing species and varieties, and from the principle of general gradation, that species have been modified; and he attributes the modification to the change of circumstances.

The author (1855) has also treated Psychology on the principle of the necessary acquirement of each mental power and capacity by gradation.e

In 1852 M. Naudin, a distinguished botanist, expressly stated, in an admirable paper on the Origin of Species ('Revue Horticole,' p. 102; since partly republished in the 'Nouvelles Archives du Muséum,' tom. i. p. 171), his belief that species are formed in an analogous manner as varieties are under cultivation; and the latter process he attributes to man's power of selection. But he does not show how selection acts under nature. He believes, like Dean Herbert, that species, when nascent, were more plastic than at present. He lays weight on what he calls the principle of finality, "puissance mystérieuse, indéterminée; fatalité pour les uns; pour les autres, volonté providentielle, dont l'action incessante sur les êtres vivants détermine, à toutes les époques de l'existence du monde, la forme, le volume, et la durée de chacun d'eux, en raison de sa destinée dans l'ordre de choses dont il fait partie. C'est cette puissance qui harmonise chaque membre à l'ensemble en l'appropriant à la fonction qu'il doit remplir dans l'organisme général de la nature, fonction qui est pour lui sa raison d'être."[c]

In 1853 a celebrated geologist, Count Keyserling ('Bulletin de la Soc. Géolog.,' 2nd Ser., tom. x. p. 357), suggested that as new diseases, supposed to have been caused by some miasma, have arisen and spread over the world, so at certain periods the germs of existing species may have been chemically affected by circumambient molecules of a particular nature, and thus have given rise to new forms.

In this same year, 1853, Dr. Schaaffhausen published an excellent pamphlet ('Verhand. des Naturhist. Vereins der Preuss. Rheinlands,' &c.), in which he maintains the progressive development of organic forms on the earth. He infers that many species have kept true for long periods, whereas a few have become modified. The distinction of species he explains by the destruction of intermediate graduated forms. "Thus living plants and animals are not separated from the extinct by new creations, but are to be regarded as their descendants through continued reproduction."

A well-known French botanist, M. Lecoq, writes in 1854 ('Etudes sur Géograph. Bot.,' tom. i. p. 250), "On voit que nos recherches sur la fixité ou la variation de l'espéce, nous conduisent directement aux idées émises, par deux hommes justement célèbres, Geoffroy Saint-Hilaire et Goethe." Some other passages scattered through M. Lecoq's large work, make it a little doubtful how far he extends his views on the modification of species.

The 'Philosophy of Creation' has been treated in a masterly manner by the Rev. Baden Powell, in his 'Essays on the Unity of Worlds,' 1855. Nothing can be more striking than the manner in which he shows that the introduction of new species is "a regular, not a casual phenomenon," or, as Sir John Herschel expresses it, "a natural in contradistinction to a miraculous process."

The third volume of the 'Journal of the Linnean Society' contains papers, read July 1st, 1858, by Mr. Wallace and myself, in which, as stated in the introductory remarks to this volume, the theory of Natural Selection is promulgated by Mr. Wallace with admirable force and clearness.

Von Baer, towards whom all zoologists feel so profound a respect, expressed about the year 1859 (see Prof. Rudolph Wagner, "Zoologisch-Anthropologische Untersuchugen,' 1861, s. 51) his conviction, chiefly grounded on the laws of geographical distribution, that forms now perfectly distinct have descended from a single parent-form.

In June, 1859, Professor Huxley gave a lecture before the Royal Institution on the 'Persistent Types of Animal Life.' Referring to such cases, he remarks, "It is difficult to comprehend the meaning of such facts as these, if we suppose that each species of animal and plant, or each great type of organisation, was formed and placed upon the surface of the globe at long intervals by a distinct act of creative power; and it is well to recollect that such an assumption is as unsupported by tradition or revelation as it is opposed to the general analogy of nature. If, on the other hand, we view 'Persistent Types' in relation to that hypothesis which supposes the species living at any time to be the result of the gradual modification of pre-existing species – a hypothesis which, though unproven, and sadly damaged by some of its supporters, is yet the only one to which physiology lends any countenance; their existence would seem to show that the amount of modification which living beings have undergone during geological time is but very small in relation to the whole series of changes which they have suffered."

In December, 1859, Dr. Hooker published his 'Introduction to the Australian Flora.' In the first part of this great work he admits the truth of the descent and modification of species, and supports this doctrine by many original observations.

The first edition of this work was published on November 24th, 1859, and the second edition on January 7th, 1860.

Notes

a Aristotle, in his 'Physicæ Auscultationes' (lib. 2, cap. 8, s. 2), after remarking that rain does not fall in order to make the corn grow, any more than it falls to spoil the farmer's corn when threshed out of doors, applies the same argument to organisation: and adds (as translated by Mr. Clair Grece, who first pointed out the passage to me), "So what hinders the different parts [of the body] from having this merely accidental relation in nature? as the teeth, for example, grow by necessity, the front ones sharp, adapted for dividing, and the grinders flat, and serviceable for masticating the food; since they were not made for the sake of this, but it was the result of accident. And in like manner as to the other parts in which there appears to exist an adaptation to an end. Wheresoever, therefore, all things together (that is all the parts of one whole) happened like as if they were made for the sake of something, these were preserved, having been appropriately constituted by an internal spontaneity, and whatsoever things were not thus constituted, perished, and still perish." We here see the principle of natural selection shadowed forth, but how little Aristotle fully comprehended the principle, is shown by his remarks on the formation of the teeth.

b I have taken the date of the first publication of Lamarck from Isid. Geoffroy Saint-Hilaire's ('Hist. Nat. Générale,' tom. ii. p. 405, 1859) excellent history of opinion on this subject. In this work a full account is given of Buffon's conclusions on the same subject.

It is curious how largely my grandfather, Dr. Erasmus Darwin, anticipated the views and erroneous grounds of opinion of Lamarck in his 'Zoonomia' (vol. i. pp. 500–510), published in 1794. According to Isid. Geoffroy there is no doubt that Goethe was an extreme partisan of similar views, as shown in the Introduction to a work written in 1794 and 1795, but not published till long afterwards: he has pointedly remarked ('Goethe als Naturforscher,' von Dr. Karl Meding, s. 34) that the future question for naturalists will be how, for instance, cattle got their horns, and not for what they are used. It is rather a singular instance of the manner in which similar views arise at about the same time, that Goethe in Germany, Dr. Darwin in England, and Geoffroy Saint-Hilaire (as we shall immediately see) in France, came to the same conclusion on the origin of species, in the years 1794–5.

c From references in Bronn's 'Untersuchungen über die Entwickelungs-Gesetze,' it appears that the celebrated botanist and palæontologist Unger published, in 1852, his belief that species undergo development and modification. D'Alton, likewise, in Pander and Dalton's work on Fossil Sloths, expressed, in 1821, a similar belief. Similar views have, as is well known, been maintained by Oken in his mystical 'Natur-Philosophie.' From other references in Godron's work 'Sur l'Espèce,' it seems that Bory St. Vincent, Burdach, Poiret, and Fries, have all admitted that new species are continually being produced.

I may add, that of the thirty-four authors named in this Historical Sketch, who believe in the modification of species, or at least disbelieve in separate acts of creation, twenty-seven have written on special branches of natural history or geology.

Editorial Headnote

31. Matthew, P. 1860. "Nature's law of selection." *Gardeners' Chronicle and Agricultural Gazette* (7 April): 312–3.

Darwin acknowledged Patrick Matthew (1790–1874) as one of the several people who anticipated him, Matthew having described a process very much akin to natural selection in the appendix of his large volume *On Naval Timber and Arboriculture: With Critical Notes on Authors who Have Recently Treated the Subject of Planting*, which was published in 1831.[1] Rather than quote the original here, we provide the letter to the *Gardener's Chronicle and Agricultural Gazette* in which Matthew drew attention to his earlier work. The letter was published under the title "Nature's Law of Selection." Matthew had been prompted to write after having read a review of *Origin* by Thomas Huxley, which had appeared in the March 3rd edition of the same publication. It is worth noting, however, that the extracts that Matthew highlighted had originally appeared in a somewhat disorganized appendix to the volume. As the historian Curtis Johnson has shown, no reviewers of *On Naval Timber* drew attention to this aspect of Matthew's book, nor did Matthew see his own work as a contribution to the species question, at least not until Darwin's work became famous.[2] Indeed, and as Darwin pointed out in his reply to the *Gardener's Chronicle*, it seems that no other naturalists had been aware of Matthew's work either.

Despite this, Darwin responded with the following letter and later included an acknowledgement of Matthew's priority in the "Historical Sketch," with which he prefaced *Origin* from the third edition of 1861 onwards.

> *Natural Selection.* – I have been much interested by Mr. Patrick Matthew's communication in the Number of your Paper, dated April 7th. I freely acknowledge that Mr. Matthew has anticipated by many years the explanation which I have offered of the origin of species, under the name of natural selection. I think that no one will feel surprised that neither I, nor apparently any other naturalist, had heard of Mr. Matthew's views, considering how briefly they are given, and that they appeared in the appendix to a work on Naval Timber and Arboriculture. I can do no more than offer my apologies to Mr. Matthew for my entire ignorance of his publication. If another edition of my work is called for, I will insert a notice to the foregoing effect. *Charles Darwin, Down, Bromley, Kent.*[3]

Notes

1 Patrick Matthew, *On naval timber and arboriculture; with critical notes on authors who have recently treated the subject of planting* (London: Longman, Rees, Orme, Brown, and Green; and Edinburgh: Adam Black, 1831.)

2 Curtis N. Johnson, *Darwin's Historical Sketch. An Examination of the 'Preface' to the Origin of Species*, (Oxford: Oxford University Press 2020), pp. 110–24.
3 Charles Darwin, "Natural selection," *Gardeners' Chronicle and Agricultural Gazette* no. 16 (21st April 1860): 362–363.

31

"NATURE'S LAW OF SELECTION"

P. Matthew

Source: P. Matthew, "Nature's Law of Selection", *Gardeners' Chronicle and Agricultural Gazette* (7 April 1860), pp. 312–3

Matthew, P. 1860. Nature's law of selection. *Gardeners' Chronicle and Agricultural Gazette* (7 April): 312–13.

NATURE'S LAW OF SELECTION.

TRUSTING to your desire that every man should have his own, I hope you will give place to the following communication.

In your Number of March 3d I observe a long quotation from the *Times,* stating that Mr. Darwin "professes to have discovered the existence and *modus operandi* of the natural law of selection," that is, "the power in nature which takes the place of man and performs a selection, *sua sponte,*" in organic life. This discovery recently published as "the results of 20 years' investigation and reflection" by Mr. Darwin turns out to be what I published very fully and brought to apply practically to forestry in my work "Naval Timber and Arboriculture," published as far back as January 1, 1831, by Adam & Charles Black, Edinburgh, and Longman & Co., London, and reviewed in numerous periodicals, so as to have full publicity in the "Metropolitan Magazine," the "Quarterly Review," the "Gardeners' Magazine," by Loudon, who spoke of it as *the* book, and repeatedly in the "United Service Magazine" for 1831, &c. The following is an extract from this volume, which clearly proves a prior claim. The same volume contains the first proposal of the steam ram (also claimed since by several others, English, French, and Americans,) and a navy of steam gun-boats as requisite in future maritime war, and which, like the organic selection law, are only as yet making way: –

"There is a law universal in nature, tending to render every reproductive being the best possibly suited to its condition that its kind, or that organised matter, is susceptible of, which appears intended to model the physical and mental or instinctive powers, to their highest perfection, and to continue them so. This law sustains the lion in his strength, the hare in her swiftness, and the fox in his wiles. As nature, in all her modifications of life, has a power of increase far beyond what is needed to supply the place of what falls by Time's decay, those individuals who possess not the requisite strength, swiftness, hardihood, or cunning, fall prematurely without reproducing – either a prey to their natural devourers, or sinking under disease, generally induced by want of nourishment, their place being occupied by the more perfect of their own kind, who are pressing on the means of subsistence."

"Throughout this volume, we have felt considerable inconvenience, from the adopted dogmatical classification of plants, and have all along been floundering between species and variety, which certainly under culture soften into each other. A particular conformity, each after its own kind, when in a state of nature, termed species, no doubt exists to a considerable degree. This conformity has existed during the last 40 centuries. Geologists discover a like particular conformity – fossil species – through the deep deposition of each great epoch, but they also discover an almost complete difference to exist between the species or stamp of life on one epoch from that of every other. We are therefore led to admit, either of a repeated miraculous creation; or of a power of change, under a change of circumstances, to belong to living organised matter, or rather to the congeries of inferior life, which appears to form superior. The derangements and changes in organised existence, induced by a change of circumstance from the interference of man, affording us proof of the plastic quality of superior life, and the likelihood that circumstances have been very different in the different epochs, though steady in each, tend strongly to heighten the probability of the latter theory.

"When we view the immense calcareous and bituminous formations, principally from the waters and atmosphere, and consider the oxidations and depositions which have taken place, either gradually, or during some of the great convulsions, it appears at least probable, that the liquid elements containing life have varied considerably at different times in composition and weight; that our atmosphere has contained a much greater proportion of carbonic acid or oxygen; and our waters aided by excess of carbonic acid, and greater heat resulting from greater density of atmosphere, have contained a greater quantity of lime and other mineral solutions. Is the inference then unphilosophic, that living things which are proved to have a circumstance-suiting power – a very slight change of circumstance by culture inducing a corresponding change of character – may have gradually accommodated themselves to the variations of the elements containing them, and, without new creation, have presented the diverging changeable phenomena of past and present organised existence?

"The destructive liquid currents, before which the hardest mountains have been swept and comminuted into gravel, sand, and mud, which intervened between and divided these epochs, probably extending over the whole surface of the globe, and destroying nearly all living things, must have reduced existence so much, that an unoccupied field would be formed for new diverging ramifications of life, which, from the connected sexual system of vegetables, and the natural instincts of animals to herd and combine with their own kind, would fall into specific groups, these remnants, in the course of time, moulding and accommodating their being anew to the change of circumstances, and to every possible means of subsistence, and the millions of ages of regularity which appear to have followed between the epochs, probably after this accommodation was completed, affording fossil deposit of regular specific character.

"There are only two probable ways of change – the above, and the still wider deviation from present occurrence – of indestructible or molecular life (which seems to resolve itself into powers of attraction and repulsion under mathematical figure and regulation, bearing a slight systematic similitude to the great aggregations of matter), gradually uniting and developing itself into new circumstance-suited living aggregates, without the presence of any mould or germ of former aggregates, but this scarcely differs from new creation, only it forms a portion of a continued scheme or system.

"In endeavouring to trace, in the former way, the principle of these changes of fashion which have taken place in the domiciles of life, the following questions occur: – Do they arise from admixture of species nearly allied producing intermediate species? Are they the diverging ramifications of the living principle under modification of circumstance? Or have they resulted from the combined agency of both? Is there only one living principle? Does organised existence, and perhaps all material existence, consist of one Proteus principle of life capable of gradual circumstance-suited modifications and aggregations, without bound under the solvent or motion-giving principle, heat or light? There is more beauty and unity of design in this continual balancing of life to circumstance, and greater conformity to those dispositions of nature which are manifest to us, than in total destruction and new creation. It is improbable that much of this diversification is owing to commixture of species nearly allied, all change by this appears very limited, and confined within the bounds of what is called species; the progeny of the same parents, under great difference of circumstance, might, in several generations, even become distinct species, incapable of co-reproduction.

"The self-regulating adaptive disposition of organised life, may, in part, be traced to the extreme fecundity of Nature, who, as before stated, has, in all the varieties of her offspring, a prolific power much beyond (in many cases a thousandfold) what is necessary to fill up the vacancies caused by senile decay. As the field of existence is limited and pre-occupied, it is only the hardier, more robust, better suited to circumstance individuals, who are able to struggle forward to maturity, these inhabiting only the situations to which they have superior adaptation and greater power of occupancy than any other kind; the weaker, less circumstance-suited, being prematurely destroyed. This principle is in constant action, it regulates the colour, the figure, the capacities, and instincts; those individuals of each species, whose colour and covering are best suited to concealment or protection from enemies, or defence from vicissitude and inclemencies of climate, whose figure is best accommodated to health, strength, defence, and support; whose capacities and instincts can best regulate the physical energies to self-advantage according to circumstances – in such immense waste of primary and youthful life, those only come forward to maturity from the strict ordeal by which Nature tests their adaptation to her standard of perfection and fitness to continue their kind by reproduction.

"From the unremitting operation of this law acting in concert with the tendency which the progeny have to take the more particular qualities of the parents, together with the connected sexual system in vegetables, and instinctive limitation to its own kind in animals, a considerable uniformity of figure, colour, and character, is induced, constituting species; the breed gradually acquiring the very best possible adaptation of these to its condition which it is susceptible of, and when alteration of circumstance occurs, thus changing in character to suit these as far as its nature is susceptible of change.

"This circumstance-adaptive law, operating upon the slight but continued natural disposition to sport in the progeny (seedling variety), does not preclude the supposed influence which volition or sensation may have over the configuration of the body. To examine into the disposition to sport in the progeny, even when there is only one parent, as in many vegetables, and to investigate how much variation is modified by the mind or nervous sensation of the parents, or of the living thing itself during its progress to maturity; how far it depends upon external circumstance, and how far on the will, irritability, and muscular exertion, is open to examination and experiment. In the first place, we ought to investigate its dependency upon the preceding links of the particular chain of life, variety being often merely types of approximations of former parentage; thence the variation of the family, as well as of the individual, must be embraced by our experiments.

"This continuation of family type, not broken by casual particular aberration, is mental as well as corporeal, and is exemplified in many of the dispositions or instincts of particular races of men. These innate or continuous ideas or habits seem proportionally greater in the insect tribes, those especially of shorter revolution; and forming an abiding memory, may resolve much of the enigma of instinct, and the foreknowledge which these tribes have of what is necessary to completing their round of life, reducing this to knowledge, or impressions and habits, acquired by a long experience. This greater continuity of existence, or rather continuity of perceptions and impressions, in insects, is highly probable; it is even difficult in some to ascertain the particular stops when each individuality commences, under the different phases of egg, larva, pupa, or if much consciousness of individuality exists. The continuation of reproduction for several generations by the females alone in some of these tribes, tends to the probability of the greater continuity of existence, and the subdivisions of life by cuttings (even in animal life) at any rate must stagger the advocate of individuality.

"Among the millions of specific varieties of living things which occupy the humid portion of the surface of our planet, as far back as can be traced, there does not appear, with the exception of man, to have been any particular engrossing race, but a pretty fair balance of powers of occupancy – or rather, most wonderful variation of circumstance parallel to the nature of every species, as if circumstance and species had grown up together. There are indeed several races which have threatened ascendancy in some particular regions, but it is man alone from whom any general imminent danger to the existence of his brethren is to be dreaded.

"As far back as history reaches, man had already had considerable influence, and had made encroachments upon his fellow denizens, probably occasioning the destruction of many species, and the production and continuation of a number of varieties or even species, which he found more suited to supply his wants, but which, from the infirmity of their condition – not having undergone selection by the law of nature, of which we have spoken, cannot maintain their ground without its culture and protection.

"It is, however, only in the present age that man has begun to reap the fruits of his tedious education, and has proven how much 'knowledge is power.' He has now acquired a dominion over the material world, and a consequent power of increase, so as to render it probable that the whole surface of the earth may soon be overrun by this engrossing anomaly, to the annihilation of every wonderful and beautiful variety of animated existence, which does not administer to his wants principally as laboratories of preparation to befit cruder elemental matter for assimiliation by his organs."

"Much of the luxuriance and size of timber depending upon the particular variety of the species, upon the treatment of the seed before sowing, and upon the treatment of the young plant, and as this fundamental subject is neither much attended to nor generally understood, we shall take it up *ab initio*.

"The consequences are now being developed of our deplorable ignorance of, or inattention to, one of the most evident traits of natural history, that vegetables as well as animals are generally liable to an almost unlimited diversification, regulated by climate, soil, nourishment, and new commixture of already formed varieties. In those with which man is most intimate, and where his agency in throwing them from their natural locality and dispositions has brought out this power of diversification in stronger shades, it has been forced upon his notice, as in man himself, in the dog, horse, cow, sheep, poultry – in the Apple, Pear, Plum, Gooseberry, Potato, Pea, which sport in infinite varieties, differing considerably in size, colour, taste, firmness of texture, period of growth, almost in every recognisable quality. In all these kinds man is influential in preventing deterioration, by careful selection of the largest or most valuable as breeders; but in timber trees the opposite course has been pursued. The large growing varieties being so long of coming to produce seed, that many plantations are cut down before they reach this maturity, the small growing and weakly varieties, known by early and extreme seeding, have been continually selected as reproductive stock, from the ease and conveniency with which their seed could be procured; and the husks of several kinds of these invariably kiln-dried, in order that the seeds might be the more easily extracted. May we, then, wonder that our plantations are occupied by a sickly short-lived puny race, incapable of supporting existence in situations where their own kind had formerly flourished – particularly evinced in the genus Pinus, more particularly in the species Scots Fir; so much inferior to those of Nature's own rearing, where only the stronger, more hardy, soil-suited varieties can struggle forward to maturity and reproduction?

"We say that the rural economist should pay as much regard to the breed or particular variety of his forest trees, as he does to that of his live stock of horses, cows, and sheep. That nurserymen should attest the variety of their timber plants, sowing no seeds but those gathered from the largest, most healthy, and luxuriant growing trees, abstaining from the seed of the prematurely productive, and also from that of the very aged and over-mature; as they, from animal analogy, may be expected to give an infirm progeny, subject to premature decay." – See "Naval Timber and Arboriculture," pages 364 and 365, 381 to 388; also 106 to 108. *Patrick Matthew, Gourdee Hill, Errol N.B., March 7.*

Editorial Headnote

32. William Charles Wells, *Two essays: upon a single vision with two eyes, the other on dew*. (Constable, London. 1818). This includes the appendix: "*An account of a female of the white race of mankind, part of whose skin resembles that of a negro, with some observations on the cause of the differences in colour and form between the white and negro races of man,*" pp. 431–8.

William Charles Wells (1757–1818) was an American-Scottish physician who received his medical education in Edinburgh and practiced in London. His "Two Essays Upon Dew and Single Vision," published in 1818, were papers he had originally read to the Royal Society of London in 1813. For his *Essay on Dew,* he had received the Royal Society's Rumford Medal. It was in the expanded 1818 publication that he included the appendix included here: "An account of a female of the white race of mankind, part of whose skin resembles that of a negro, with some observations on the cause of the differences in colour and form between the white and negro races of man." Darwin acknowledged this work in the preface or "Historical Sketch" he wrote for the third and subsequent editions of *On the Origin of Species*. Darwin read Wells's work in 1865, and thus he included mention of Wells only from the 4th edition of *Origin* onwards:

> In this paper he [Wells] distinctly recognizes the principle of natural selection, and this is the first recognition which has been indicated; but he applies it only to man, and to certain characters alone. After remarking that negroes and mulattoes enjoy an immunity from certain tropical diseases, he observes, firstly, that all animals tend to vary in some degree, and, secondly, that agriculturalists improve their domesticated animals by selection; and then he adds, but what is done in this latter case by art, seems to be done with equal efficacy, though more slowly, in the formation of varieties of mankind, fitted for the country which they inhabit.[1]

Note

1 Charles Darwin, *On the Origin of Species, by means of Natural Selection, Or, the Preservation of the Favoured Races in the Struggle for Life* (London: John Murray, 1866), p. xiv. For a more detailed consideration of Wells see Curtis N. Johnson, *Darwin's Historical Sketch. An Examination of the 'Preface' to the Origin of Species*, (Oxford: Oxford University Press 2020), pp. 78–82.

32

TWO ESSAYS: UPON A SINGLE VISION WITH TWO EYES, THE OTHER ON DEW

William Charles Wells

Source: William Charles Wells, *Two Essays: upon a single vision with two eyes, the other on dew*, including appendix, *An account of a female of the white race of mankind, part of whose skin resembles that of a negro, with some observations on the cause of the differences in colour and form between the white and negro races of man* (Constable, London. 1818), pp. 431–8

Two inferences may, I think, be made from what has been related respecting Hannah West.

The first is, that the blackness of the skin in negroes is no proof of their forming a different species of men from the white race.

When a white man is much exposed to the action of the sun, his skin becomes more or less brown, and as the intensity of this colour, after equal degrees of exposure, is generally proportional to the heat of the climate, it has hence been supposed, that the colour of negroes is derived from a very great degree of the same cause. But this conclusion seems to me very faulty. For, setting aside that a white man, rendered brown by the sun's rays, begets as white children as those of another of the same race, the colour of whose skin had never been altered, it appears to me probable, from observations lately made on two negroes, that the action of the sun tends rather to diminish than augment the colour of their race. Both of those persons were born in European settlements, and had been accustomed to have their bodies clothed, yet, in both, the trunk, arms, and lower extremities, were considerably darker than the face, and in one, were somewhat darker than the hands. But admitting this observation to be of no force, still it must be granted, in consequence of what has been said upon the state of part of West's skin, – that great heat is not indispensably necessary to render the human colour black; which is the second conclusion to be drawn from the account which has been given of her.

On considering the difference of colour between Europeans and Africans, a view has occurred to me of this subject, which has not been given by any author, whose works have fallen into my hands. I shall, therefore, venture to mention it here, though at the hazard of its being thought rather fanciful than just.

DOI: 10.4324/9781003490548-36

There is no circumstance, perhaps, in which these two races differ so much, as in their capacity to bear, with impunity, the action of the causes of many diseases. The fatality to Europeans of the climate of the middle parts of Africa, which are, however, inhabited by negroes without injury to their health, is well known. Let it then be supposed, that any number of Europeans were to be sent to that country, and that they were to subsist themselves by their bodily labour; it seems certain, that the whole colony would soon become extinct. On the other hand, the greater liability of negroes in Europe to be attacked with fatal diseases is equally well established. If, therefore, a colony of the former race were brought to Europe, and forced to labour in the open air for their subsistence, many of them would quickly die, and the remainder, from their inability to make great bodily exertions in cold weather, and their being frequently diseased, would be prevented from working an equal number of days in the year with the whites. The consequence would be, that without taking farther into account the unfriendliness of the climate to them, their gains would be inadequate to the maintenance of themselves and their families. They would thence become feeble, and be rendered still more incapable of supporting life by their labour. In the mean time, their children would die from want, or diseases induced by deficient or improper nourishment, and in this way, a colony of the negro race in a cold country would quickly cease to exist.

This difference in the capacity of the two races to resist the operation of the causes of many diseases, I assume as a fact, though I am utterly unable to explain it. I do not, however, suppose, that their different susceptibility of diseases depends, properly, on their difference of colour. On the contrary, I think it probable, that this is only a sign of some difference in them, which, though strongly manifested by its effects in life, is yet too subtle to be discovered by an anatomist after death; in like manner as a human body, which is incapable of receiving the small-pox, differs in no observable thing from another, which is still liable to be affected with that disease.

Regarding then as certain, that the negro race are better fitted to resist the attacks of the diseases of hot climates than the white, it is reasonable to infer, that those, who only approach the black race, will be likewise better fitted to do so, than others who are entirely white. This is, in fact, found to be true, with regard to the mixture of the two races; since mulattoes are much more healthy in hot climates than whites. But amongst men, as well as among other animals, varieties of a greater or less magnitude are constantly occurring. In a civilized country, which has been long peopled, those varieties, for the most part, quickly disappear, from the intermarriages of different families. Thus, if a very tall man be produced, he very commonly marries a woman much less than himself, and their progeny scarcely differs in size from their countrymen. In districts, however, of very small extent, and having little intercourse with other countries, an accidental difference in the appearance of the inhabitants will often descend to their late posterity. The clan of the Macras, for instance, possess both sides of Loch-Duich in Scotland; but those who inhabit one side of the loch are called the black Macras, and the others the white, from a difference which has always been observed in their complexions. Again, those who attend to the improvement of domestic animals, when they find individuals possessing, in a greater degree than common, the qualities they desire, couple a male and female of these together, then take the best of their offspring as a new stock, and in this way proceed, till they approach as near the point in view, as the nature of things will permit. But, what is here done by art, seems to be done, with equal efficacy, though more slowly, by nature, in the formation of varieties of mankind, fitted for the country which they inhabit. Of the accidental varieties of man, which would occur among the first few and scattered inhabitants of the middle regions of Africa, some one would be better fitted than the others to bear the diseases of the country. This race would consequently multiply, while the others would decrease, not only from their inability to sustain the attacks of disease, but from their incapacity of contending with their more vigorous neighbours. The colour of this vigorous race I take for granted, from what has been already said, would be dark. But the same disposition to form varieties still existing, a darker and a darker race would in the course of time occur, and as the darkest would be the best fitted for the climate, this would at length become the most prevalent, if not the only race, in the particular country in which it had originated.

In like manner, that part of the original stock of the human race, which proceeded to the colder regions of the earth, would in process of time become white, if they were not originally so, from persons of this colour being better fitted to resist the diseases of such climates, than others of a dark skin.

The cause which I have stated, as likely to have influence on the colour of the human race, would necessarily operate chiefly during its infancy, when a few wandering savages, from ignorance and improvidence, must have found it difficult to subsist throughout the various seasons of the year, even in countries the most favourable to their health. But, when men have acquired the knowledge of agriculture, and other arts, and in consequence adopt a more refined mode of life, it has been found, that an adherence to their ancient customs and practices will preserve them long as a distinct race from the original inhabitants of the country to which they had emigrated. Examples of this kind are frequent in the islands in the eastern seas in the torrid zone, where the inhabitants of the sea-coast, evidently strangers, are in some degree polished, and of a brown colour, while the ancient natives, who live in the interior parts, are savage and black. Similar facts occur in respect to other species of animals. It seems certain, for instance, that fine woolled sheep, like the Spanish, never both arose and sustained their breed in the northern parts of Europe; yet, by care, this feeble race, after being formed in Spain, has been propagated and preserved in very cold countries. Thus the late Mr. Dryander, the learned librarian of the Royal Society, informed me, that the breed of fine woolled Spanish sheep had been kept perfect in Sweden during a very long term of years, I think he said a century. If, then, my memory be accurate upon this point, we have here an example of a variety of animals, much more liable to be affected by external circumstances than the human race, being preserved without change, in a country very different from their own, by assimilating their new state as much as possible to their old, during at least fifty generations, that is, during a period equivalent to 1500 years in the history of man.

Editorial Headnotes

33. William Herbert, "On the Production of Hybrid Vegetables; With the Results of Many Experiments Made in the Investigation of the Subject. In a Letter to the Secretary," *Transactions of the Horticultural Society,* (1822), v.4:15–22, 47.

34. William Herbert, *Amaryllidaceae* (London: J. Ridgeway, 1837), p. 19–20, p. 337–40.

William Herbert (1778–1847) was a botanist, notable poet, botanical illustrator, and plant breeder who wrote extensively on hybridization. He enjoyed a brief political career as a Member of Parliament in the first decade of the nineteenth century, but later, in 1814, he took Holy Orders and was promoted to the position of Dean of Manchester. Darwin first read Herbert's works on botany in the 1830s and noted in chapter three of *On the Origin of Species* that he had acknowledged the importance of competition among plant species.[1] While Darwin acknowledged the importance of many plant and animal breeders in the development of his theory, he later included acknowledgement of Herbert in the "historical sketch" that he included as a preface to *Origin* from the 1861 third edition specifically because, where many horticulturalists had commented on the ability of breeders to produce varieties, Herbert was unique in stating clearly that "new species might also be made through hybridization in nature." As the historian of science Curtis N. Johnson has shown in his detailed study of Darwin's "historical sketch," Darwin was indebted to two of Herbert's works in particular: an essay that had appeared in the fourth volume of the *Horticultural Transactions* in 1822, and his major work *Amaryllidaceae* (1837). Darwin first read Herbert's 1822 essay in 1837 and the latter early in 1839.[2] Johnson makes a convincing case that Darwin was likely alerted to the existence of Herbert's work by his reading of the second volume of Charles Lyell's *Principles of Geology*. In that work, Lyell had rejected the transmutationist ideas of the French naturalist Jean-Baptiste Lamarck and doubted Herbert's contention that different species could produce permanent hybrids in nature.[3] There Lyell had written:

> Mr. Herbert, in one of his ingenious papers on mule plants . . . says, whenever species, having a certain degree of affinity to each other, are transported from different countries, and brought for the first time into contact, they give rise to hybrid species.

However, he had concluded that

> Until some fact of this kind is fairly established, and a new species, capable of perpetuating itself in a state of perfect independence of man,

> can be pointed out, we think it reasonable to call in question entirely this hypothetical source of new species.[4]

Darwin and Herbert also corresponded on this issue in 1839, and they met in person prior to Herbert's death in 1847. It is clear that Darwin was impressed not only with Herbert's careful hybridization experiments and with his belief that hybrids might form new species in nature but also with Herbert's clear belief that a species was not fundamentally different from a variety, just one that was well marked and reproduced its kind. Furthermore, Herbert also talked about chance or accidental variations. Given the importance of these ideas for Darwin's later formulation of natural selection, it is likely for these reasons, as well as his personal and cordial relationship with Herbert, that he highlighted Herbert's work in the historical sketch.

Notes

1 Charles Darwin, *On the Origin of Species by Means of Natural Selection, Or, the Preservation of the Favoured Races in the Struggle for Life* (London: John Murray, 1859), p. 62.
2 Curtis N. Johnson, *Darwin's Historical Sketch. An Examination of the 'Preface' to the Origin of Species*, (Oxford: Oxford University Press 2020), pp. 85–6.
3 Johnson, *Darwin's Historical Sketch,* p. 87.
4 Charles Lyell, *Principles of Geology*, volume 2, (London: John Murray, 1832), pp. 225–6; quoted in Johnson, *Darwin's Historical Sketch*, p. 87–8.

33

"ON THE PRODUCTION OF HYBRID VEGETABLES; WITH THE RESULTS OF MANY EXPERIMENTS MADE IN THE INVESTIGATION OF THE SUBJECT. IN A LETTER TO THE SECRETARY"

William Herbert

Source: William Herbert, "On the Production of Hybrid Vegetables; With the Results of Many Experiments Made in the Investigation of the Subject. In a Letter to the Secretary", *Transactions of the Horticultural Society*, 4 (1822), pp. 15–22, 47

V. *On the Production of* Hybrid Vegetables; *with the Result of many Experiments made in the Investigation of the Subject. In a Letter to the Secretary. By the Hon. and Rev.* WILLIAM HERBERT, *D. C. L. F. H. S. &c.*

Read December 21, 1819.

MY DEAR SIR,

IT was your wish, that I should address some further communications to the Horticultural Society, concerning the production of *Hybrid Vegetables*. It would, perhaps, have been more advisable, to have waited till points, that are still doubtful, could have been accurately investigated; but, according to your request, I will report the progress of my experiments, and explain my ideas on the subject. It opens a wide and interesting field: the vegetable kingdom may certainly be greatly enriched by artificial intermixtures, and I think that new plants may be so formed, which will be capable of reproducing themselves, as distinct generations, by seed, in the same manner as natural species. I am aware of the opinion of our worthy President, who conceives that the production of a fertile offspring, from two plants of species supposed to be distinct, is in itself a decisive proof, that the parents are really of one species, and that Botanists have been in error concerning them; or, in other words, that fertility is evidence, that the plant is not really hybrid, but produced by the intermixture of two varieties of one species. I am,

DOI: 10.4324/9781003490548-37

however, satisfied, from the progress I have already made, that several plants, which I have raised, are not only, in the fair sense of the word, hybrid, but also fertile; and if they should perpetuate themselves by seed, without reverting to the form of either parent, they will be entitled to be considered by the Botanist as distinct species.

If it is meant only that a fertile offspring may be supposed to intimate, that the two parent plants have branched out from one common stock since the creation of the world, I am fully disposed to admit the truth of that position; but I should go much further, considering that many species, which we cannot now, by artificial means, prevail upon to intermix, have also descended from one original; and I doubt very much whether such a multiplication of distinct species may not also have taken place in the animal and insect tribes; but, to produce an intermixture between species that may have so diverged, the will of the animal must consent, while that of the plant need not be consulted.

I suspect that in the early periods of the world, there existed only the distinct genera of plants, or heads of families, not, however, exactly according to the present divisions of Botanists; who, indeed, are perpetually at variance with each other, as to the features which are sufficient to constitute a variety, a species, or a distinct genus. The lapse of centuries and diversity of soil and climate have probably wrought the most wide and permanent distinctions between vegetables, that have originated from a common stock, possibly even between the arborescent Ferns of the Andes, and the herbaceous inhabitants of our forests; but I should neither decide, if I found it impossible to produce a fertile offspring from the intermixture of any two plants, that they must have been distinct from the commencement of the world; nor, if I did succeed in procuring it, that they must not at present rank as distinct species of vegetables; I should merely think, as the Botanist would probably have already decided, that they were referable to one genus, or family. I apprehend the whole mystery to be this; that, in the progress of the distension of genera into various species, some species have retained such affinity as to admit of easy artificial intermixture; others have preserved so little affinity as to render their union more difficult; and others have departed so widely from their prototype, that the art of man cannot now reconcile them, and that, to bring them together again, it would be necessary to tread back or reverse the process of the centuries that have worked their difference. To which may be added, that in some plants the variation may be striking in many external appearances, and yet less considerable in the parts of fructification than in other species, of which the leaf or corolla may be more similar. If it be admitted, that diversity of species could have been produced by variations of soil, temperature, or humidity, it will be readily understood that such diversity might have been further multiplied by hybrid intermixture, as the species were brought together by the natural progress of their diffusion.

Mr. Knight has raised fertile Strawberries from the mixture of the Chili, the Carolina, and the Scarlet Strawberries, which some persons had deemed to be distinct. But this division was probably erroneous, for we find that they will intermingle naturally, and become confounded by seminal variations, without any artificial impregnation, and therefore it was a pretty clear case that they had been improperly separated. On the other hand, Mr. KNIGHT has failed, as yet, of producing a fertile plant between the American and European Strawberries: but are we to conclude from that circumstance, that plants, so nearly allied as the Scarlet and Alpine Strawberries, were from the creation of the world distinct, and at the same time hold that all the African Gladioli, between which (permanently dissimilar as they are in appearance) I can raise fertile intermixtures, were originally one species? Such an opinion would be so paradoxical, that it would require to be supported by very clear proof; and yet it would be difficult, by experiments, in any manner to confirm it.

It is not even true that all mules amongst animals are entirely sterile. There are well attested instances of the fertility of the mule between the ass and the mare; and mules between the gold-finch and canary-bird, the distinction of which no naturalist could deny, are frequently known to breed; and I understand that all the more beautiful mules are produced by a second cross with the canary-bird. I have known the mule siskin couple with a canary-bird; and I understand that the mules between two very distinct birds, the common pheasant and the silver pheasant, have proved fertile. The analogy, therefore, of animals, if such could be relied on, in considering the generation of vegetables, would not completely establish fertility as the test of a common origin. The organs in mule animals are said to be perfect, and the cause of their rarely breeding is not ascertained, though it is perhaps a sluggishness of constitution, as the equine mule is said to be more fertile in warmer countries. I suspect also, that the plants which thrive in a high temperature, will intermix more readily than those of colder regions. Strawberries, especially Hautbois, in their natural state, are often sterile.

Many plants, which Botanists have considered distinct, are certainly not so; as, for instance, Ixia (or Tritonia) crocata, of which seminal varieties have been erroneously named I. squalida, I. miniata, I. fenestrata, and I. deusta; Ixia flexuosa and I. polystachya, are the same; Babiana stricta, B. villosa, B. sulphurea, and rubro-cyanea, are not distinct; I have had a natural seedling from Babiana sulphurea with a pale eye like that of B. rubro-cyanea. I raised from the natural seed of one umbel of an highly manured Red Cowslip, a Primrose, a Cowslip, and Oxlips of the usual and other colours, a Black Polyanthus, a Hose-in-Hose Cowslip, and a natural Primrose bearing its flower on a Polyanthus stalk. From the seed of that very Hose-in-Hose Cowslip, I have since raised a Hose-in-Hose Primrose. I therefore consider all these to be only local varieties, depending upon soil and situation. I have raised a powdered Auricula and a Primula Helvetica from the seed of P. nivalis; and I have raised a Primula Helvetica also from P. Viscosa. I therefore esteem these Swiss Primulas to be local varieties of one species. The Violas are proved by cultivation to have been too much divided. The great Heartsease, which adorns Covent Garden market, under the name of Viola grandiflora, is found all yellow in Craven in Yorkshire, under the name of Viola lutea; with large dark purple flowers, without any yellow, in the neighbourhood of Moor-rig, above the falls of the Tees, in the county of Durham; and with mixed purple and yellow flowers, under the name of Viola amœna, in Weardale, a few miles from the last named place. The seeds gathered in Teesdale from the dark purple Hearts-ease once produced a dirty purple and yellow flower in my garden. These are therefore only local varieties, which by their uniformity in their natural abodes have misled the Botanist.

I believe the Orange, Citron, Lime, Lemon, and Shaddock, to be varieties of one plant. I do not, however, consider that Mr. KNIGHT's experiment[a] has proved the Almond and the Peach to be one species. The Peach is extremely similar to the Almond, with the exception of the sweet pulp, which may be, very probably, the effect of cultivation; and, if any amelioration of the pulp could be produced in seedling Almonds, I should incline to think that a long course of cultivation might have improved the Almond into a Peach. But the production of a fruit resembling a Peach, from an impregnation of the Almond with a plant so very similar, only shews that in an intermixture between two plants, which have such close affinity, the type of the male (as is frequently the case) has been very conspicuous; and this, even if the Peach had been known to have grown wild, with a sweet pulp, before the deluge, would not have surprised me.

The science of the Botanist, at the best, is very unstable, because it is entirely a science of conjecture, liable, at all times, to be overset by the test of cultivation. He carefully observes in plants the features that are least liable to variation; and by their means is enabled to subdivide the Classes, Genera, and Species of vegetables; but experience sometimes shews that the features, on which he relies, are very variable. Rhododendron and Azalea belong to two classes, widely separated by the number of the anthers, which is the characteristic feature of those classes; yet they are found to breed so freely together, and accord so exactly in the seed and capsule, that it can scarcely be doubted that they have branched from one original stock. Indeed, the Azalea is reported to have been occasionally seen with ten anthers, instead of five. At the same time, I doubt the possibility of intermixing the Alpine Rhododendrons with those of Pontus, India, and America, although united in one genus.

The most distinguished Botanists are perpetually at variance with each other, as to the subdivisions of the vegetable system; nor has any precise meaning been affixed to the terms by which they are known, as a guide to their labours; and, if we ask what is an Order? What is a Genus? What is a Species? What is a Variety? We shall find the answers very unsatisfactory. The most rational interpretation of the terms, I think, will be found as follows: *An Order*, – all the genera or original stocks which have general affinities to each other, though not such as to warrant a belief that they have branched from one stock. *A Genus*, all the species which have peculiar affinities, distinguishing them from all others; and which, I think, render it probable that they have branched, since the creation of the world, from one original. *A Species*, – a race of plants that will, in the present state of the world, perpetuate itself without varying in essential particulars, so as to confound itself with any other. *A permanent* or *local Variety*, – that which will perpetuate itself in a particular form, if kept in its native soil or situation, or at a distance from all other varieties; but which would, otherwise, confound itself with them. *An accidental Variety*, – that which cannot with certainty be perpetuated by seed in any situation. I do not believe that a better definition for the purposes of science can be given; and, if Botanists attended to it, their classifications would not be liable to such perpetual variety and contradiction: but, at all events, the experience of the Cultivator must always have weight to supersede the conjectural decisions of the Botanist.

[...]

I hope that this report of the progress of my experiments, however imperfect, may throw some light upon the mysteries of the vegetable kingdom. I must refer those who may wish to intermingle species of Amaryllideæ to my division of the genera, in the forty-seventh Volume of the *Botanical Magazine*, now publishing.

I am, dear Sir,
Yours very truly,
WILLIAM HERBERT.
Spofforth, November 29, 1819.

Note

a Horticultural Transactions, Vol. III. page 1.

34

AMARYLLIDACEAE

William Herbert

Source: William Herbert, *Amaryllidaceae* (London: J. Ridgeway, 1837), p. 19–20, 337–40

Jussieu has stated, in the perspicuous Latin Introduction to his History of Plants, that the foundation of botanical science rests on the distinction of species, which he defines to be a perennial succession of like individuals renewed by continual generations, and adds that although they may be sometimes a little diversified by climate, disease, or culture, the seminal produce will return to the original type. He then considers the union of species in genera to be a discretionary work for the assistance of the memory, attempted at first without rule, and latterly constructed with more propriety by reference to the fructification. Upon the same premises I should have come to the same conclusion; but that statement is contradicted by the complete confusion of vegetable species which may be produced by reciprocal intermixture; and horticultural experiments have established, beyond the possibility of refutation, that botanical species are only a higher and more permanent class of varieties; and that even the local and cultivated varieties to which he alludes, if they do not come in contact with other varieties, instead of reverting, as he supposed, to an original type, are renewed with considerable uniformity in their generations. It is now certain that individuality does not, as Jussieu had imagined, coincide with the species of botanists, but with a higher and more comprehensive grade, which seems to accord with such genera as are truly defined, and would agree with all if properly reformed; and if he had known, what our later experiments have established, his logical mind would certainly have assented to my proposition, which is in truth but the application of his own doctrine to facts which have since come to light. The labours of the many distinguished cultivators of the science have been daily tending to place the genera on a correct footing, but it will never be perfectly accomplished till the erroneous notion of the original diversity of all vegetable species is thoroughly discarded. If the original constructors of the system had known all that is at present ascertained, they would perhaps have called the botanical genera species, and the species varieties; but it does not appear to me advisable at present to make such an alteration, on account of the extreme inconvenience of such an extensive change. We are not certain that there is an exact analogy as to the diversification of individuals in their generation, between animals and vegetables, the latter having branched more extensively into

 DOI: 10.4324/9781003490548-38

different grades of variety, some of which can re-unite with facility, while others seem so constitutionally altered, as to be almost incapable of now mixing with the other grades; so that the arrangement and nomenclature is not objectionable, if we only bear in mind that the botanic genera define the individuality of kind. If any botanist should tell me, that in framing a generic character he was not seeking signs whereby to come at the knowledge of a fact, namely, the individuality of the thing described, I should answer, that if such is not his object, whatever name he may assign to the character, he has nothing in his science real or tangible; that it is the vague offspring of variable opinion; that he is amusing himself with child's play, and might as well build castles with the sand. If, instead of proceeding by single steps from the first point of general agreement to the last important point of generic identity, we throw together a number of kinds, and include them under the character of an order framed to cover their diversities, we are arranging them in groups to which no such limits were assigned by the Creator; and our classifications, by whatever name we may call them, are artificial; and we must be liable to find our labours, on further investigation, defeated by the reality. If we divide all vegetables, according to their agreement, by single points of difference, till we reach that which separates them from their nearest kindred, every plant, when rightly known, will have its proper place, and the whole system will be built on truth, and not the offspring of opinion.

[...]

Having, in fact, the same fundamental opinion, that the production of a fertile intermixture, designated the common origin of the parents, I held also, what experience has since in a great measure confirmed, that the production of any intermixture amongst vegetables, whether fertile or not, gave reason to suspect that the parents were descended from one common stock, and shewed that they were referable to one genus; but that there was no substantial and natural difference between what botanists had called species, and what they had termed varieties; the distinction being merely in degree, and not absolute; so that, without first reforming the terms used in botany, and ascertaining more precisely what was meant by a species, those who argued on the subject were fighting the air; and I suggested, as my view, that the birth of an intermixture afforded presumptive evidence that the parents were of one genus, meaning thereby kind or descent, and implying such an affinity as to enable them to breed together, and to induce a probability that they had diverged from one original created type. The real point in discussion at that period was, whether there did exist a positive and invariable line of fertility or sterility in all mixed vegetable productions, founded upon an original identity or diversity in the parental stocks; and whether it was possible for two plants, which were considered according to the general system of botanists to be distinct species, to produce a fertile cross, without proving an error of the subdivision in that particular case. Further experiments have shewn, that the sterility or fertility of the offspring does not depend upon original diversity of stock; and that, if two species are to be united in a scientific arrangement on account of a fertile issue, the botanist must give up his specific distinctions generally, and entrench himself within the genera. It has been objected that if any plants, now different, had descended from one original type, we might expect to find new forms and combinations daily arising round us by the process of nature, as well as by artificial agency; whereas the catalogue of European vegetables does not appear to be increased by the production of new plants in a wild[a] state; but it is most probable, that if the Almighty created the original types capable of permanent variations under different circumstances, perhaps of soil or climate, those variations were worked at a very early period, on the first diffusion of seeds into every different portion of the world, especially by the operation of the flood, and may have in part resulted from the changes of climate which accompanied it and shortened the life of man. We must recollect, that although the different races of dogs, which all freely interbreed, are universally admitted to have come from one type, though now outwardly more unlike to each other than numberless distinct species of other animals, we know not what the similitude of that type was; we have no record concerning the original wild dog, nor whether there existed immediately before or after the deluge any dogs in an undomesticated state; nor have we any knowledge of the time or place when any one of the several races, as greyhound, terrier, spaniel, bull-dog, &c. took its birth; nor is there a single known instance of two parent dogs of the same race, giving birth to individuals of a new race, or materially dissimilar to themselves, except where they are mongrels, and one of the ancestral types reappears more strongly than the other. Neither have we any information concerning the origin of the different races of mankind, which are as different in appearance as the species of vegetables; we have not seen any new race arise within the period of historical certainty; and whatever we do know concerning them, refers the time of their branching out from the common stock to very remote antiquity, at a period antecedent to or coeval with the dispersion of mankind over the globe. If it had been otherwise, the various races would have been blended, instead of occupying different localities. It is probable that the various races of dogs owe their origin to a very early period; to the days, when the effects arising from change of situation, were first experienced by the several created members of the animal and vegetable kingdom: and it is no more essential to believe that individuals of every one of the present species of fox, or antelope, or finch (many of which are more like to each other than the greyhound is to the terrier, though they do not intermingle), entered with their present respective aspects into the ark, than that all the calceolarias on the mountains of Chili, or all the mezembryanthemums on the wastes of Southern Africa, exhibited their present peculiarities in the days of the patriarch. It was perhaps part of the wise scheme of Providence, for the purpose of peopling the world with the immense diversity of forms that occupy it, to give each created race a disposition to branch into diversities, acquiring constitutional peculiarities, which should keep them more or less separated; and the same phænomenon is observable in the languages of man, which are infinitely numerous; yet there is no reason to believe that many languages were given to man on the confusion of tongues; on the contrary, the cloven tongues that gave back the power of universal speech, imply that they were few; but from these have branched out innumerable languages, which cannot be reunited, and no person can show when or how any one of them arose, though we may trace the mingling of one with another in the later years of the world. One thing seems pretty certain, amongst the mysteries in which this subject is enveloped, that the differences worked, whether in plants or animals, in a state of domesticity, do not effect so great a constitutional separation inducing an indisposition to reunite and produce a prolific offspring, as the changes which have been wrought by nature in the wilderness.

I have said in the preliminary observations on Amaryllidaceæ, that a perfect analogy between animals and vegetables in their generations is not apparent; but I do not mean to assert, that, if this subject can ever be thoroughly bottomed, it may not be found to exist. A reformation of Zoology is in progress; for example, in Ornithology, the Linnæan genus Motacilla was after a time confined to the wagtails, a large group being detached as Silviæ; but later observers found that group to consist of several families, and have since correctly distinguished at least the robins, the red-starts, the nightingales, the hedge warblers, the fruit-eating warblers, the sedge warblers, the chats, the troglodyte wrens, and the greenish wrens, as separate genera with their respective diversities; and within those generic limits I suspect that the power of crossing may be confined, and their several species, however now immutably distinct, may have respectively branched out from one stock since the period of the deluge. I have lately had under my observation a dog, whose father was a fox in an innyard at Ripon, and it has singularly the manners as well as the voice of a fox, but it is the parent of many families of puppies: and I feel satisfied that the fox and the dog are of one origin, and suspect the wolf and jackall to be of the same; nor could I ever contemplate the black line down the back of a dun pony without entertaining a suspicion that the horse, unknown in a wild state except where it has escaped from domesticity, may be a magnificent improvement of the wild ass in the very earliest age of the world: bearing in mind, that both in the animal and vegetable creation, the diversities arising from inscrutable causes in the wild races of the forest, are of a more unalterable character than those which spring up under the care and cultivation of man. With respect to animals in their wild state, their union with their own species seems to be mainly guided by voice and smell, and in domesticity that instinctive preference is evidently much weaker, and the will to keep themselves distinct is therefore lessened. The various species of greenish wrens are so similar in plumage, that it requires nice examination to distinguish them, yet they have different notes, manners, and habits of building their nests, even when in the same locality; but we have no certainty that if their predilection for the voice and smell of their own race was weakened, they would not be capable of producing a fertile cross; and we draw our conclusions from a few instances of domestic mules between species which happen to be widely removed from each other, as the pheasant and fowl, the goldfinch and canary bird; whereas we should apply to this subject, concerning which our knowledge is very limited, the consideration of the fact ascertained concerning vegetables, which have no will to interfere with our experiments, that some crosses are sterile and some quite fertile, without any apparent reason, except the greater or less approximation of constitution in the parents: and that the cross-bred plant, which has seemed for a long course of years to be absolutely sterile, becomes under some circumstances productive.

In accordance with the principle above stated, when it was shown that the botanic subdivisions of Rhododendron, Azalea, and Rhodora, comprehended plants which were capable of intermixing, I asserted that the botanist must reconsider and alter his subdivisions, and acknowledge that, notwithstanding their peculiarities, they constituted but one genus or kind.

Note

a Ranunculus, Anemone, Hypericum, Scleranthus, Drosera, Potentilla, Geum, Medicago, Galium, Centaurca, Stachys, Rhinanthus, Digitalis, Verbascum, Gentiana, Mentha, Quercus, Salix, and Narcissus, are however a long list of Genera enumerated by Schiede, 1825, and Lasch Linn. 1829, as having produced spontaneous hybrids, to which Crinum may be added.

BIBLIOGRAPHY

Barbour, T. and Wheeler, W. M. (eds.), *The Lamarck Manuscripts at Harvard* (Cambridge, MA: Harvard University Press, 1933), p. xiv.

Barlow, N. (ed.), *The Autobiography of Charles Darwin 1809–1882* (London: W. W. Norton & Co., 1958).

de Blainville, H. M. D., 'Analyse des principaux travaux dans les sciences physiques, publiés dans l'année 1821', in *Journal de physique, de chimie, d'histoire naturelle et des arts*, 94 (janvier 1822), p. 54.

Blumenbach, J. F., 'Über den Bildungstrieb (Nisus formativus) und seinen Einfluß auf die Generation und Reproduction', in *Göttingisches Magazin der Wissenschaften und Litteratur*, 1 (1780), pp. 247–266.

Blumenbach, J. F., *A Manual of Comparative Anatomy Translated from the German of J.F. Blumenbach with Additional Notes by William Lawrence Esq. FRS. Second Edition, Revised and Augmented, by William Coulson, Demonstrator of Anatomy at the Medical School, Aldersgate Street, and Member of the Zoological Society* (London: W. Simpkin & R. Marshall, 1827).

Bowler, P. J., *The Non-Darwinian Revolution: Reinterpreting a Historical Myth* (Baltimore: Johns Hopkins University Press, 1988), p. ix.

Bowler, P. J., *Evolution. The History of an Idea* (Berkley: University of California Press, 1989).

Brown, A. W., *The Metaphysical Society. Victorian Minds in Crisis, 1869–1880* (New York: Columbia University Press, 1947).

Brown, B. R., *Until Darwin. Science, Human Variety, and the Origins of Race* (London: Pickering and Chatto, 2010).

Browne, J., 'Botany of Gentlemen: Erasmus Darwin and "The Loves of Plants"', *Isis*, 80:4 (1989), pp. 593–621.

Browne, J., *Charles Darwin. Voyaging. A Biography* (Princeton N.J., Princeton University Press, 1995).

Browne, J., *Charles Darwin. The Power of Place: A Biography* (New York: Alfred A. Knopf, 2002).

Burkhardt, F., et. al. (eds.), *The Correspondence of Charles Darwin*, 30 vols (Cambridge: Cambridge University Press, 1985–2023).

Burkhardt, R., *The Spirit of System: Lamarck and Evolutionary Biology* (Cambridge, MA: Harvard University Press, 1997).

Carpenter, W. B., 'Inaugural Address of Dr. William Carpenter, F.R.S., President', *Nature* (15 August 1872), pp. 206–12.

Carpenter, W. B., *Principles of Mental Physiology, with their applications to the training and discipline of the mind and the study of its morbid conditions* (New York: D. Appleton & Co, 1874).

[Chambers, R.], *The Vestiges of the Natural History of Creation* (London: John Churchill, 1844).

[Chambers, R.], *The Vestiges of the Natural History of Creation* (London: George Routledge & Sons, 1887).

[Chilton, W.], 'Theory of Gradation. XLV', *Oracle of Reason*, September 30, (1843), pp. 332–34.

Claeys, G., 'Malthus and Godwin: rights, utility, and productivity', in Robert J. Mayhew (ed.), *New Perspectives on Malthus* (Cambridge: Cambridge University Press, 2016), pp. 52–73.

Cobbe, F. P., 'Darwinism in Morals', *Theological Review: A Quarterly Journal of Religious Thought and Life*, 8:33 (1871), pp. 167–192.

Cooter, R., *The Cultural Meaning of Popular Science. Phrenology and the Organization of Consent in Nineteenth-Century Britain* (Cambridge: Cambridge University Press, 1985).

Cooter, R. and Pumfrey, S., 'Separate Spheres and Public Places. Reflections on the History of Science Popularization and Science in Popular Culture', *History of Science*, 32:3 (1994), pp. 237–267.

Corsi, P., *The Age of Lamarck: Evolutionary Theories in France 1790–1830*, (Berkeley: University of California Press, 1988).

Corsi, P., 'Edinburgh Lamarckians? The Authorship of Three Anonymous Papers (1826–1829)', *Journal of the History of Biology* 54 (2021), pp. 345–374.

Cuvier, G., *Essay on The Theory of the Earth* (Edinburgh: William Blackwood, 1827).

Darwin, C., *Journal of Researches into the Natural History and Geology of the Countries Visited During the Voyage of HMS Beagle Round the World, Under the Command of Capt. Fitz Roy, R.N.* (New York: Harper & Brothers, 1846).

Darwin, C., *On the Origin of Species by Means of Natural Selection; or the Preservation of the Favoured Races in the Struggle for Life* (London: John Murray, 1859), p. 62.

Darwin, C., 'Natural selection', *Gardeners' Chronicle and Agricultural Gazette*, 16 (21 April 1860), pp. 362–363.

Darwin, C., *On the Origin of Species by Means of Natural Selection, or the Preservation of the Favored Races in the Struggle for Life* (London: John Murray, 1860), p. 490.

Darwin, C., *On the Origin of Species by Means of Natural Selection, or the Preservation of the Favoured Races in the Struggle for Life* (London: John Murray, 1861).

Darwin, C., *On the Origin of Species, by means of Natural Selection, Or, the Preservation of the Favoured Races in the Struggle for Life* (London: John Murray, 1866), p. xiv.

Darwin, C., *The Descent of Man and Selection in Relation to Sex* (London: John Murray, 1871).

Delrue, V., 'Le spectacle du ciel: nut en schronheid als drijfveren van de meteorologie (1799–1816)', *Jaarboek de Achttiende Eeuw*, 54:1 (2022), pp. 148–165

Desmond, A., 'Robert E. Grant: The social predicament of a pre-Darwinian transmutationist', *Journal of the History of Biology*, 17:2 (1984), pp. 189–223.

Desmond, A., *The Politics of Evolution. Medicine, Morphology, and Reform in Radical London* (Chicago: University of Chicago Press, 1989).

Desmond, A., 'Artisan Resistance and Evolution in Britain, 1819–1848', *Osiris*, 3 (1987), pp. 77–110.

Desmond, A., and Moore, J., *Darwin. The Life of a Tormented Evolutionist* (London & New York: W. W. Norton & Co., 1991).

Desmond, A. and Moore, J., *Darwin's Sacred Cause: How a Hatred of Slavery Shaped Darwin's Views on Human Evolution* (New York: Houghton Mifflin Harcourt, 2009).

Dewitt, A., 'Evolution and Political Revolution in Blackwood's Periodical Poetry', *Victorian Periodicals Review*, 56, 2 (Summer 2023), pp. 158–181.

Elliott, P. A., 'Erasmus Darwin, Herbert Spencer, and the Origins of the Evolutionary Worldview in British Provincial Scientific Culture, 1770–1850', *Isis*, 94:1 (2003), pp. 1–29.

Elliott, P. A., '"The Derbyshire Darwinians": The Persistence of Erasmus Darwin's Influence on a British provincial Library and Scientific Community, c.1780–1850', in R. Arnot and C.U.M. Smith (eds.), *The Genius of Erasmus Darwin* (Aldershot: Ashgate, 2005).

Engels, F. and Marx, K., *Karl Marx and Friedrich Engels: Collected Works*, 50 vols (New York: International Publishers, 1975–2004), 41, p. 381.

England, R.; Lightman, B.; and Marshall, C. (eds.), *The Metaphysical Society (1869–1880). Intellectual Life in Mid-Victorian England* (Oxford: Oxford University Press, 2019).

Evans, E. J., *The Great Reform Act of 1832* (London: Routledge, 1994).

Fichman, M., *An Elusive Victorian. The Evolution of Alfred Russel Wallace* (Chicago: University of Chicago Press, 2004).

Galera, A., 'The Impact of Lamarck's Theory of Evolution Before Darwin's Theory', *Journal of the History of Biology*, 50:1 (Feb. 2017), pp. 53–70.

Godwin, W., *An Enquiry Concerning Political Justice and Its Influence on General Virtue and Happiness*, 2 vols (Dublin: Luke White, 1793).

Habermas, J., *The Structural Transformation of the Public Sphere: An Inquiry in a Category of Bourgeois Society*, tr. Thomas Burger and Frederick Lawrence (Cambridge, MA: MIT Press, 1991 [1962]).

Hale, P. J., 'Darwin's Other Bulldog: Charles Kingsley and the Popularisation of Evolution in Victorian Britain', *Science and Education*, 21 (2012), pp. 977–1013.

Hale, P. J., *Political Descent. Malthus, Mutualism and the Politics of Evolution in Victorian England* (Chicago: University of Chicago Press, 2014).

Hale, P. J., 'Rejecting the Myth of the Non-Darwinian Revolution', *Victorian Review*, 41:2 (2015), pp. 13–18.

Hale, P. J., 'Between Intuition and Empiricism. William Benjamin Carpenter on Man, Mind, and Moral Responsibility', in R. England, B. Lightman, and C. Marshall (eds.), *The Metaphysical Society (1869–1880). Intellectual Life in Mid-Victorian England* (Oxford: Oxford University Press, 2019), pp. 205–27.

Hale, P. J., 'Darwin, Design, and the Honey Bee. The Morality of the Hive', *American Bee Journal* (October 2020), pp. 1103–1107.

Hale, P. J., 'Charles Darwin, sexual selection, and the evolution of other-regarding ethics', *British Journal of the History of Science, Themes*, 6 (2021), pp. 157–177.

Hale, P. J. and Testa, C. C. (eds.), *Evolution Before Darwin* (London: Routledge, 2025).

Hale, P. J. and Testa, C. C. (eds.), *Evolution, Race, and Colonialism in Nineteenth-Century Britain* (London: Routledge, 2025).

Hale, P. J. and Testa, C. C. (eds.), *Evolution and Religion in Nineteenth-Century Britain* (London: Routledge, 2025).

Hale, P. J. and Testa, C. C. (eds.), *Evolution, Sex, Gender, and the Woman Question in Nineteenth-Century Britain* (London: Routledge, 2025).

Hale, P. J. and Testa, C. C. (eds.), *Evolution and Socialism in Nineteenth-Century Britain* (London: Routledge, 2025).

Hale, P. J. and Testa, C. C., 'Mapping evolutionary ideas in Britain in the first half of the nineteenth century', in P. J. Hale and C. C. Testa (eds.), *Evolution Before Darwin* (London: Routledge, 2025.

Harvey, D. A., *The French Enlightenment and Its Others* (New York: Palgrave Macmillan, 2012), pp. 138–146.

Hetherington, H., *The Free-thinker's Information for the People* 1, no.1–2 (1842), pp. 1–16.

Hofstadter, R., *Social Darwinism in American Thought* (Boston: Beacon Press, 1992).

Huxley, T. H., 'Darwin On the Origin of Species', *Westminster Review*, 73:144 (1860), pp. 541–70.

Jacyna, L. S., 'Immanence or Transcendence: Theories of Life and Organization in Britain 1790–1835', *Isis*, 74 (1983), pp. 311–29.

Jenkins, B., 'Race before Darwin: Variation, adaptation and the natural history of man in post-Enlightenment Edinburgh, 1790–1835', *British Journal for the History of Science*, 53:3 (2020), pp. 333–350.

Jenkins, B., *Evolution Before Darwin: Theories of the Transmutation of Species in Edinburgh, 1804–1834* (Edinburgh: University of Edinburgh Press, 2021).

Johnson, C. N., *Darwin's Historical Sketch. An Examination of the 'Preface' to the Origin of Species* (Oxford: Oxford University Press, 2020), p. 6.

Jones, G. and Peel, R. A. (eds.), *Herbert Spencer. The Intellectual Legacy* (London: Galton Institute, 2004).

Joyce, P., *Visions of the People. Industrial England and the Question of Class, 1848–1914* (Cambridge: Cambridge University Press, 1994).

Junker, T., 'Blumenbach's theory of human races and the natural unity of mankind', in G. Lauer and N. Rupke (eds.), *Johann Friedrich Blumenbach: Race and Natural History, 1750–1850* (London: Routledge, 2019), pp. 96–112.

Kingsley, C., 'President's Address', *Report and Transactions of the Devonshire Association for the Advancement of Science, Literature and the Arts*, 4:1 (1871), p. 383.

Knox, R., 'Inquiry into the Origin and Characteristic Difference of the Native Races Inhabiting the Extra-Tropical part of Southern Africa', *Memoirs of the Wernerian Natural History Society*, 5 (1824), pp. 206–219.

Knox, R., *The Races of Men. A Fragment* (Philadelphia: Lea & Blanchard, 1850).

Knox, R., 'Contributions to the Philosophy of Zoology with Special Reference to the Natural History of Man', *The Lancet* (14 July 1855), pp. 24–26.

Kropotkin, P., *Mutual Aid. A Factor in Evolution* (London: Freedom Press, 1902).

Lamarck, J. B., *Philosophie Zoologique* (Paris: Musée d'Histoire Naturelle, 1809).

Lamarck, J. B., *Zoological Philosophy*, tr. Hugh Elliot (London: Macmillan and Co., 1914 [1809]), pp. 368–371.

Largent, M., 'The So-Called Eclipse of Darwinism', *Transactions of the American Philosophical Society*, 99:1 (2009), pp. 3–21.

Lawrence, W., *Lectures on Physiology, Zoology, and the Natural History of Man* (London: Smith, 1822).

Lenoir, T., *The Strategy of Life. Teleology and Mechanics in Nineteenth Century German Biology* (Chicago: University of Chicago Press, 1982).

Lightman, B., 'Darwin and the Popularization of Evolution', *Notes and Records of the Royal Society*, 64 (2010), pp. 5–24.

Lightman, B. (ed.), *Global Spencerism. The Communication and Appropriation of a British Evolutionist* (Leiden: Brill, 2015).

Lightman, B., 'The "Greatest Living Philosopher" and the Useful Biologist. How Spencer and Darwin Viewed Each Other's Contributions to Evolutionary Theory', in Ian Hesketh (ed.), *Imagining the Darwinian Revolution. Historical Narratives of Evolution from the Nineteenth-Century to the Present*, (Pittsburgh: University of Pittsburgh Press, 2022), pp. 37–57.

Lightman, B. and Zon, B. (eds.), *Evolution and Victorian Cultur*e (Cambridge: Cambridge University Press, 2014).

Lovejoy, A. O., 'Buffon and the Problem of Species', in *Forerunners of Darwin: 1745–1859*, ed. B. Glass, W. L. Straus, and O. Temkin (Baltimore: Johns Hopkins University Press,1959), pp. 84–113.

Lyell, C., *Principles of Geology*, 3 vols (London: John Murray, 1830–33).

Lyell, C., *Principles of Geology*, 3 vols, facsimile reprint (Chicago: University of Chicago Press, 1990; 1991).

Matthew, P., 'Nature's law of selection', *Gardeners' Chronicle and Agricultural Gazette* (7 April 1860), pp. 312–13.

Mayhew, R. J., *Malthus. The Life and Legacies of an Untimely Prophet* (Harvard: Belknap, 2014).

Mill, J. S., *Autobiography of John Stuart Mill* (New York: New American Library, 1964), p. xii.

Nelson, W. M., 'Making Men: Enlightenment Ideas of Racial Engineering', *American Historical Review*, 15:5 (Dec. 2010), pp. 1364–1394.

Pick, D., *Faces of Degeneration. A European Disorder, c. 1848–1918* (Cambridge: Cambridge University Press, 1989).

Pritchard, J. C., *A Review of the Doctrine of a Vital Principle: As Maintained by Some Writers on Physiology with Observations on the Causes of Physical & Animal Life* (London: John and Arthur Arch, 1829).

Richards, E., 'The "Moral Anatomy" of Robert Knox. The Interplay Between Biological and Social Thought in Victorian Scientific Naturalism', *Journal of the History of Biology*, 22:3 (1989), pp. 373–436.

Richards, E., *Darwin and the Making of Sexual Selection* (Chicago: University of Chicago Press, 2017).

Richards, R. J., *Darwin and the Emergence of Evolutionary Theories of Mind and Behavior* (Chicago: University of Chicago Press, 1987).

Richards, R. J., *The Meaning of Evolution: The Morphological Construction and Ideological Reconstruction of Darwin's Theory* (Chicago: University of Chicago Press, 1992).

Richards, R. J., 'Kant and Blumenbach on the Bildungstrieb: A historical misunderstanding', *Studies in the History and Philosophy of the Biological and Biomedical Sciences*, 31:1 (2000), pp. 11–32.

Richards, R. J., 'The Relation of Spencer's Evolutionary Theory to Darwin's', in G. Jones and R. A. Peel (eds.), *Herbert Spencer. The Intellectual Legacy* (London: The Galton Institute, 2004), pp. 17–36.

Rudwick, M. J. S., *Bursting the Limits of Time: The Reconstruction of Geohistory in the Age of Revolution* (Chicago: University of Chicago Press, 2005).

Rudwick, M. J. S., *Worlds Before Adam. The Reconstruction of Geohistory in the Age of Reform* (Chicago: University of Chicago Press, 2008).

Ruse, M., *The Darwinian Revolution. Science Red in Tooth and Claw* (Chicago: University of Chicago Press, 1977).

Ruse, M., *Darwinism as Religion. What Literature Tells Us About Evolution* (Oxford: Oxford University Press, 2017).

Schwarz, L., Infidel Feminism. Secularism, Religion, and Women's Emancipation, England 1830–1914 (Manchester: Manchester University Press).

Secord, J. A. 'Edinburgh Lamarckians: Robert Jameson and Robert E. Grant', *Journal of the History of Biology*, 24:1 (1991), pp. 1–18.

Secord, J. A., *Victorian Sensation. The Extraordinary Publication, Reception, and Secret Authorship of The Vestiges of the Natural History of Creation* (Chicago: University of Chicago Press, 2000).

Secord, J. A., *Visions of Science. Books and Readers at the Dawn of the Victorian Age* (Chicago: University of Chicago Press, 2015), pp. xiii.

Shaw, G. B., 'Preface' to *Back to Methuselah: A Metabiological Pentateuch* (London: Constable and Company Ltd., 1921).

Showalter, E., *The Female Malady. Women, Madness and English Culture 1830–1980* (London: Virago, 1987).

Spary, E., *Utopia's Garden. French Natural History from the Old Regime to Revolution*, (Chicago: University of Chicago Press, 2000).

Spencer, H., *Social Statics; or, the Conditions Essential to Human Happiness Specified, and the First of Them Developed* (London: John Chapman, 1851).

Spencer, H., *The Principles of Psychology* (London: Longman, Brown, Green and Longman, 1855).

Spencer, H., *An Autobiography*, 2 vols (New York: D. Appleton, 1904).

St. Clair, W., *The Godwins and the Shelleys: The Biography of a Family* (London: Faber & Faber, 1989).

Stark, L., 'Mutual Aid: The Workers' History of Science', *ISIS*, vol. 114, Iss.4 (Dec.2023), pp. 841–849.

Taylor, B., *Eve and the New Jerusalem. Socialism and Feminism in the Nineteenth* Century (London: Virago, 1991).

Testa, C. C., 'Species Transformation and Social Reform: The Role of the Will in Jean-Baptiste Lamarck's Transformist Theory', *Journal of the History of Biology*, 56:1 (March 2023), pp. 125–151.

Thompson, E. P., *The Making of the English Working Class* (London: Penguin, 1963).

Wallace, A. R., 'On the Tendency of Varieties to Depart Indefinitely from the Original Type', *Proceedings of the Linnean Society* (August 1858).

Wallace, A. R., 'On the Law Which has Regulated the Introduction of New Species', *Annals and Magazine of Natural History* (September 1855), reprinted in A. R. Wallace, *Contributions to the Theory of Natural Selection. A Series of Essays* (London: Macmillan & Co., 1871), pp. 1–25.

Winter, A., 'The Construction of Orthodoxies and Heterodoxies in the Early Victorian Life Sciences', in B. Lightman (ed.), *Victorian Science in Context* (Chicago: University of Chicago Press, 1997), pp. 24–50.

Wollstonecraft, M., *A Vindication of the Rights of Woman with Strictures on Political and Moral Subjects* (Boston: Peter Edes, 1792).

van Wyhe, J. (ed.), *Combe's Constitution of Man and Nineteenth Century Responses*, 3 vols (London: Thoemmes Press, 2004).

van Wyhe, J., *Phrenology and the Origins of Scientific Naturalism* (London: Routledge, 2004).

van Wyhe, J., 'Mind the Gap: Did Darwin Avoid Publishing his Theory for Many Years?', *Notes and Records of the Royal Society*, 61:2 (May 2007), pp. 177–205.

van Wyhe, J., 'The Impact of A. R. Wallace's Sarawak Law paper reassessed', *Studies in the History and Philosophy of the Biological and Biomedical Sciences*, 60 (2016), pp. 55–66.

van Wyhe, J., 'History of Phrenology on the Web', www.historyofphrenology.org.uk/texts/gordon.htm. Accessed 26th November 2023.

Zammito, J. H., *The Gestation of German Biology. Philosophy and Physiology from Stahl to Schelling* (Chicago: University of Chicago Press, 2018).

INDEX

Note: Page numbers in *italics* indicate a figure and page numbers in **bold** indicate a table on the corresponding page.